Ageing and Nutrition through Lifespan

Ageing and Nutrition through Lifespan

Special Issue Editor

Stefanos Tyrovolas

MDPI • Basel • Beijing • Wuhan • Barcelona • Belgrade

Special Issue Editor
Stefanos Tyrovolas
Parc Sanitari Sant Joan de Deu,
Fundacio Sant Joan de Deu,
CIBRESAM, Sant Boi de
Llobregat (Barcelona)
Spain

Editorial Office
MDPI
St. Alban-Anlage 66
4052 Basel, Switzerland

This is a reprint of articles from the Special Issue published online in the open access journal *Nutrients* (ISSN 2072-6643) from 2018 to 2019 (available at: https://www.mdpi.com/journal/nutrients/special_issues/Ageing_Nutrition_Lifespan).

For citation purposes, cite each article independently as indicated on the article page online and as indicated below:

LastName, A.A.; LastName, B.B.; LastName, C.C. Article Title. *Journal Name* **Year**, *Article Number*, Page Range.

ISBN 978-3-03928-945-5 (Hbk)
ISBN 978-3-03928-946-2 (PDF)

Contents

About the Special Issue Editor

Stefano Tyrovolas has a B.S. in Dietetics and Nutrition, M.S. in Applied Dietetics, and Ph.D. in Nutrition (Public Health Nutrition) from Harokopio University of Athens, Greece. During 2012–2013, he was a visiting lecturer at the University of Kalamata in the topic of Human Nutrition. Since 2010 Dr Tyrovolas is working in the research field of Ageing and Nutrition Epidemiology. Dr. Stefanos Tyrovolas is the Principal Investigator of the Research Line on Frailty, Sarcopenia and Nutrition in the older adults at Fundació Sant Joan de Déu (FSJD) and Parc Sanitari Sant Joan de Deu (PSSJD) and a Miguel Servet Researcher from Instituto de Salud Carlos III (ISCIII). Dr. Tyrovolas has contributed in more than 100 articles published in peer-reviewed international journals and has numerous presentations in national and international conferences. His research interests include nutrition, healthy ageing, frailty, sarcopenia and public health prevention for the elderly.

Preface to "Ageing and Nutrition through Lifespan"

Population is ageing at an unprecedented speed globally. As concept, ageing is considered a continuous process starting from birth and is accompanied by various physiological changes and a number of chronic diseases that affect health and quality of life. Ageing as a continuous process is depending on life course exposures to health risks, lifestyle and nutrition, socioeconomic background, and other factors.

There is considerable interest among scientists regarding the direct and indirect effect of nutrition in optimal ageing. Nutrition has a beneficial effect in a variety of chronic disease that impact the process of ageing. Given the importance of this issue, the journal *Nutrients* is planning a Special Issue on "Ageing and Nutrition through Lifespan" with the aim of providing a source for accurate, up-to-date scientific information on this topic.

We invite you and your co-workers to consider submission of your original research findings or a review article on the topic. Manuscripts should focus on the direct impact of specific food components, dietary patterns, energy intake, macro-, micro- nutrients, alcohol intake, food insecurity as well as malnourishment and appetite to the ageing process (healthy, active, successful ageing, frailty and other similar indices) across lifespan. In a similar way, we also welcome manuscripts that focus on the indirect effect of nutrition to the ageing process throughout the pathway of chronic disease (i.e., obesity, diabetes, depression and mental diseases).

This book is dedicated to my beloved father, Athanasio Tyrovola.

Stefanos Tyrovolas
Special Issue Editor

Article

Prevalence of Medication-Dietary Supplement Combined Use and Associated Factors

Ignacio Aznar-Lou [1,2,*], Cristina Carbonell-Duacastella [1], Ana Rodriguez [3], Inés Mera [3] and Maria Rubio-Valera [1,2,4]

[1] Teaching, Research & Innovation Unit, Institut de Recerca Sant Joan de Déu, Esplugues de Llobregat, Parc Sanitari Sant Joan de Déu, 08830 Sant Boi de Llobregat (Barcelona), Spain
[2] Centro de Investigación Biomédica en Red en Epidemiología y Salud Pública, CIBERESP, 28029 Madrid, Spain
[3] Spanish Society of Community and Family Pharmacy (SEFAC), 28045 Madrid, Spain
[4] School of Pharmacy, University of Barcelona, 08028 Barcelona, Spain
* Correspondence: i.aznar@pssjd.org; Tel.: +34-936406350 (ext. 12543)

Received: 27 August 2019; Accepted: 5 October 2019; Published: 15 October 2019

Abstract: Introduction: The use of medication has increased in recent years in the US while the use of dietary supplements has remained stable but high. Interactions between these two kinds of products may have important consequences, especially in the case of widely used medications such as antihypertensives and antibiotics. The aim of this paper is to estimate the prevalence of potentially serious drug–dietary supplement interactions among tetracyclines, thiazides, and angiotensin II receptor blocker users by means of the NHANES 2013–2014 dataset. Methods: Data from 2013–2014 NHANES were obtained. Potential interactions analysed were tetracyclines with calcium, magnesium, and zinc, thiazides with vitamin D, and angiotensin II receptors blockers with potassium. Prevalence was calculated for each potential interaction. Logistic regression was used to assess associated factors. Results: 864 prescriptions issued to 820 patients were analysed. Overall prevalence of potential interaction was 49%. Older age and higher educational level were strongly associated with being at risk of a potential interaction. Factors such as age, race, civil status, citizenship, country of birth, BMI, and physical activity did not show notable associations. Conclusions: Healthcare professionals should be aware of other medical products when they prescribe or dispense a medication or a dietary supplement, especially to the older population and people with a higher educational level.

Keywords: prevalence; interactions; dietary supplements; antibiotics, antihypertensive medication

---

1. Introduction

The use of prescription medicines has increased recently in the United States (US) [1]. This general increase is not homogenous across all drug classes. While important increases are noted in the consumption of some prescription drugs such as anti-hypertensive agents (i.e., angiotensin II receptor blockers and thiazides), consumption of other drug classes, for instance oral antibiotics, has decreased. The use of dietary supplements is high and has remained stable over recent years in the general population in the US. The latest figures point to a prevalence in the use of dietary supplements of around 50% [2]. This use is higher in the older population, and among females, non-Hispanic whites, and people with a higher level of education.

One of the main health concerns related to the use of prescription medicines is the potential risk of adverse events and interactions. Interactions may occur between medications but also between medications and dietary supplements [3]. Qato et al. reported a range of prevalences of concomitant use of medication prescription and dietary supplements of between 0.2% and 2% of the general population [4]. Drug–drug interactions are usually checked by prescription and dispensing systems, which will prompt a warning message to the physician or pharmacist if a drug–drug interaction is detected. However, these systems have no information on patients' use of dietary supplements, as most of these supplements are obtained over the counter [3].

Interactions between prescription medication and dietary supplements may occur with widely used medications, such as antihypertensive agents and antibiotics [5,6]. The consequences of these interactions may vary between drugs and dietary supplements. Certain minerals including calcium, magnesium, and zinc may interact with tetracycline. The effects of these interactions could cause a reduction in tetracycline absorption (decreasing or eliminating the therapeutic effect) [3,5]. Other minerals, such as potassium, can cause interactions with anti-hypertensive drugs, including angiotensin II receptor blockers, provoking hyperkalaemia [5]. The use of thiazides has to be especially controlled in patients with osteoporosis because they could also interact with vitamin D and/or calcium causing hypercalcaemia and a potential metabolic alkalosis [5].

Since consumption of dietary supplements is rarely supervised by healthcare professionals, this information may help these professionals in their prescription and medication-counselling practice [3]. Specifically, although studies regarding consequences of prescription medication and dietary supplement interaction do exist, only a few of them have focused on American populations [4], revealing the lack of information. Thus, the aim of this paper is to estimate the prevalence of potentially serious drug–dietary supplement interactions among tetracyclines, thiazides, and angiotensin II receptor blocker users by means of the NHANES 2013–2014 dataset. The secondary aim of this paper is to outline the profile of a patient at risk of interactions between medication and dietary supplements. The NHANES survey is among the few large population-based nationally representative health and nutrition studies to apply standard design, and it includes the most detailed information regarding dietary habits and dietary supplement intake, as well as medication prescription. Due to the potential consequences of prescription medication and dietary supplement interaction, it is important to discern what patient profile is at the greatest risk of consuming dietary supplements that cause serious interactions. Thus, the information derived from our study will be important for effective US and international public health planning.

2. Methods

2.1. Study Design

We obtained data from the 2013–2014 National Health and Nutrition Examination Survey (NHANES). NHNAES is a nationally representative cross-sectional survey conducted by the National Center for Health Statistics (NCHS) and Centers for Disease Control and Prevention. NHANES includes information about the non-institutionalised US population. This database is used worldwide and has produced satisfactory results [6,7].

2.2. Ethics

The NCHS Research Ethics Review Board (ERB) approved the 2013–2014 study protocol (protocol 2011-17). NHANES and all participants provided written informed consent.

2.3. Interactions under Study

Table 1 shows the drug–dietary supplement interactions studied. We considered medication groups if they represented a prevalence of use higher than 2% of the population included in the database.

Table 1. Drug–dietary supplement interactions and their consequences.

Prescription Medicine	Dietary Supplement	Potential Clinical Consequences of the Interaction	Prevalence of Potential Interactions
Tetracyclines	Calcium	Decreased therapeutic effect as a consequence of a reduction in the absorption of tetracycline [3].	44.3%
	Magnesium	Decreased therapeutic effect as a consequence of a reduction in the absorption of tetracycline [3,5].	26.9%
	Zinc	Decreased therapeutic effect as a consequence of a reduction in the absorption of tetracycline [3,5].	37.4%
Thiazides	Calcium Vitamin D	Hypercalcemia and metabolic alkalosis. Thiazides reduce the urinary excretion of calcium. Vitamin D increases the absorption of calcium [5].	53.5% 52.1%
Angiotensin II receptor blockers	Potassium	Higher risk of hyperkalaemia, especially in patients with decreased renal function, heart failure, or diabetes [5].	28.8%

Tetracycline interacts with divalent ions such as calcium, magnesium, and zinc, forming a relatively stable and poorly absorbed chelate, preventing absorption of the antibiotic due to a lower amount of calcium in the gut available to be absorbed. This interaction may reduce or even abolish the therapeutic effect of the antibiotic, thereby diminishing anti-infectious efficiency. For this reason, tetracycline should be taken one hour before or two hours after meals [3,5].

Thiazide diuretics can cause increased calcium reabsorption in distal tubules of the kidneys, which contributes to hypercalcemia. Another cause of hypercalcemia is the excess of vitamin D, for example, through high doses of oral supplements, which increases the absorption of calcium in the gut. Due to the retention of calcium in the body, metabolic alkalosis may be developed [5].

Finally, angiotensin II receptor blockers are potassium-sparing and can, therefore, have additional hyperkalaemic effects if combined with potassium supplements or salt substitutes containing potassium. The use of potassium supplements is the main risk factor for developing hyperkalaemia, as this causes a rapid rate of increase in serum potassium levels. Other contributory risk factors such as poor renal function, heart failure, and diabetes should also be considered, as they are associated with a faster rate of hyperkalaemia progress [5,8].

3. Population and Prescription Medication Information

The sample was composed of tetracyclines, thiazides, and/or angiotensin II receptor blocker users. These drugs were chosen due to the potential severity of their interactions and their high prevalence of use in the American population. Prescription medication information was obtained through the Prescription Medication subsection included in The Dietary Supplement and Prescription Medication section of the Sample Person Questionnaire. This section provides personal information on the use of prescription medication in the month prior to the participant's interview. The name of the medication was provided by the participant to the interviewer, who entered it into the computer where it was automatically matched to a generic drug name and code. Medication is presented following the WHO Drug Statistics Methodology of ATC index [9].

Tetracyclines: We considered a patient to be a tetracycline consumer if he/she reported having taken a medication with a generic drug name included in the ATC group J01AA.

Thiazides: We considered a patient to be a thiazide consumer if he/she reported having taken a medication with a generic drug name included in the ATC group C03AA.

Angiotensin II receptor blockers: We considered a patient to be an angiotensin II receptor blocker consumer if he/she reported having taken a medication with a generic drug name included in the ATC group C09CA.

4. Dietary Supplement Information

Dietary supplement information was obtained through the dietary supplement subsection also included in The Dietary Supplement and Prescription Medication section of the Sample Person Questionnaire. This subsection allows for collection of personal data on the use of dietary supplements in the month prior to the participant's interview. Interviewers reported the supplement product name, which was automatically disaggregated to up to 34 nutrients.

We considered a participant to be a nutrient supplement consumer if he/she had taken any supplement containing at least one of the nutrients under study (Table 1).

5. Other Covariates

Demographic covariates were sex, age, race/ethnicity (Hispanic origin: Mexican-American, other Hispanic, non-Hispanic white, non-Hispanic black, other race), educational level (primary, secondary, university), civil status (married/with partner, widow/er, divorced, single), citizenship (American/non-American), and country of birth (U.S./Other country).

We also considered body mass index (BMI) and physical activity. BMI was determined from height and weight measured by health technicians previously trained by an expert anthropometrist. We categorised this variable as follows: underweight (<18.5), normal weight (18.5–24.9), overweight (25.0–29.9), and obesity type I (30.0–34.9) and obesity type II and type III (≥35.0).

Physical activity was self-reported and measured through a question asking if participants did any moderate-intensity sports in a typical week. The answer was dichotomic (yes/no).

6. Statistical Analysis

Prevalence rates were calculated for each potential interaction described in Table 1. The reference population consisted of those who reported having taken one of the medications under study.

A multivariate logistic regression analysis was conducted to determine the factors associated with a higher probability of having a potential interaction using the presence/absence of the interaction as the dependent variable and demographic and clinical variables as independent variables. The regression model provided odds ratios (OR) and 95% confidence intervals for the associations between the dependent variable and each of the independent variables. Crude associations obtained from bivariate logistic regressions were also presented.

To test potential interactions between demographic and clinical variables, such as age and physical exercise or gender and BMI, we tested the association between the probability of having a potential interaction and the interacting term of the dependent variables. The interacting terms were not statistically significant and, therefore, were not included in the final model.

All analyses were performed taking into account the appropriate weights. This procedure was developed with the goal of obtaining nationally representative estimates and accounting for unequal probability of selection derived from study design and non-response.

STATA 13.1MP was used to perform the statistical analyses.

7. Results

7.1. Sample Demographic Characteristics

The total sample was made up of 864 prescriptions issued to 820 individuals. Five per cent of those prescriptions involved tetracyclines, 61% involved thiazides, and 46% angiotensin II receptor blockers. Some of these prescriptions consisted of a combination of a thiazide and an angiotensin II inhibitor.

Sociodemographic characteristics are shown in Table 2. The proportion of women was 57%. The majority of the sample was non-Hispanic white (70%) and had secondary-level education (57%). Regarding civil status, 66% of the sample was defined as married or with a partner. Some 88% had been born in the U.S. and a higher percentage (97%) had American citizenship. The main BMI category was type II or III obesity (32%). Only 30% of the sample did moderate-intensity sport.

Table 2. Sociodemographic characteristics of the sample (n = 820).

	% or Mean	95% CI
Gender,% (n)		
Men	42.6 (350)	38.7; 46.7
Women	57.3 (470)	53.3; 61.3
Age, mean (range)	61.4 (20–80)	60.3; 62.6
Age group,% (n)		
20–39	6.4 (41)	4.6; 8.8
40–59	34.1 (249)	29.4; 39.2
60–79	50.0 (470)	45.6; 54.4
≥80	9.5 (104)	7.6; 11.9
Race,% (n)		
Mexican-American	3.9 (63)	2.0; 7.7
Other Hispanic	3.5 (62)	2.3; 5.5
Non-Hispanic white	70.4 (357)	65.5; 75.0
Non-Hispanic black	15.6 (237)	11.4; 21.0
Other race	6.4 (101)	4.4; 9.2
Education,% (n)		
Primary	15.7 (191)	11.9; 20.5
Secondary	57.4 (450)	53.6; 61.1
University	26.9 (179)	22.7; 31.7
Civil status *,% (n)		
Married/with partner	65.5 (482)	62.0; 68.9
Widow/er	12.9 (131)	10.4; 15.8
Divorced	14.4 (131)	12.3; 16.9
Single	7.1 (75)	5.2; 9.5
Citizenship *,% (n)		
American	88.3 (633)	84.4; 91.3
Non-American	11.7 (186)	8.7; 15.6
Country of birth *,% (n)		
U.S.	97.7 (779)	96.1; 98.6
Other country	2.3 (40)	1.4; 3.9
Body mass index (categories),% (n)		
Underweight	0.5 (5)	0.1; 2.2
Normal weight	10.6 (109)	7.6; 14.5
Overweight	30.7 (247)	26.4; 35.3
Obesity type I	26.5 (206)	22.2; 31.4
Obesity types II and III	31.6 (253)	28.3; 35.2
Physical activity in a typical week,% (n)		
Yes	40.4 (313)	35.9; 45.0
No	59.6 (507)	55.0; 64.1

* The following variables contain one missing value: civil status, citizenship and country of birth.

7.2. Prevalence of Potential Interactions

Table 1 shows the prevalence of potential interactions among the sample. Forty-four percent of the people using tetracyclines were consuming calcium and 26% and 37% of them were consuming magnesium and zinc, respectively. Among the users of thiazides, 54% and 52% used calcium and vitamin D, respectively. Finally, 26% of consumers of antagonist II receptor blockers presented a potential interaction due to the concomitant use of potassium. Overall, 49% of the participants were at risk of at least one of the studied interactions.

7.3. Factors Associated with Potential Interactions

Table 3 shows the factors associated with suffering a potential interaction between medications and dietary supplements. According to the adjusted analysis, age and educational level were strongly associated with the probability of a potential interaction. Also, compared to the non-Hispanic white population, the non-Hispanic black population presented a lower probability of a potential interaction.

The remaining variables (sex, civil status, citizenship, country of birth, BMI category, and physical activity) showed no statistically significant association with the risk of a potential interaction.

Table 3. Factors associated with potential interactions based on the multivariate weighted logistic regression model [$].

	Bivariate Analysis		Multivariate Analysis	
	OR	95% CI	OR	95% CI
Gender				
Men	ref		ref	-
Women	1.08	0.79; 1.49	1.23	0.87; 1.75
Age (1 year increase)	1.02	1.01; 1.03	1.02	1.01; 1.03
Race				
Non-Hispanic white	ref		ref	-
Other Hispanic	0.42	0.17; 1.03	0.44	0.14; 1.33
Mexican-American	0.44	0.24; 0.80	0.55	0.25; 1.20
Non-Hispanic black	0.42	0.29; 0.60	0.45	0.30; 0.66
Other race	0.58	0.83; 1.78	0.60	0.29; 1.26
Education				
Primary	ref		ref	-
Secondary	2.10	1.31; 3.40	1.95	1.18; 3.23
University	2.03	1.24; 3.31	1.63	1.01; 2.63
Civil status *				
Married/with partner	ref		ref	-
Widow/er	0.98	0.66; 1.46	0.84	0.50; 1.43
Divorced	0.92	0.65; 1.32	1.05	0.71; 1.56
Single	0.56	0.25; 1.26	0.95	0.39; 2.34
Citizenship *				
American	ref		ref	-
Non-American	0.40	0.16; 0.99	0.86	0.27; 2.72
Country of birth *				
U.S.	ref		ref	-
Other country	0.65	0.41; 1.02	1.12	0.57; 2.20
Body mass index (categories)				
Underweight *				
Normal weight	ref		ref	-
Overweight	0.90	0.53; 1.51	0.77	0.45; 1.29
Obesity type I	1.23	0.77; 1.97	1.00	0.59; 1.71
Obesity types II and III	1.17	0.83; 1.67	1.04	0.74; 1.48
Physical activity in a typical week				
Yes	ref		ref	-
No	0.69	0.41; 1.17	0.71	0.41; 1.24

[$] Potential interactions include the interaction of tetracycines with calcium, magnesium or zinc, thiazides with vitamin D and Angiotensin II receptor blockers with potassium; * Few values to be considered. CI = Confidence interval; OR = Odds ratio.

Older people had a higher risk of using a prescription medication and a dietary product with a potential interaction effect (OR = 1.02 (95% CI 1.01, 1.03)) per year, i.e., OR is 1.22 in patients 10 years older). People with a higher educational level (secondary or university) showed a higher risk of using a dietary product with a potential interaction with one prescription medicine (OR = 2.0 (95%CI 1.18; 3.23) and OR = 1.6 (95% CI 1.01; 2.63), respectively).

8. Discussion

One in every two people who take one of the considered medications is at risk of a potential interaction. Specifically, older people and the population with a higher educational level represent a profile at risk of a potential interaction between medications and nutritional supplements. Older people are also more likely to use both drugs and supplements because of a higher potential to get sick [10]. This is an important issue from a public health perspective, as there are population groups, such as the older population, who are high consumers of these two kinds of health products. This study is one of the first to examine factors associated with specific medication-dietary supplement interactions [11,12].

Qato et al. already showed that more than two-thirds of older adults used prescription medication with OTC medication or dietary supplements [4]. This is in line with our results indicating that the older population has greater probability of suffering a potential interaction. According to Kantor et al. supplement use in the US showed a downward trend among young adults aged 20 to 39 years, stable use among middle-aged adults aged 40–64 years, and an increase among adults over 65 years of age; this last population has a greater probability of being under pharmacotherapeutic treatment due to their clinical status [2]. In this population, the intake of both products is essential. Polypharmacy is a well-known phenomenon that is mainly observed in older populations [13]. The appropriateness of these medications is questioned in some cases [14]; however, in other situations such as antihypertensive or diuretic medication, the need is beyond doubt as cardiovascular illnesses are among the most important causes of death and disability in the US [15]. A similar scenario is observed with dietary supplements, where the effect of the supplementation may reduce the risk of several chronic diseases [16,17].

People with a primary level of education showed lower likelihood of being at risk of a potential interaction. This group also has a lower likelihood of taking dietary supplements. An explanation for this might be that people with a higher educational level might be over-concerned due to a flood of health information about health, and they might be taking dietary supplements when they do not need them. However, this result may also indicate a higher concern regarding the health of people with a higher education; higher education mediates the impact on health outcomes through health literacy [18]. Furthermore, people with a higher education level, which are likely to have a higher socioeconomic status, may have more resources to access dietary products.

These results are especially important for healthcare professionals. Determining whether patients at risk of suffering a potential interaction are taking a dietary supplement that might interact with their prescribed medication is important. In the case of tetracyclines, the effect of the drug may be decreased and the patient may be uncovered for a potential infection; in the case of thiazides, patients may suffer metabolic alkalosis; and in the case of angiotensin II receptor blockers, cardiac function may be affected. This information should be considered at the time of prescription or dispensation. A secondary assessment could be made to analyze whether this supplement is really needed, and if so, to try to adapt it to the pharmacologic treatment. In addition, the assistance of a specialist, such as a nutritionist, is recommendable.

In this line, public policies designed to inform healthcare professionals how to detect potential interactions and tools to help them identify them are highly recommended. These tools could be incorporated in the respective electronic tools of prescription and dispensing, and they might not only remind healthcare professionals to ask about the use of those dietary supplements that may generate an interaction, but also offer alternatives to avoid the interaction in the event that the supplement is recommended. In addition, it is important to inform citizens that, before taking a supplement, it is desirable to consult their doctors to verify that the supplement is necessary and safe taking into account their prescription drugs.

This study has several strengths. It is among the first to evaluate medication use with specific dietary supplements in a representative sample of the American population, providing useful information for targeted public health planning. In the NHANES, medication and dietary supplement use were assessed through in-home interviews, and boxes were seen by the interviewers in most participants. This reduces the recall bias, which is especially notable in medication [1]. However, the present study also has several limitations. First, there is no certainty that the interactions detected occurred. Furthermore, some of these interactions were dose-dependent, and information on dose was not available. Second, information only showed self-reported recent consumption, so it was impossible to discern whether the consumption was concurrent. Finally, other important medication groups that might generate serious interactions, such as quinolones, were not assessed due to their limited representation in the sample. These limitations are not restricted to the present study, as the methodology and analysis strategy followed were similar to those of previous studies based on the NHANES database [4].

9. Conclusions

There are two main population groups at risk of potential interactions: older people and the population with a higher educational level. With respect to other races, non-Hispanic whites present a higher risk of potential interactions. These results add important information about ways of approaching patients when they receive a prescription or ask for a medication. Health policy should take this information into account so as to inform healthcare professionals, and electronic tools should be developed or adapted to help in the re-assessment of their pharmacotherapeutic planning.

Author Contributions: Conceptualization, I.A.-L. and M.R.-V.; methodology, I.A.-L. and M.R.-V.; formal analysis, I.A.-L., M.R.-V. and C.C.-D. investigation; investigation, C.C.-D., I.M. and A.R.; writing-original draft, I.A.-L. and C.C.-D.; writing—review and editing, I.M., A.R., supervision M.R.-V.

Funding: This research received no external funding.

Acknowledgments: I.A.-L. and M.R.-V. have a research contract with the CIBERESP (CB16/02/00429) funded by the Instituto de Salud Carlos III, and CCD has a research contract with the Instituto de Salud Carlos III (PI15/00114). Both contracts are co-funded by the European regional development fund (ERDF). We are grateful to Stephen Kelly and Tom Yohannan for their contribution to editing the English of the article.

Conflicts of Interest: The authors declare no conflict of interest.

References

1. Kantor, E.D.; Rehm, C.D.; Haas, J.S.; Chan, A.T.; Giovannucci, E.L. Trends in prescription drug use among adults in the United States from 1999–2012. *JAMA* **2015**, *314*, 1818–1830. [CrossRef] [PubMed]
2. Kantor, E.D.; Rehm, C.D.; Du, M.; White, E.; Giovannucci, E.L. Trends in dietary supplement use among US adults from 1999–2012. *JAMA* **2016**, *316*, 1464–1474. [CrossRef] [PubMed]
3. Bushra, R.; Aslam, N.; Khan, A.Y. Food-drug interactions. *Oman Med. J.* **2011**, *26*, 77–83. [CrossRef] [PubMed]
4. Qato, D.M.; Wilder, J.; Schumm, L.P.; Gillet, V.; Alexander, G.C. Changes in prescription and over-the-counter medication and dietary supplement use among older adults in the United States, 2005 vs. 2011. *JAMA Intern. Med.* **2016**, *176*, 473–482. [CrossRef] [PubMed]
5. Baxter, K. *Stockley Interacciones Farmacológicas (Stockley's Drug Interactions)*, 3rd ed.; Pharma, E., Ed.; Pharmaceutical Press: London, UK, 2009.
6. Tyrovolas, S.; Koyanagi, A.; Kotsakis, G.A.; Panagiotakos, D.; Shivappa, N.; Wirth, M.D.; Hébert, J.R.; Haro, J.M. Dietary inflammatory potential is linked to cardiovascular disease risk burden in the US adult population. *Int. J. Cardiol.* **2017**, *240*, 409–413. [CrossRef] [PubMed]
7. Malek, A.M.; Newman, J.C.; Hunt, K.J.; Marriott, B.P. Race/Ethnicity, Enrichment/Fortification, and Dietary Supplementation in the U.S. Population, NHANES 2009–2012. *Nutrients* **2019**, *11*, 1005. [CrossRef] [PubMed]
8. Indermitte, J.; Burkolter, S.; Drewe, J.; Krähenbühl, S.; Hersberger, K.E. Risk factors associated with a high velocity of the development of hyperkalaemia in hospitalised patients. *Drug Saf.* **2007**, *30*, 71–80. [CrossRef] [PubMed]
9. Norwegian Insitute of Public Health WHO Collaborating Centre for Drug Statistics Methodology. International Language for Drug Utilization Research. Available online: https://www.whocc.no/ (accessed on 27 August 2019).
10. Kuerbis, A.; Sacco, P.; Blazer, D.G.; Moore, A.A. Substance abuse among older adults. *Clin. Geriatr. Med.* **2014**, *30*, 629–654. [CrossRef] [PubMed]
11. Nahin, R.L.; Pecha, M.; Welmerink, D.B.; Sink, K.; Dekosky, S.T.; Fitzpatrick, A.L. Concomitant use of prescription drugs and dietary supplements in ambulatory elderly people: Clinical investigations. *J. Am. Geriatr. Soc.* **2009**, *57*, 1197–1205. [CrossRef] [PubMed]
12. Qato, D.M.; Alexander, G.C.; Conti, R.M.; Johnson, M.; Schumm, P.; Lindau, S.T. Use of prescription and over-the-counter medications and dietary supplements among older adults in the United States. *JAMA J. Am. Med. Assoc.* **2008**, *300*, 2867–2878.

13. Oktora, M.P.; Denig, P.; Bos, J.H.J.; Schuiling-Veninga, C.C.M.; Hak, E. Trends in polypharmacy and dispensed drugs among adults in the Netherlands as compared to the United States. *PLoS ONE* **2019**, *14*, e0214240. [CrossRef] [PubMed]

14. Vozoris, N.T. Benzodiazepine and opioid co-usage in the US population, 1999–2014: An exploratory analysis. *Sleep* **2019**, *42*, zsy264. [CrossRef] [PubMed]

15. Dwyer-Lindgren, L.; Bertozzi-Villa, A.; Stubbs, R.W.; Morozoff, C.; Kutz, M.J.; Huynh, C.; Barber, R.M.; Shackelford, K.A.; Mackenbach, J.P.; van Lenthe, F.J.; et al. US county-level trends in mortality rates for major causes of death, 1980–2014. *JAMA J. Am. Med. Assoc.* **2016**, *316*, 2385–2401. [CrossRef] [PubMed]

16. Parva, N.R.; Tadepalli, S.; Singh, P.; Qian, A.; Joshi, R.; Kandala, H.; Nookala, V.K.; Cheriyath, P. Prevalence of Vitamin D Deficiency and Associated Risk Factors in the US Population (2011–2012). *Cureus* **2018**, *10*, e2741. [CrossRef] [PubMed]

17. McKay, D.L.; Perrone, G.; Rasmussen, H.; Dallal, G.; Hartman, W.; Cao, G.; Prior, R.L.; Roubenoff, R.; Blumberg, J.B. The Effects of a Multivitamin/Mineral Supplement on Micronutrient Status, Antioxidant Capacity and Cytokine Production in Healthy Older Adults Consuming a Fortified Diet. *J. Am. Coll. Nutr.* **2000**, *19*, 613–621. [CrossRef] [PubMed]

18. Van Der Heide, I.; Wang, J.; Droomers, M.; Spreeuwenberg, P.; Rademakers, J.; Uiters, E. The relationship between health, education, and health literacy: Results from the dutch adult literacy and life skills survey. *J. Health Commun.* **2013**, *18*, 172–184. [CrossRef] [PubMed]

 *nutrients* 

Article

The Effect of Maternal Obesity on Breast Milk Fatty Acids and Its Association with Infant Growth and Cognition—The PREOBE Follow-Up

Andrea de la Garza Puentes [1,2,3,*], Adrià Martí Alemany [1], Aida Maribel Chisaguano [4], Rosa Montes Goyanes [5], Ana I. Castellote [1,2,6], Franscisco J. Torres-Espínola [7,8], Luz García-Valdés [7,8], Mireia Escudero-Marín [7,8], Maria Teresa Segura [7,8], Cristina Campoy [7,8,9] and M. Carmen López-Sabater [1,2,6,*]

[1]   Department of Nutrition, Food Sciences and Gastronomy, Faculty of Pharmacy and Food Sciences, University of Barcelona, 08028 Barcelona, Spain
[2]   Institut de Recerca en Nutrició i Seguretat Alimentària UB (INSA-UB), 08921 Barcelona, Spain
[3]   Teaching, Research & Innovation Unit, Parc Sanitari Sant Joan de Déu, 08830 Sant Boi, Spain
[4]   Nutrition, Faculty of Health Sciences, University of San Francisco de Quito, 170157 Quito, Ecuador
[5]   Food Research and Analysis Institute, University of Santiago de Compostela, 15705 Santiago de Compostela, Spain
[6]   CIBER Physiopathology of Obesity and Nutrition CIBERobn, Institute of Health Carlos III, 28029 Madrid, Spain
[7]   Centre of Excellence for Paediatric Research EURISTIKOS, University of Granada, 18071 Granada, Spain
[8]   Department of Paediatrics, University of Granada, 18071 Granada, Spain
[9]   CIBER Epidemiology and Public Health CIBEResp, Institute of Health Carlos III, 28029 Madrid, Spain
*   Correspondence: adelagarza@ub.edu (A.d.l.G.P.); mclopez@ub.edu (M.C.L.-S.); Tel.: +39-934-024-512 (A.d.l.G.P. & M.C.L.-S.)

Received: 6 August 2019; Accepted: 24 August 2019; Published: 9 September 2019

Abstract: This study analyzed how maternal obesity affected fatty acids (FAs) in breast milk and their association with infant growth and cognition to raise awareness about the programming effect of maternal health and to promote a healthy prenatal weight. Mother–child pairs ($n = 78$) were grouped per maternal pre-pregnancy body mass index (BMI): normal-weight (BMI = 18.5–24.99), overweight (BMI = 25–29.99) and obese (BMI > 30). Colostrum and mature milk FAs were determined. Infant anthropometry at 6, 18 and 36 months of age and cognition at 18 were analyzed. Mature milk exhibited lower arachidonic acid (AA) and docosahexaenoic acid (DHA), among others, than colostrum. Breast milk of non-normal weight mothers presented increased saturated FAs and n6:n3 ratio and decreased α-linolenic acid (ALA), DHA and monounsaturated FAs. Infant BMI-for-age at 6 months of age was inversely associated with colostrum n6 (e.g., AA) and n3 (e.g., DHA) FAs and positively associated with n6:n3 ratio. Depending on the maternal weight, infant cognition was positively influenced by breast milk linoleic acid, n6 PUFAs, ALA, DHA and n3 LC-PUFAs, and negatively affected by n6:n3 ratio. In conclusion, this study shows that maternal pre-pregnancy BMI can influence breast milk FAs and infant growth and cognition, endorsing the importance of a healthy weight in future generations.

Keywords: maternal obesity; breastfeeding; breast milk; colostrum; mature milk; fatty acids; LC-PUFA; omega-3; omega-6; DHA; AA; children; growth; cognition; early life nutrition; programming

---

1. Introduction

In spite of efforts made, as well as existing evidence-based information, for tackling obesity and the burden of the disease, obesity is a societal challenge that is still on the rise, including in women of reproductive age, and this is affecting the health of future generations [1]. Early-life nutrition plays a key role in infant growth and development and has a programming effect related to the appearance of future non-communicable diseases, such as obesity, diabetes and others [2,3]. Breast milk composition and breastfeeding practice are some of the most influential factors of child outcomes [4–6]. Even though lactation comprises a relatively short period in the average person's lifespan, the exposure to breast milk in the first months of life occurs during a very critical period of rapid growth and development [2,7–9]. Maternal obesity influences the nutritional status of the child through different mechanisms, breastfeeding being one of them. If the mother of the child has obesity, the fatty acid (FA) profile in breast milk can be different, with a prevalence of pro-inflammatory FAs beyond those critical for neurodevelopment [10]. Thus, the early nutritional status and future health of the child can be affected.

Breast milk contains long-chain (LC) polyunsaturated fatty acids (PUFAs), which are crucial nutrients—especially docosahexaenoic (DHA) and arachidonic acid (AA)—involved in growth, the immune system, vision, and cognitive and motor development [11]. These nutrients are associated with the prevention of obesity [12,13] and other infectious and chronic diseases in the future life [14]. However, maternal characteristics, such as diet [15] or obesity [10], may alter the FA content in human milk. Studies have shown that the breast milk of mothers with overweight and obesity have higher levels of n6 FAs and lower levels of n3 FAs than the breast milk of normal-weight mothers [16–18], and a high ratio of n6:n3 LC-PUFAs in red blood cells membrane phospholipids has been reported as a risk factor for obesity [19]. In fact, in high-fat rodent models of maternal obesity, lowering the maternal n6:n3 ratio using a novel genetic model or supplemental fish oil has been shown to prevent offspring obesity [20]. Nevertheless, the results appear to be inconsistent [18,21].

The direct impact of maternal weight on the infant cognition has also been studied [21–24]. Mostly, observational, prospective and longitudinal studies correlate a high pre-pregnancy maternal body mass index (BMI) with poorer cognitive performance [24]. High gestational weight gain (GWG) seems to augment this correlation, as well [25]. However, three studies have failed to find an association between maternal obesity and cognitive infant deficits [26–28].

Although there are studies that have analyzed the influence of maternal weight on breast milk FA composition [10,18,29–35], none of these studies have further assessed its effect on infant cognition and growth. Furthermore, there is a lot of variability regarding the timing of breastmilk collection in the existing studies, and most of them focus on the analysis of mature breastmilk, without considering the evolution of the different FAs from colostrum to mature milk. Therefore, the current study aims to analyze the implications of maternal obesity on FA levels in colostrum and mature milk and their association with infant growth and cognition, to raise awareness about the programming effect of maternal nutrition and promote a healthy weight in women.

2. Materials and Methods

2.1. Statement of Ethics

This study was carried out in accordance with the ethical standards recognized by the Declaration of Helsinki (2004), the EEC Good Clinical Practice guidelines (document 111/3976/88 of July 1990) and current Spanish legislation governing clinical research in humans (Royal Decree 561/1993 on clinical trials). Additionally, the study was approved by San Cecilio University Hospital Ethics Committee and the Faculty of Medicine at the University of Granada. Written informed consent was obtained from all participants at the beginning of the study.

2.2. Study Population and Design

For the present study, a subsample of mother–child pairs ($n = 78$) from the PREOBE cohort was selected and classified according to maternal pre-pregnancy BMI: normal-weight (BMI = 18.5–24.99 Kg/m^2, $n = 34$), overweight (BMI = 25–29.99 Kg/m^2, $n = 27$) and obese (BMI > 30 Kg/m^2, $n = 17$).

The PREOBE study (Role of Nutrition and Maternal Genetics on the Programming of Development of Fetal Adipose Tissue) is an observational cohort study of a total of 331 pregnant women that analyzes the impact of maternal obesity and gestational diabetes. The information regarding the PREOBE study has been published elsewhere [34] and was registered at www.ClinicalTrials.gov (NCT01634464). Figure 1 presents the study design and information of the PREOBE study.

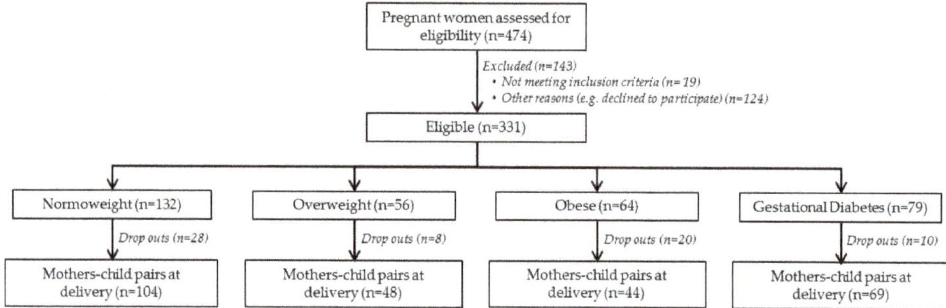

Figure 1. Participants in the PREOBE cohort and classification following BMI and gestational diabetes criteria.

Briefly, the study and recruitment of participants were carried out at San Cecilio University Hospital and the Mother-Infant Hospital in the city of Granada, Spain. The inclusion criteria were: singleton pregnancy, gestation between 12 and 20 weeks at enrollment, and an intention to deliver in one of the two obstetrics centers mentioned above. Women were excluded if they were participating in other research studies, receiving drug treatment or supplements of DHA or folate for more than the first three months of pregnancy, suffering from disorders such as hypertension, pre-eclampsia, fetal intrauterine growth retardation, infections, hypo- or hyperthyroidism and hepatic renal diseases, or following an unusual or vegan diet. Maternal age, pre-pregnancy BMI, parity, smoking status, diet, alcohol habits, socio-demographic information, education, gestational weight gain, infant anthropometry, gender and feeding practices were recorded. After birth, the women were encouraged to breastfeed their infants.

2.3. Breast Milk Sample Collection

Colostrum and mature milk were collected at 2–4 and 28–32 days postpartum, respectively, by an experienced nurse at the hospitals or by the mother at home (after receiving training by the nurse). Samples were collected over the course of an entire day (24 h) from both breasts before and after each feed. Milk samples were gathered in sterile polypropylene tubes by mechanically expressing each breast with a breast pump. Mothers were given 14 tubes with a capacity of 5 mL and the total volume obtained from each mother ranged from 45 to 70 mL. The samples collected at each time were frozen at −20 °C at home, and mothers brought them to the 3-month offspring follow-up visit. Each time, the samples were transported in ice boxes to the laboratory, where they were stored at −80 °C until analysis. All samples from each woman were mixed and aliquoted prior to analysis.

2.4. Fatty Acid Analysis of Breast Milk

The FA composition of breast milk was determined according to the method described by Chisaguano et al. [35]. 50 µL human milk samples were used for the analysis. FA methyl esters (FAMEs) were prepared with sodium methylate in methanol (0.5 M) and boron trifluoride methanol solution (14% v/v). They were then separated and quantified by fast gas chromatography (GC)using a HP-6890 Series GC System (Hewlett-Packard, Waldbronn, Germany) equipped with a flame ionization detector (FID), a split/splitless injector, a HP-7683B Series autoinjector, and a fused-silica SP-2560 capillary column (75 m 0.18 mm internal diameter, 0.14 µm thickness) coated with a 100% bis-cyanopropyl polysiloxane stationary phase (Supelco, Saunderton, UK). The chromatographic conditions used were: hydrogen as the carrier gas at a constant linear velocity of 22 cm/s (which gave an initial pressure of 39 psi). The detector and injector temperatures were set at 300 °C and 250 °C, respectively; the split ratio was at 1:50 and the injection volume was 1 µL. Oven temperatures were programmed as follows: the initial temperature was set at 120 °C, which was increased at a rate of 25 °C min^{-1} to 180 °C. This temperature was held for 6 min and finally increased to 240 °C at a rate of 25 °C min^{-1}, and held for 9 min.

FAs were identified by a comparison of the peak retention times of those of the standard solution Supelco 37-component FAME mix (Sigma-Aldrich, St. Louis, MO, USA). FAs were then quantified by standard normalization (% total fatty acids), and they are therefore expressed as a percentage of the total amount of FAs. FA summatories were derived by adding the corresponding single FAs to saturated FAs (SFAs), monounsaturated FAs (MUFAs), PUFAs, n6 PUFAs, n3 PUFAs, n6 LC-PUFAs and n3 LC-PUFAs. Moreover, n6 to n3 ratios were created for analysis.

2.5. Assessment of Anthropometric Infant Outcomes

After birth, the infants received a medical examination during which anthropometric measurements were recorded. Data, such as weight, length and BMI at 6, 18 and 36 months of age were included in the present study. Length and weight (with light clothing and no shoes) were recorded using a Harpenden Infantometer (Model 702) calibrated stadiometer (Holtain, Wales, United Kingdom) and a Multina Comfort calibrated balance scale (SOEHNLE, Backnang, Germany), respectively. Weight, length and BMI measurements were ultimately converted to weight-for-age z-scores (WAZ), length-for-age z-scores (LAZ) and BMI-for-age z-scores (BMIZ) (SD scores), according to World Health Organization (WHO) child growth standards [36,37].

2.6. Assessment of Infant Cognitive Development

Infant cognitive development was assessed at 18 months of age using the Bayley Scales of Infant Development III (BSID III) [38], by trained psychologists in the presence of the mother of the child. These scales measure the level of motor, language and cognitive or mental development. The present study uses the Cognitive Composite score, which is the global score of the scales and represents the overall cognitive development of the children.

2.7. Statistical Analysis

Statistical analyses were performed using the SPSS statistical software package for Windows (version 23.0; SPSS Inc., Chicago, IL, USA). The Kolmogorov-Smirnov test was used to study the normal distribution of the data and non-normally distributed data were natural log-transformed. Means and standard deviations (SD) were used to describe continuous variables. The characteristics of the population were analyzed using the ANOVA and Bonferroni post-hoc test. To analyze the FA evolution from colostrum to mature milk, a paired Student's *t*-test was used. The independent Student's *t*-test was used to compare the breast milk FA composition between maternal weight groups. The associations between breast milk FAs and child anthropometric measurements and cognitive scores were determined using linear regression analyses and corrected for potential confounders such as maternal BMI, smoking, education, GWG and parity, and infant characteristics, such as gender and feeding practices. The Bonferroni correction (0.05/(48 FAs × 3 study groups = 144 analyses)) was applied to take multiple testing into account and *p*-value thresholds were set at 0.002. In the tables, *p*-values ≤ 0.05 are highlighted in bold, while those ≤0.002 are additionally marked by stars.

3. Results

3.1. Characteristics of the Population

The characteristics of the population are shown in Table 1. Normal-weight women presented the highest GWG, followed by overweight and finally mothers with obesity. The latter group had the highest n6:n3 ratio in dietary intake, while normal-weight mothers had the lowest intake of AA. No significant differences were found in infant characteristics according to maternal BMI.

Table 1. Characteristics of the population.

Characteristic	Normal-Weight		Overweight		Obesity		p
		Mean (SD)		Mean (SD)		Mean (SD)	
Maternal characteristics	n		n		n		
Age (years)	34	31 (4)	27	32 (4)	17	32 (4)	0.492
Pre-pregnancy BMI (kg/m^2)	34	22.14 (1.54) [a]	27	27.59 (1.35) [b]	17	33.40 (2.65) [c]	**<0.001 ***
Weight Gain (kg)	25	13.17 (3.55)	23	10.32 (5.20)	15	9.14 (7.06)	**0.042**
Education (%)							0.660
<High school	26	14.71	19	11.11	11	23.53	
High school	3	8.82	5	18.52	2	11.76	
>High school	5	76.47	3	70.37	4	64.71	
Smoking during pregnancy (%)							0.415
No, never	17	77.27	15	71.43	10	73.68	
Yes	3	13.64	5	23.81	1	7.14	
Quit	2	9.09	1	4.76	3	21.43	
Maternal dietary intake							
Energy (Kcal/day)	27	2066.37 (261.93)	23	2089.59 (542.07)	12	2058.08 (469.97)	0.961
Lipids (g)	27	86.89 (17.29)	23	85.25 (25.86)	12	93.23 (20.95)	0.468
Lipids (%)	27	37.76 (5.26)	23	39.04 (7.73)	12	41.17 (5.38)	0.307
SFA(g/d)	27	30.81 (6.36)	23	30.05 (8.67)	12	34.19 (5.87)	0.203
MUFA (g/d)	27	36.53 (10.92)	23	39.33 (17.63)	12	36.32 (12.08)	0.926
PUFA(g/d)	27	12.10 (3.17)	23	13.34 (6.59)	12	14.63 (3.95)	0.285
n6 PUFA (g/d)	27	2.48 (1.95)	23	2.89 (2.53)	12	3.44 (1.50)	0.122
n3 PUFA (g/d)	27	0.18 (0.11)	23	0.21 (0.13)	12	0.20 (0.09)	0.427
n-3 from fish (g/d)	27	0.36 (0.31)	23	0.28 (0.34)	12	0.45 (0.34)	0.690
AA (g/d)	27	0.11 (0.06) [a]	23	0.17 (0.08) [b]	12	0.16 (0.08) [ab]	**0.005**
EPA(g/d)	27	0.12 (0.11)	23	0.09 (0.11)	12	0.16 (0.12)	0.213
DHA (g/d)	27	0.24 (0.18)	23	0.22 (0.21)	12	0.31 (0.21)	0.269
n6:n3	27	12.99 (2.98) [a]	23	13.53 (3.40) [a]	12	19.12 (8.94) [b]	**0.004**
Infant characteristics							
Sex, male (%)	14	41.18	11	40.74	7	41.18	0.999
Birth weight (g)	32	3359.06 (352.35)	27	3340.37 (511.85)	16	3532.35 (389.61)	0.277
Birth length (cm)	31	50.52 (1.57)	27	50.30 (1.88)	16	51.22 (1.80)	0.245
Birth head Circumference (cm)	26	34.31 (1.36)	22	34.36 (1.39)	16	34.69 (1.40)	0.674
Placenta (g)	30	496.67 (144.11)	25	509.20 (130.25)	16	568.13 (143.17)	0.332
Newborn according Lubchenco curves [#] (%)							0.627
SGA	0	0.00	1	4.55	0	0.00	
AGA	26	81.25	16	72.73	11	73.33	
LGA	6	18.75	5	22.73	4	26.67	
Breastfeeding [†] (%)							0.290
Exclusive	16	53.33	16	66.67	8	50.00	
Mixt	10	33.33	3	12.50	3	18.75	
Artificial	4	13.33	5	20.83	5	31.25	

Different superscript letters indicate differences among BMI groups, according to ANOVA and the Bonferroni post-hoc test. Chi-square test was applied to qualitative variables. p-values ≤ 0.05 are highlighted in bold and those ≤0.002 are additionally marked by stars. [#] The newborns were divided into three groups according to the Lubchenco curves: SGA: Small for Gestational Age; AGA: Appropriate for Gestational Age; LGA: Large for Gestational Age (LGA). [†] Breastfeeding practice information was collected at 3 months of age of the child. SFA: Saturated Fatty Acids; MUFA: Monounsaturated Fatty Acids; PUFA: Polyunsaturated Fatty Acids; AA: Arachidonic Acid; EPA: Eicosapentaenoic Acid; DHA: Docosahexaenoic Acid.

3.2. Breast Milk Fatty Acid Evolution

The FA evolution from colostrum to mature milk is shown in Table 2. In spite of maternal pre-pregnancy BMI, mature breast milk presented lower levels of C16:1n9, C20:1n9, AA, C22:1n9, C22:4n6, C22:5n6, C22:5n3, DHA, C24:0, C24:1n6 and n3 LC-PUFAs, and higher levels of C8:0, C10:0, medium-chain FAs (MCFAs), eicosapentaenoic acid (EPA):AA and DHA:AA ratios than those found in colostrum.

Table 2. Human milk fatty acids profile according to maternal pre-pregnancy BMI.

	Normal-Weight			Overweight			Obesity		
	Colostrum (n = 26)	Mature Milk (n = 20)	p	Colostrum (n = 21)	Mature Milk (n = 23)	p	Colostrum (n = 16)	Mature Milk (n = 14)	p
	Mean (SD)	Mean (SD)		Mean (SD)	Mean (SD)		Mean (SD)	Mean (SD)	
C6:0	0.06 (0.04)	0.08 (0.03)	0.19	0.05 (0.03)	0.10 (0.04)	<0.001 *	0.05 (0.03)	0.08 (0.07)	0.002 *
C8:0	0.13 (0.09)	0.23 (0.07)	0.002 *	0.13 (0.08)	0.24 (0.09)	<0.001 *	0.08 (0.03) [#]	0.24 (0.10)	<0.001 *
C10:0	0.86 (0.57)	1.29 (0.37)	0.010	0.86 (0.47)	1.36 (0.33)	0.002 *	0.70 (0.57)	1.47 (0.32)	<0.001 *
C12:0	4.10 (2.07)	4.86 (1.80)	0.039	4.18 (1.97)	4.86 (1.92)	0.14	3.67 (2.49)	5.32 (1.65)	0.029
C14:0	5.15 (1.41)	4.79 (1.71)	0.88	5.38 (1.92)	4.67 (1.66)	0.38	5.04 (2.31)	5.23 (1.22)	0.32
C14:1	0.08 (0.02)	0.10 (0.03)	0.20	0.11 (0.03) [†]	0.11 (0.04)	0.51	0.10 (0.05)	0.12 (0.06)	0.10
C15:0	0.21 (0.03)	0.21 (0.06)	0.80	0.24 (0.04) [†]	0.20 (0.05)	0.016	0.19 (0.04) [#]	0.22 (0.08)	0.32
C16:0	21.13 (2.32)	19.56 (2.29)	0.006	21.24 (1.46)	19.53 (2.02)	0.001 *	21.96 (2.11)	21.20 (2.55) [#]	0.42
C16:1t	0.12 (0.03)	0.13 (0.04)	0.59	0.14 (0.04)	0.11 (0.05)	0.06	0.13 (0.04)	0.13 (0.06)	0.98
C16:1n9	0.50 (0.05)	0.43 (0.06)	<0.001 *	0.54 (0.13)	0.45 (0.07)	0.010	0.54 (0.10)	0.46 (0.10)	0.024
C16:1n7	1.48 (0.45)	1.83 (0.62)	0.11	1.63 (0.36)	1.80 (0.49)	0.014	1.78 (0.60)	1.88 (0.58)	0.61
C17:0	0.30 (0.04)	0.29 (0.08)	0.74	0.34 (0.03) [†]	0.29 (0.06)	0.026	0.29 (0.05) [#]	0.32 (0.07)	0.28
C17:1	0.15 (0.03)	0.17 (0.03)	0.28	0.18 (0.04) [†]	0.18 (0.04)	0.88	0.17 (0.07)	0.17 (0.05)	0.83
C18:0	5.74 (0.67)	5.74 (0.41)	0.20	5.58 (0.85)	5.88 (0.67)	0.12	5.44 (0.81) [†]	6.03 (0.49)	<0.001 *
C18:1n9	38.47 (3.51)	39.69 (4.23)	0.41	38.16 (4.57)	38.73 (5.53)	0.87	37.36 (3.11)	36.63 (0.96) [†]	0.30
C18:1n9t	0.33 (0.14)	0.28 (0.09)	0.12	0.31 (0.09)	0.26 (0.08)	0.24	0.24 (0.05) [†#]	0.22 (0.04) [†]	0.15
C18:1n7	1.69 (0.22)	1.58 (0.24)	0.02	1.69 (0.28)	1.63 (0.23)	0.81	1.89 (0.42)	1.59 (0.32)	0.022
C18:1n-7t11	0.28 (0.12)	0.30 (0.12)	0.95	0.28 (0.09)	0.26 (0.11)	0.15	0.27 (0.14)	0.24 (0.08)	0.72
C18:2n6 (LA)	13.03 (2.42)	13.60 (3.21)	0.31	13.31 (3.20)	15.20 (3.86)	0.050	12.98 (2.74)	13.90 (3.14)	0.49
C18:2n7c9t11	0.12 (0.03)	0.13 (0.05)	0.69	0.14 (0.04)	0.12 (0.04)	0.32	0.13 (0.05)	0.14 (0.06)	0.67
C18:3n6	0.09 (0.04)	0.17 (0.06)	0.010	0.11 (0.07)	0.18 (0.05)	<0.001 *	0.13 (0.09)	0.17 (0.05)	0.15
C18:3n3 (ALA)	0.54 (0.18)	0.59 (0.21)	0.61	0.53 (0.13)	0.58 (0.16)	0.14	0.41 (0.04) [†#]	0.46 (0.08) [†#]	0.16
C20:0	0.20 (0.03)	0.18 (0.01)	0.012	0.18 (0.02)	0.17 (0.03)	0.005	0.19 (0.04)	0.18 (0.04)	0.75
C20:1n9	0.77 (0.20)	0.50 (0.07)	<0.001 *	0.74 (0.26)	0.46 (0.05)	<0.001 *	0.77 (0.24)	0.48 (0.06)	<0.001 *
C20:3n6	0.60 (0.15)	0.47 (0.11)	<0.001 *	0.65 (0.21)	0.48 (0.06)	0.004	0.68 (0.24)	0.52 (0.14)	0.05
C20:4n6 (AA)	0.67 (0.19)	0.49 (0.05)	0.005	0.66 (0.13)	0.49 (0.12)	<0.001 *	0.67 (0.23)	0.47 (0.10)	0.004
C20:5n3 (EPA)	0.05 (0.02)	0.07 (0.03)	0.07	0.04 (0.02)	0.06 (0.02)	<0.001 *	0.05 (0.03)	0.07 (0.02)	0.10

Table 2. *Cont.*

	Normal-Weight			Overweight			Obesity		
	Colostrum (n = 26)	Mature Milk (n = 20)	p	Colostrum (n = 21)	Mature Milk (n = 23)	p	Colostrum (n = 16)	Mature Milk (n = 14)	p
	Mean (SD)	Mean (SD)		Mean (SD)	Mean (SD)		Mean (SD)	Mean (SD)	
C22:0	0.08 (0.03)	0.06 (0.02)	**0.007**	0.09 (0.03)	0.08 (0.03)	0.20	0.08 (0.02)	0.07 (0.02)	**0.039**
C22:1n9	0.18 (0.06)	0.09 (0.02)	**<0.001 ***	0.19 (0.08)	0.09 (0.02)	**<0.001 ***	0.18 (0.06)	0.09 (0.01)	**<0.001 ***
C22:2n6	0.07 (0.03)	0.04 (0.01)	**0.003**	0.08 (0.03)	0.06 (0.03)	0.22	0.06 (0.02)	**0.06 (0.02)** [†]	0.22
C22:4n6	0.25 (0.13)	0.10 (0.02)	**<0.001 ***	0.28 (0.15)	0.12 (0.04)	**<0.001 ***	0.37 (0.27)	0.12 (0.03)	**<0.001 ***
C22:5n6	0.12 (0.04)	0.05 (0.01)	**<0.001 ***	0.12 (0.05)	**0.07 (0.03)** [†]	**0.048**	0.12 (0.05)	**0.09 (0.05)** [†]	**0.040**
C22:5n3	0.16 (0.08)	0.11 (0.03)	**0.012**	0.14 (0.05)	0.11 (0.03)	**0.002 ***	0.19 (0.09)	0.12 (0.04)	**0.007**
C22:6n3 (DHA)	0.41 (0.14)	0.28 (0.11)	**<0.001 ***	0.35 (0.08)	**0.22 (0.06** [†]	**<0.001 ***	0.39 (0.13)	0.25 (0.07)	**0.002 ***
C23:0	0.12 (0.05)	0.05 (0.02)	**<0.001 ***	0.12 (0.05)	0.06 (0.03)	**<0.001 ***	0.13 (0.05)	0.09 (0.07) [†]	0.06
C24:0	0.10 (0.03)	0.05 (0.02)	**0.002 ***	0.10 (0.04)	0.06 (0.03)	**0.007**	0.09 (0.03)	0.06 (0.03)	**0.031**
C24:1	0.20 (0.11)	0.06 (0.02)	**<0.001 ***	0.19 (0.13)	0.06 (0.04)	**<0.001 ***	0.19 (0.09)	0.07 (0.03)	**<0.001 ***
EPA:AA	0.07 (0.03)	0.12 (0.03)	**0.005**	0.06 (0.03)	**0.09 (0.04)** [†]	**<0.001 ***	0.07 (0.03)	0.12 (0.07)	**0.004**
DHA:AA	0.63 (0.22)	0.77 (0.19)	**0.005**	0.55 (0.13)	0.76 (0.24)	**<0.001 ***	0.61 (0.21)	0.90 (0.34)	**0.004**
SFA	27.45 (2.66)	25.83 (2.48)	**0.008**	27.43 (1.93)	25.95 (2.49)	**0.004**	27.96 (2.34)	**27.80 (2.62)** [†,#]	0.96
MCFA	5.14 (2.69)	6.46 (2.19)	**0.021**	5.23 (2.46)	6.56 (2.27)	**0.037**	4.50 (3.06)	7.11 (1.99)	**0.010**
MUFA	42.80 (3.51)	43.73 (4.17)	0.87	42.61 (4.91)	42.76 (5.70)	0.74	42.17 (3.56)	**40.74 (1.37)** [†]	0.12
n6 PUFA	15.41 (2.53)	15.24 (3.31)	0.77	15.79 (3.23)	16.89 (3.93)	0.29	15.67 (3.01)	15.64 (3.19)	0.93
n3 PUFA	1.16 (0.30)	1.05 (0.24)	**0.018**	1.07 (0.15)	0.96 (0.21)	0.12	1.04 (0.23)	0.90 (0.13)	0.06
n6 LC-PUFA	2.30 (0.59)	1.48 (0.22)	**<0.001 ***	2.36 (0.63)	1.52 (0.24)	**<0.001 ***	2.56 (0.93)	1.58 (0.34)	**<0.001 ***
n3 LC-PUFA	0.62 (0.22)	0.45 (0.16)	**<0.001 ***	0.54 (0.12)	0.38 (0.11)	**<0.001 ***	0.63 (0.23)	0.44 (0.09)	**0.008**
n6:n3 PUFA	14.28 (4.76)	15.08 (3.91)	0.06	15.13 (4.37)	**18.28 (5.33)** [†]	**0.033**	15.67 (4.22)	17.80 (4.97)	0.15
n6:n3 LC-PUFA	4.00 (1.16)	3.62 (1.18)	0.08	4.49 (0.95)	4.28 (1.24)	0.41	4.23 (1.05)	3.73 (0.92)	0.14

Means are expressed as percentages of total FAs. The *p*-value shown refers to the breast milk evolution within each group of weight, according to Student's paired *t*-test. *p*-values ≤ 0.05 are highlighted in bold and those ≤0.002 are additionally marked by stars. [†] Indicates differences ($p \leq 0.05$) compared with normal-weight group and its corresponding breast milk, according to Student's independent *t*-test. [#] Indicates differences ($p \leq 0.05$) between overweight and obese groups and corresponding breast milk, according to Student's independent *t*-test; LA: Linoleic Acid; AA: Arachidonic Acid; ALA: α-linolenic Acid; EPA: Eicosapentaenoic Acid; DHA: Docosahexaenoic Acid; SFA: Saturated Fatty Acids; MCFA: Medium-chain Fatty Acids; MUFA: Monounsaturated Fatty Acids; PUFA: Polyunsaturated Fatty Acids; LC-PUFA: Long-Chain Polyunsaturated Fatty Acids.

Regarding other biologically important FAs, and always compared to colostrum levels, the mature milk of normal-weight mothers showed higher levels of C12:0 and C18:3n6, and lower levels of C16:0, C20:0, C20:3n6, C22:0, C22:2 n6, C23:0, saturated fatty acids (SFAs) and n3 PUFAs; the mature milk of overweight mothers showed increased levels of C6:0, C16:1n7, linoleic acid (LA), C18:3n6, EPA and n6:n3 ratio, and decreased concentrations of C15:0, C16:0, C17:0, C20:0, C23:0 and SFAs; and finally, the mature milk of mothers with obesity had higher levels of C6:0, C12:0 and C18:0 and lower levels of C18:1n7 and C22:0.

3.3. Breast Milk FAs According to Maternal Weight Group

Table 2 also shows the differences in breast milk FAs between weight groups. Compared to normal-weight women, the overweight group had higher levels of C14:1, C15:0, C17:0 and C17:1 in colostrum; and higher levels of C22:5n6 and n6:n3 ratio and lower levels of DHA and EPA:AA in mature milk.

On the other hand, compared to normal-weight mothers, mothers with obesity had lower levels of C18:0, C18:1n9t and ALA in colostrum; and lower levels of C18:1n9, C18:1n9t, ALA and MUFAs and higher levels of C22:2n6, C22:5n6, C23:0 and SFAs in mature milk.

We also compared overweight with mothers with obesity and found that the group with obesity had lower concentrations of C8:0, C15:0, C17:0, C18:1n9t, ALA in colostrum; and higher levels of C16:0 and SFAs and lower levels of ALA in mature milk.

3.4. Association of Breast Milk FAs with Infant Growth

Table 3 shows the associations between breast milk FAs and infant growth. All associations were observed after adjusting for potential confounders, which included maternal pre-pregnancy BMI, maternal smoking, weight gain during pregnancy, maternal education, gender of the child and type of infant feeding practice.

Table 3. Associations between breast milk PUFA levels and anthropometric measurements in infants.

Fatty Acid	BMIZ				WAZ				LAZ			
	Colostrum		Mature Milk		Colostrum		Mature Milk		Colostrum		Mature Milk	
	6mo $n = 37$ 18mo $n = 38$ 36mo $n = 16$		6mo $n = 39$ 18mo $n = 37$ 36mo $n = 13$		6mo $n = 37$ 18mo $n = 38$ 36mo $n = 16$		6mo $n = 39$ 18mo $n = 38$ 36mo $n = 13$		6mo $n = 38$ 18mo $n = 38$ 36mo $n = 18$		6mo $n = 39$ 18mo $n = 38$ 36mo $n = 14$	
	β	*p*	β	*p*	β	*p*	β	*p*	β	*p*	β	*p*
	C18:3n3 (ALA)											
6mo	−0.11	0.63	−0.15	0.40	−0.06	0.77	−0.22	0.24	0.05	0.80	−0.12	0.49
18mo	0.32	0.13	−0.11	0.57	0.11	0.63	−0.13	0.51	−0.20	0.26	−0.06	0.69
36mo	−0.44	0.21	0.01	0.99	−0.27	0.52	0.06	0.95	−0.17	0.63	−0.86	0.22
	C18:2n6 (LA)											
6mo	0.33	0.12	−0.13	0.43	**0.42**	**0.027**	0.16	0.36	0.15	0.40	0.09	0.60
18mo	−0.19	0.34	0.04	0.85	0.18	0.37	0.05	0.79	0.06	0.73	0.17	0.26
36mo	0.12	0.75	−0.22	0.50	0.02	0.96	−0.136	0.75	0.01	0.98	−0.33	0.41
	C20:4n6 (AA)											
6mo	**−0.44**	**0.016**	0.02	0.91	−0.20	0.26	0.04	0.84	0.25	0.13	0.03	0.86
18mo	−0.03	0.89	−0.12	0.57	0.06	0.77	−0.13	0.52	0.10	0.50	0.15	0.39
36mo	0.30	0.30	0.09	0.81	0.42	0.20	0.07	0.86	0.20	0.48	0.12	0.75
	C20:5n3 (EPA)											
6mo	**−0.51**	**0.012**	0.00	0.99	−0.36	0.07	−0.13	0.49	0.18	0.31	−0.18	0.29
18mo	−0.30	0.12	0.00	0.98	−0.13	0.51	0.08	0.66	0.15	0.35	0.09	0.62
36mo	−0.74	0.155	−1.08	0.30	−0.58	0.34	−1.13	0.28	0.14	0.69	−0.59	0.61

Table 3. *Cont.*

Fatty Acid	BMIZ				WAZ				LAZ			
	Colostrum		Mature Milk		Colostrum		Mature Milk		Colostrum		Mature Milk	
	6mo *n* = 37 18mo *n* = 38 36mo *n* = 16		6mo *n* = 39 18mo *n* = 37 36mo *n* = 13		6mo *n* = 37 18mo *n* = 38 36mo *n* = 16		6mo *n* = 39 18mo *n* = 38 36mo *n* = 13		6mo *n* = 38 18mo *n* = 38 36mo *n* = 18		6mo *n* = 39 18mo *n* = 38 36mo *n* = 14	
	β	*p*	β	*p*	β	*p*	β	*p*	β	*p*	β	*p*
C22:6n3 (DHA)												
6mo	**−0.37**	**0.043**	−0.16	0.38	−0.31	0.07	−0.29	0.10	0.00	0.99	−0.23	0.17
18mo	0.14	0.42	0.03	0.88	0.08	0.66	0.00	0.99	−0.05	0.74	−0.03	0.84
36mo	0.42	0.29	0.33	0.46	0.65	0.13	0.38	0.39	0.46	0.22	0.58	0.14
n6 PUFA												
6mo	0.21	0.32	0.13	0.45	0.34	0.07	0.16	0.35	0.20	0.27	0.10	0.55
18mo	0.16	0.41	−0.06	0.776	0.18	0.40	0.04	0.83	0.08	0.65	0.18	0.23
36mo	0.20	0.587	−0.19	0.64	0.15	0.72	−0.11	0.79	0.07	0.84	−0.31	0.44
n3 PUFA												
6mo	**−0.38**	**0.047**	−0.19	0.27	−0.33	0.07	−0.32	0.07	−0.00	0.991	−0.22	0.18
18mo	0.16	0.38	−0.11	0.56	0.04	0.84	−0.11	0.53	−0.12	0.427	−0.05	0.75
36mo	−0.20	0.60	0.17	0.78	0.05	0.90	0.21	0.74	0.05	0.897	−0.18	0.77
n6 LC−PUFA												
6mo	**−0.38**	**0.047**	−0.06	0.77	−0.17	0.36	0.00	0.98	0.19	0.253	0.09	0.65
18mo	−0.05	0.77	−0.27	0.19	0.03	0.88	−0.17	0.41	0.10	0.508	0.19	0.25
36mo	0.40	0.22	0.11	0.78	0.60	0.09	0.12	0.76	0.25	0.390	−0.03	0.95
n3 LC−PUFA												
6mo	**−0.43**	**0.020**	−0.19	0.28	−0.34	0.05	−0.33	0.06	0.03	0.866	−0.24	0.16
18mo	0.07	0.70	−0.04	0.82	0.05	0.78	−0.02	0.90	−0.01	0.955	0.02	0.89
36mo	0.28	0.44	0.19	0.63	0.53	0.18	0.21	0.59	0.42	0.211	0.40	0.31
n6:n3												
6mo	**0.42**	**0.031**	0.30	0.10	**0.45**	**0.011**	**0.45**	**0.013**	0.11	0.519	0.30	0.08
18mo	−0.04	0.82	0.05	0.78	0.06	0.74	0.14	0.47	0.14	0.369	0.21	0.19
36mo	0.30	0.34	−0.23	0.55	0.05	0.88	−0.18	0.65	0.01	0.978	−0.26	0.58
LC n6:n3												
6mo	0.12	0.56	0.14	0.41	0.22	0.24	0.29	0.09	0.15	0.373	0.24	0.13
18mo	−0.13	0.48	−0.08	0.66	−0.04	0.86	−0.05	0.77	0.11	0.490	0.07	0.65
36mo	0.35	0.54	−0.07	0.86	0.16	0.81	−0.08	0.85	−0.31	0.516	−0.30	0.46

Associations were evaluated using lineal regression analyses. β and *p* are corrected values after adjustment for potential confounders: maternal pre-pregnancy BMI, maternal smoking, weight gain during pregnancy, maternal education, sex of the child and type of infant feeding practice. *p*-values ≤ 0.05 are highlighted in bold and those ≤0.002 are additionally marked by stars. mo: month; LA: Linoleic Acid; AA: Arachidonic Acid; ALA: α-linolenic Acid; EPA: Eicosapentaenoic Acid; DHA: Docosahexaenoic Acid; PUFA: Polyunsaturated Fatty Acids; LC-PUFA: Long chain Polyunsaturated Fatty Acids.

At 6 months of age, we found that colostrum levels of AA, EPA, DHA, n3 PUFAs, n6 LC-PUFAs and n3 LC-PUFAs were inversely associated with infant BMIZ, while the n6:n3 ratio was positively associated with it. Also, at 6 months of age, LA and the n6:n3 ratio in both colostrum and mature milk were positively associated with WAZ. No associations were found between mature milk and any variable at 1.5 or 3 years of age.

3.5. Associations of Breast Milk FAs with Infant Cognition

Table 4 presents the associations between breast milk PUFAs and infant cognition at 18 months of age. When the whole population was analyzed, no associations were found. However, infants born to normal-weight mothers presented a positive association between cognition scores and LA and n6 PUFA levels in colostrum. On the other hand, the infants of overweight mothers presented a direct association of DHA and n3 LC-PUFA levels in colostrum with cognitive score, while the n6:n3 ratio in colostrum was inversely associated with it. With respect to infants born to mothers with obesity, a positive association was found between ALA levels in mature milk and cognition.

Table 4. Associations between breast milk PUFA levels and infant cognition score at 18 months of age, according to maternal pre-pregnancy BMI.

| Fatty Acid | All | | | | Normal-Weight | | | | Overweight | | | | Obesity | | | |
| | Colostrum (n = 75) | | Mature Milk (n = 77) | | Colostrum (n = 14) | | Mature Milk (n = 12) | | Colostrum (n = 11) | | Mature Milk (n = 15) | | Colostrum (n = 12) | | Mature Milk (n = 11) | |
	β	p	β	p	β	p	β	p	β	p	β	p	β	p	β	p
C18:3n3 (ALA)	0.08	0.718	0.01	0.942	0.29	0.581	0.82	0.212	0.44	0.393	−0.15	0.662	0.55	0.468	**2.34**	**0.008**
C18:2n6 (LA)	0.20	0.339	0.12	0.527	**0.84**	**<0.001 ***	0.88	0.069	−0.95	0.061	−0.16	0.687	0.12	0.869	0.53	0.470
C20:4n6 (AA)	0.03	0.889	−0.20	0.333	−1.23	0.136	−0.56	0.270	0.32	0.594	−0.28	0.544	0.48	0.416	−0.87	0.405
C20:5n3 (EPA)	0.31	0.133	−0.11	0.580	0.00	0.996	0.59	0.428	−0.07	0.932	−0.18	0.635	0.80	0.233	−0.29	0.658
C22:6n3 (DHA)	−0.16	0.396	−0.27	0.161	−0.73	0.124	−0.19	0.720	**0.88**	**0.045**	0.33	0.362	−0.03	0.954	−1.16	0.104
n6 PUFA	0.24	0.251	0.13	0.495	**0.81**	**0.002 ***	0.88	0.064	−0.97	0.111	−0.15	0.713	0.16	0.803	0.45	0.520
n3 PUFA	−0.01	0.966	−0.12	0.541	−0.23	0.687	0.37	0.512	0.70	0.057	−0.04	0.922	0.13	0.829	−0.78	0.489
n6 LC-PUFA	0.14	0.446	0.18	0.407	0.01	0.984	0.31	0.536	0.53	0.227	0.36	0.481	0.28	0.650	−0.09	0.924
n3 LC-PUFA	−0.11	0.563	−0.25	0.190	−0.71	0.113	−0.08	0.873	**1.01**	**0.004**	0.21	0.585	0.04	0.949	−0.89	0.189
n6:n3	0.13	0.489	0.22	0.255	0.74	0.067	0.29	0.639	**−0.97**	**0.002 ***	−0.10	0.805	0.01	0.985	0.57	0.426
LC n6:n3	0.27	0.166	0.29	0.105	0.52	0.172	0.16	0.752	0.05	0.952	−0.08	0.830	0.22	0.696	0.83	0.286

Associations were evaluated using lineal regression analyses. β and p are corrected values after adjustment for potential confounders: maternal smoking, weight gain during pregnancy, maternal education, sex of the child and type of infant feeding practice. Cognition performance was analyzed at 1.5 years of age. p-values ≤ 0.05 are highlighted in bold and those ≤0.002 are additionally marked by stars. LA: Linoleic Acid; AA: Arachidonic Acid; ALA: α-linolenic Acid; EPA: Eicosapentaenoic Acid; DHA: Docosahexaenoic Acid; PUFA: Polyunsaturated Fatty Acids; LC-PUFA: Long chain Polyunsaturated Fatty Acids.

4. Discussion

The present study offers the evaluation of human breast milk FA composition during the first month postpartum according to maternal weight, and the impact on child outcomes from 6 months to 3 years of age. This is one of the very few studies analyzing the influence of maternal weight on breast milk FA composition and, to our knowledge, the second one to assess this parameter in both colostrum and mature breast milk. Moreover, we believe this is the first study to address its effect on both infant cognitive developmental parameters and growth all in one study.

Upon analysis of the population characteristics, we observed that women with obesity had the lowest GWG, even though no nutritional intervention was carried out. This finding is in line with the results of a systematic review about GWG in women with obesity where they concluded that GWG decreased with each higher BMI classification [39]. In fact, weight loss during pregnancy is more common in women with obesity than non-obese women [40] and GWG decreases as the severity of obesity increases [41]. Nonetheless, we observed that women with obesity had the highest n6:n3 ratio in their dietary intake, which suggests they had the lowest-quality dietary intake, as similarly demonstrated by other studies where a high weight status is related to a high dietary intake of n6 FAs and a low intake of n3 FAs [42].

We did not observe any differences in infant characteristics according to maternal BMI. Evidence suggests that infants born to mothers with obesity have an increased risk of having a higher weight and length at birth [1], but this was not the case in our population. Since GWG is directly associated with birth weight [43], a possible explanation for this finding is that women with obesity showed the lowest GWG, and therefore their offspring did not present increased weight at birth. This is in the line with the conclusions of the systematic review of Faucher and Barger, in which several studies reported a linear decrease in the prevalence of being large for gestational age (LGA) and less GWG [39], suggesting that women with obesity and low GWG would have some benefits on fetal growth. This could be the reason why in our study children from women with obesity did not present higher weight or length. Nevertheless, our data still showed a tendency in which formula-fed infants and those considered LGA represented a higher percentage in the groups of women with overweight and obesity, according to Lubchenco's curves [44].

Regardless of maternal weight, our data showed that when breast milk transitioned from colostrum (2–4 days postpartum) to mature milk (28–32 days postpartum), the levels of crucial FAs such as AA, DHA, and n3 LC-PUFA were decreased. Other authors have also demonstrated this finding [45,46]. The high content of crucial FAs in colostrum has biological relevance because it is highly associated with child outcomes, possibly because of the nutrient supply during the first few days of life, which are critical for infant health [47]. Nonetheless, we also found higher levels of EPA:AA and DHA:AA in mature milk, which are positively associated with health outcomes as well [48]. Analyzing the breast milk evolution within each weight group, we found that SFA concentrations decreased in the mature milk of women with normal-weight and overweight, but not in the mature milk of women with obesity. As described in other studies, the human milk of these women could present higher levels of SFA [17]. The factors attributed to this increased amount of SFA in breast milk of women with obesity could be the metabolic status and diet. It is well known that obesity is intrinsically a pro-inflammatory state influenced by dietary intake [49], where the ratio of n6:n3 PUFA is a clear factor affecting inflammation and obesity development [50]. In our study, the higher dietary intake of n6:n3 PUFA ratio in women with obesity, might be enhancing the pro-inflammatory state, and thereby affecting the levels of breast milk SFAs. Moreover, the fact that women with an increased BMI may have an increased intake of n6:n3 PUFA might also explain the higher n6:n3 PUFA ratio found in the mature milk of overweight women compared to their colostrum, although their increased intake of dietary n6:n3 was not significant.

To the best of our knowledge, eight studies have evaluated the FA composition of breast milk according to maternal BMI, but the results available in the literature are not entirely consistent and the studies differ in terms of the weight groups tested and the timing of sample collections. Out of these eight studies, only one shares the same collection timing for colostrum that we used [32]; another one used a similar timing for both colostrum and mature milk [16], but the other 6 studies collected the milk in different times ([10,17,18,29–31]). Regarding the weight groups used, 5 did not share the same groups that we used [10,16,17,29,31], and 2 out of the 3 studies that did [18,30,31] used a different criteria to classify weight according to BMI [30,32].

Although our results and the ones available in the literature suggest that a high maternal weight status alters human milk nutrient content, there is an inconsistency regarding which FAs are the most influenced according to BMI groups. This could be attributed to numerous factors, such as sample size, population, methods, FAs included in the analysis, weight group classification and the timing of breast milk collection. However, it is important to highlight that, even without a clear consistency among studies, an increased BMI is found to alter FA concentrations in breast milk, generally increasing SFA and n6 PUFAs and decreasing FAs from the n3 series. An important factor that could explain the differences found among weight groups could be related to dietary intake during late pregnancy, since several studies have demonstrated that this affects breast milk composition [51]. This suggests that women with overweight and obesity could have an increased dietary intake of n6 FAs and SFAs and a poor intake of n3 FAs. As previously mentioned, this happened in our population, where we found that the n6:n3 ratio of dietary intake was higher in women with an increased BMI, especially those with obesity. Since a maternal pro-inflammatory diet is positively associated with increased concentrations of SFA and MUFA in breast milk [10], specific maternal metabolic markers could be an interesting approach to predicting the predominance of certain FAs in breast milk.

This study also analyzed the possible association between breast milk FA composition and infant growth and cognition. It is well known that many nutrients are critical for proper infant growth and neurodevelopment. Animal models and epidemiological studies suggest that PUFAs such as AA and DHA are particularly important [52,53]. Thus, we evaluated the association between the PUFA levels in breast milk and infant anthropometric measurements at 6, 18 and 36 months of age. For this analysis, we used the z-score values WAZ, LAZ and BMIZ to evaluate with greater accuracy which children were within or outside the normal range [1,36,54]. Our findings showed that LC-PUFAs—especially AA, EPA, DHA, n3 and n6 LC-PUFAs—in colostrum had a negative association with infant BMIZ at 6 months. In accordance with these results, a recent review that analyzed the association between n3 PUFAs and growth suggested that DHA during pregnancy, lactation and early life may be associated with significant benefits in infant growth and development [55]. Similarly, Pedersen et al. observed a negative association between DHA levels in breast milk and BMI in children from 2 to 7 years of age. They also found an overall inverse association between breast milk DHA and body fat percentage [56]. Although it is important to mention that BMI is not the best method to quantify body composition, and especially to assess body fat in children [57,58], DHA content in breast milk could have some benefits in postponing the age of adiposity rebound [56], which is the second rise in adiposity that usually occurs between 3 and 7 years of age [59]. It is known that the age that rebound occurs predicts later fatness, meaning that an earlier rebound would be a risk factor for later obesity [59]. On the other hand, our data also indicated that n6 PUFA levels may contribute to a fat mass increase in children [59], since LA in colostrum and the n6:n3 ratio in both mature milk and colostrum could influence WAZ and BMIZ at 6 months of age. Since the n6 PUFAs in mature milk were generally increased in overweight and obese mothers, their children could be more susceptible to developing obesity [17,20,50]. Indeed, children from overweight and obese mothers presented a tendency to be LGA. In contrast, Much et al. found that AA and n6 PUFAs in mature breast milk were negatively associated with infant weight and BMI (up to 4 months of age) [33], suggesting that the role of these n6 FAs (including AA) might be age-dependent and serve as important regulating factors for growth in early postnatal life. Due to the low variability of AA contents in breast milk across populations (0.24–1% of FAs) [7], a possible explanation of this discrepancy could be the quantitative amount of milk intake by the breastfed infant, meaning that depending on the daily ingested volume of milk, AA would have its growth-regulatory effects or not [33]. Further studies are needed to look into such quantitative aspects. In our study, we only found significant associations between FAs and infant growth at 6 months of life, but not at 18 nor 36. This finding may be due to the child's own diet, lifestyle and metabolism. However, the associations found at 6 months are relevant, because it is a crucial age that represents a critical period in the child's development and programming [3]. A curious result that we found is that length was not correlated to any PUFA, which again is in disagreement with Much et al. They inversely correlated DHA, EPA and n3 PUFA with length at 1 year of age. Their milk collection was at 6 weeks and 4 months postpartum [33], whereas in our study it was at 2–4 days and 28–32 days postpartum. Therefore, the possible evolution of FA species over time would be a possible explanation for the different results. From our study and the evidence gathered, we can see that PUFAs in breastmilk influence infant growth; however, there is a high variability in existing results. Further studies are needed to obtain more conclusive outcomes [2].

PUFAs are also critical for an adequate brain growth and function in aspects such as neurogenesis, nerve impulse transmission, neuronal integrity, and vitality and gene expression in the brain [52,53,60]. Thus, we explored the association between breast milk PUFA levels and cognitive score at 18 months of life. On the one hand, when we analyzed the total population, we found no association between any FA in breast milk and child development. Similarly, there have been observational studies that found no strong evidence for a beneficial role of LC-PUFAs in order to explain the positive relationship between breastfeeding and cognition [61]. This raises the question as to whether LC-PUFA levels may only be beneficial in children's mental development when breastfeeding levels are high [62]. Although we corrected the analysis by the type of breastfeeding, this information was collected at 3 months of age, so we do not know which effect could have had a longer period of exclusive breastfeeding.

We also explored the association between breast milk FA levels and infant cognition according to maternal BMI. In general, we found a direct association between n3 and n6 PUFA levels in colostrum and infant cognition at 18 months of age. The colostrum from overweight mothers was the one that presented more relevant associations, specifically, a high n6:n3 ratio was negatively associated with cognition, whereas higher DHA concentrations were directly associated with better cognitive scores, which is in line with Bernard et al. [63]. This suggests that the cognition of infants born to overweight women could be enhanced by promoting n3 FAs, more specifically DHA, in the maternal diet. These results are in line with meta-analyses, animal and epidemiologic studies [60,64,65], and support WHO recommendations on breastfeeding for the two first years of life or beyond [66]. We must consider that, in our study, the cognitive score was assessed at 1.5 years of life, and at this age, there are many factors related to the child that could influence their cognition. The potential cofounders that we have used to adjust this analysis were mainly related to the mother, and only the gender and type of feeding practice were related to the child. Important factors such as infant diet or physical activity are lacking and could have a huge influence in the results because intake of micronutrients, such as n3 FAs, vitamin B12, folic acid, zinc, iron and iodine, together with malnutrition and general dietary patterns and other lifestyle habits, influence child cognitive development as well [67–69].

Overall, our study highlights the importance of the maternal health before, during and after pregnancy, since it could have a great impact in the breast milk FA composition and, in consequence, in the offspring's growth and cognition which affects their future health. Many women start developing healthy habits when they are pregnant or planning a pregnancy. However, as presented in our study, the pre-pregnancy health status has an important effect in the quality of the human milk, consequently affecting the health of the child. Therefore, bigger efforts must be put in place to promote and guarantee a healthier lifestyle and nutritional status in the general population to pursuit healthier future generations.

We acknowledge some limitations in our study, such as the small sample size. However, it is important to understand that, even though PREOBE is a larger cohort, we were not able to include all the participants in the present study due to lack of data or samples, possibly related to indisposition to participate given the complexity and sensitivity of the periods involved: childbirth and breastfeeding. Although risk factors, such as socio-demographic information and maternal diet, allowed us to adjust our statistical models for potential confounders, we cannot rule out residual confounding, especially coming from data related to the infants at 1.5 and 3 years of age because data on their dietary intake, lifestyle and other characteristics, could be greatly influencing the results. Another limitation is that women receiving supplements of DHA for over 3 months were excluded, but we do not know the possible effect of that initial supplementation in the breast milk FA profile. Moreover, recording the timing between sample collection or the last meal, collecting information on what was consumed before and after each sample was taken, and analyzing the different breast milk samples of one day without mixing them, would provide valuable data to assess the human milk nutrient content and impact. In general, further research is required to provide a better understanding of the role that FAs play in obesity development and management, paying special attention to the methods used for analysis and promoting the comparison of results between cohorts.

5. Conclusions

In conclusion, our results show that (1) the FA composition of colostrum and mature milk was different. Regardless of maternal weight, mature milk had lower levels of AA and DHA (among others) than colostrum; (2) Maternal obesity influenced the FA concentrations in breast milk. Overall, breast milk of mothers with a high BMI presented increased SFA levels and n6:n3 ratio, and decreased ALA, DHA and MUFA concentrations; and (3) The early supply of n6 and n3 PUFAs through colostrum influenced infant weight status and cognition, at 6 and 18 months of life, respectively. Infant BMIZ at 6 months of age was inversely associated with colostrum levels of n6 and n3 LC-PUFAs (e.g., AA and DHA) and positively associated with n6:n3 ratio. Depending on the maternal BMI, infant cognition may be positively affected by colostrum levels of LA, n6 PUFAs, DHA, n3 LC-PUFAs and ALA, and negatively affected by the n6:n3 ratio. Since the maternal pre-pregnancy weight can influence the breast milk FAs, which is related to the early nutritional status of the child and to health conditions throughout the life span, this study endorses the need for early preventive health care through diet and lifestyle. A healthy weight in women before, during and after pregnancy should be encouraged to promote beneficial FAs in breast milk and promote healthier future generations.

Author Contributions: The authors' responsibilities were as follows. C.C., M.C.L.-S. and A.I.C. designed the project. F.J.T.-E., L.G.-V., M.E.-M., M.T.S. were involved in participant recruitment, data and sample collection. A.I.C., A.M.A., A.d.l.G.P. and R.M.G. processed the samples. A.d.l.G.P. and A.M.A. conducted the data analysis and interpretation. A.d.l.G.P., A.M.A. and A.M.C. wrote the manuscript. All authors performed a critical review of the final manuscript.

Funding: This research was funded by the European Commission (DynaHEALTH-HORIZON 2020 GA No: 633595) and the Spanish Ministry of Economy and Competitiveness (BFU2012-40254-C03-02). Further support was obtained from, Spanish Ministry of Innovation and Science (Junta de Andalucía), Excellence Projects (P06-CTS-02341). ADLGP thanks the Mexican government and the National Council on Science and Technology (CONACYT) for her PhD grant. The funders had no role in the study design, data collection, data analysis, decision to publish, or preparation of the manuscript.

Acknowledgments: The authors sincerely thank the women and children involved in the PREOBE study for their valuable participation. We are also grateful to personnel, scientists, staff and all people involved in the PREOBE team who have made this research possible. We acknowledge members of the Autonomous University of Barcelona for their assistance in statistical analysis. Members of the PREOBE team include: University of Granada. Spain: EURISTIKOS Excellence Centre for Paediatric Research. Department of Paediatrics: Cristina Campoy (PI), Luz Mª García-Valdés, Francisco J Torres-Espínola, Mª Teresa Segura, Cristina Martínez-Zaldívar, Tania Anjos, Antonio Jerez, Daniel Campos, Rosario Moreno-Torres, Mª José Aguilar, Iryna Rusanova, Jole Martino, Signe Altmäe; Department of Obstetrics and Gynecology: Jesús Florido, Carmen Padilla; Department of Biostatistics: Mª Teresa Miranda; Mind, Brain and Behavior International Research Centre: Andrés Catena, Miguel Pérez-García; Department of Legal Medicine: Jose A. Lorente, Juan C. Alvarez; Department of Pharmacology: Ahmad Agil; ICTAN-CSIC–Madrid. Spain: Ascensión Marcos, Esther Nova, Department of Nutrition and Bromatology. University of Barcelona. Spain: Mª Carmen López-Sabater; Lorgen, S.L.: Carmen Entrala; Rowett Institute, University of Aberdeen, UK: Harry McArdle, University of Nöttingham, UK: Michael Symonds; Ludwig-Maximiliam University of Munich, Germany: Berthold Koletzko, Hans Demmelmair, Olaf Uhl; University of Graz, Austria: Gernot Desoye; Abbott Laboratories: Ricardo Rueda; University of Umeå, Sweden: Staffan K Berglund.

Conflicts of Interest: The authors declare no conflict of interest.

References

1. Poston, L.; Caleyachetty, R.; Cnattingius, S.; Corvalán, C.; Uauy, R.; Herring, S.; Gillman, M.W. Preconceptional and maternal obesity: Epidemiology and health consequences. *Lancet Diabetes Endocrinol.* **2016**, *4*, 1025–1036. [CrossRef]

2. Koletzko, B.; Godfrey, K.M.; Poston, L.; Szajewska, H.; van Goudoever, J.B.; de Waard, M.; Brands, B.; Grivell, R.M.; Deussen, A.R.; Dodd, J.M.; et al. Nutrition During Pregnancy, Lactation and Early Childhood and its Implications for Maternal and Long-Term Child Health: The Early Nutrition Project Recommendations. *Ann. Nutr. Metab.* **2019**, *74*, 93–106. [CrossRef]

3. Langley-Evans, S.C. Nutrition in early life and the programming of adult disease: A review. *J. Hum. Nutr. Diet.* **2015**, *28*, 1–14. [CrossRef] [PubMed]

4. Pérez-Escamilla, R.; Martinez, J.L.; Segura-Pérez, S. Impact of the Baby-friendly Hospital Initiative on breastfeeding and child health outcomes: A systematic review. *Matern. Child Nutr.* **2016**, *12*, 402–417. [CrossRef]

5. Mosca, F.; Giannì, M.L. Human milk: Composition and health benefits. *La Pediatr. Med. Chir.* **2017**, *39*, 155. [CrossRef]

6. Binns, C.; Lee, M.; Low, W.Y. The Long-Term Public Health Benefits of Breastfeeding. *Asia Pac. J. Public Health* **2016**, *28*, 7–14. [CrossRef] [PubMed]

7. Brenna, J.T.; Varamini, B.; Jensen, R.G.; Diersen-Schade, D.A.; Boettcher, J.A.; Arterburn, L.M. Docosahexaenoic and arachidonic acid concentrations in human breast milk worldwide. *Am. J. Clin. Nutr.* **2007**, *85*, 1457–1464. [CrossRef] [PubMed]

8. Savino, F.; Liguori, S.A.; Fissore, M.F.; Oggero, R. Breast milk hormones and their protective effect on obesity. *Int. J. Pediatr. Endocrinol.* **2009**, *2009*, 327505. [CrossRef]

9. Lauritzen, L.; Carlson, S.E. Maternal fatty acid status during pregnancy and lactation and relation to newborn and infant status. *Matern. Child Nutr.* **2011**, *7*, 41–58. [CrossRef] [PubMed]

10. Panagos, P.G.; Vishwanathan, R.; Penfield-Cyr, A.; Matthan, N.R.; Shivappa, N.; Wirth, M.D.; Hebert, J.R.; Sen, S. Breastmilk from obese mothers has pro-inflammatory properties and decreased neuroprotective factors. *J. Perinatol.* **2016**, *36*, 284–290. [CrossRef] [PubMed]

11. Martin, C.; Ling, P.R.; Blackburn, G.; Martin, C.R.; Ling, P.R.; Blackburn, G.L. Review of Infant Feeding: Key Features of Breast Milk and Infant Formula. *Nutrients* **2016**, *8*, 279. [CrossRef] [PubMed]

12. Muhlhausler, B.S.; Ailhaud, G.P. Omega-6 polyunsaturated fatty acids and the early origins of obesity. *Curr. Opin. Endocrinol. Diabetes Obes.* **2013**, *20*, 56–61. [CrossRef] [PubMed]

13. Ailhaud, G.; Guesnet, P. Fatty acid composition of fats is an early determinant of childhood obesity: A short review and an opinion. *Obes. Rev.* **2004**, *5*, 21–26. [CrossRef] [PubMed]

14. Gertosio, C.; Meazza, C.; Pagani, S.; Bozzola, M. Breastfeeding and its gamut of benefits. *Minerva Pediatr.* **2016**, *68*, 201–212. [PubMed]

15. Barrera, C.; Valenzuela, R.; Chamorro, R.; Bascuñán, K.; Sandoval, J.; Sabag, N.; Valenzuela, F.; Valencia, M.P.; Puigrredon, C.; Valenzuela, A.; et al. The Impact of Maternal Diet during Pregnancy and Lactation on the Fatty Acid Composition of Erythrocytes and Breast Milk of Chilean Women. *Nutrients* **2018**, *10*, 839. [CrossRef] [PubMed]

16. Storck Lindholm, E.; Strandvik, B.; Altman, D.; Möller, A.; Palme Kilander, C. Different fatty acid pattern in breast milk of obese compared to normal-weight mothers. *Prostaglandins Leukot. Essent. Fat. Acids* **2013**, *88*, 211–217. [CrossRef] [PubMed]

17. Mäkelä, J.; Linderborg, K.; Niinikoski, H.; Yang, B.; Lagström, H. Breast milk fatty acid composition differs between overweight and normal weight women: The STEPS Study. *Eur. J. Nutr.* **2013**, *52*, 727–735. [CrossRef] [PubMed]

18. Marín, M.C.; Sanjurjo, A.; Rodrigo, M.A.; de Alaniz, M.J.T. Long-chain polyunsaturated fatty acids in breast milk in La Plata, Argentina: Relationship with maternal nutritional status. *Prostaglandins Leukot. Essent. Fat. Acids* **2005**, *73*, 355–360. [CrossRef] [PubMed]

19. Simopoulos, A.P. An Increase in the Omega-6/Omega-3 Fatty Acid Ratio Increases the Risk for Obesity. *Nutrients* **2016**, *8*, 128. [CrossRef]

20. Heerwagen, M.J.R.; Stewart, M.S.; de la Houssaye, B.A.; Janssen, R.C.; Friedman, J.E. Transgenic increase in N-3/n-6 Fatty Acid ratio reduces maternal obesity-associated inflammation and limits adverse developmental programming in mice. *PLoS ONE* **2013**, *8*, e67791. [CrossRef]

21. Edlow, A.G. Maternal obesity and neurodevelopmental and psychiatric disorders in offspring. *Prenat. Diagn.* **2017**, *37*, 95–110. [CrossRef] [PubMed]

22. Rivera, H.M.; Christiansen, K.J.; Sullivan, E.L. The role of maternal obesity in the risk of neuropsychiatric disorders. *Front. Neurosci.* **2015**, *9*, 194. [CrossRef]

23. Veena, S.R.; Gale, C.R.; Krishnaveni, G.V.; Kehoe, S.H.; Srinivasan, K.; Fall, C.H. Association between maternal nutritional status in pregnancy and offspring cognitive function during childhood and adolescence; a systematic review. *BMC Pregnancy Childbirth* **2016**, *16*, 220. [CrossRef] [PubMed]

24. Contu, L.; Hawkes, C.A. A Review of the Impact of Maternal Obesity on the Cognitive Function and Mental Health of the Offspring. *Int. J. Mol. Sci.* **2017**, *18*, 1093. [CrossRef] [PubMed]

25. Huang, L.; Yu, X.; Keim, S.; Li, L.; Zhang, L.; Zhang, J. Maternal prepregnancy obesity and child neurodevelopment in the Collaborative Perinatal Project. *Int. J. Epidemiol.* **2014**, *43*, 783–792. [CrossRef] [PubMed]

26. Heikura, U.; Taanila, A.; Hartikainen, A.L.; Olsen, P.; Linna, S.L.; Wendt, L.; Jarvelin, M.R. Variations in Prenatal Sociodemographic Factors associated with Intellectual Disability: A Study of the 20-Year Interval between Two Birth Cohorts in Northern Finland. *Am. J. Epidemiol.* **2007**, *167*, 169–177. [CrossRef] [PubMed]

27. Brion, M.J.; Zeegers, M.; Jaddoe, V.; Verhulst, F.; Tiemeier, H.; Lawlor, D.A.; Smith, G.D. Intrauterine effects of maternal prepregnancy overweight on child cognition and behavior in 2 cohorts. *Pediatrics* **2011**, *127*, e202–e211. [CrossRef]

28. Craig, W.Y.; Palomaki, G.E.; Neveux, L.M.; Haddow, J.E. Maternal Body Mass Index during Pregnancy and Offspring Neurocognitive Development. *Obstet. Med.* **2013**, *6*, 20–25. [CrossRef]

29. Kwon, M. Nutrient Content of Human Breast Milk from Overweight and Normal Weight Caucasian Women of Northeast Tennessee. Master's Thesis, East Tennessee State University, Johnson City, TN, USA, 2017.

30. Rudolph, M.C.; Young, B.E.; Lemas, D.J.; Palmer, C.E.; Hernandez, T.L.; Barbour, L.A.; Friedman, J.E.; Krebs, N.F.; MacLean, P.S. Early infant adipose deposition is positively associated with the n-6 to n-3 fatty acid ratio in human milk independent of maternal BMI. *Int. J. Obes. (Lond.)* **2017**, *41*, 510–517. [CrossRef]

31. Kim, H.; Kang, S.; Jung, B.M.; Yi, H.; Jung, J.A.; Chang, N. Breast milk fatty acid composition and fatty acid intake of lactating mothers in South Korea. *Br. J. Nutr.* **2017**, *117*, 556–561. [CrossRef]

32. Sinanoglou, V.J.; Cavouras, D.; Boutsikou, T.; Briana, D.D.; Lantzouraki, D.Z.; Paliatsiou, S.; Volaki, P.; Bratakos, S.; Malamitsi-Puchner, A.; Zoumpoulakis, P. Factors affecting human colostrum fatty acid profile: A case study. *PLoS ONE* **2017**, *12*, e0175817. [CrossRef] [PubMed]

33. Much, D.; Brunner, S.; Vollhardt, C.; Schmid, D.; Sedlmeier, E.M.; Brüderl, M.; Heimberg, E.; Bartke, N.; Boehm, G.; Bader, B.L.; et al. Breast milk fatty acid profile in relation to infant growth and body composition: Results from the INFAT study. *Pediatr. Res.* **2013**, *74*, 230–237. [CrossRef] [PubMed]

34. Berglund, S.K.; García-Valdés, L.; Torres-Espinola, F.J.; Segura, M.T.; Martínez-Zaldívar, C.; Aguilar, M.J.; Agil, A.; Lorente, J.A.; Florido, J.; Padilla, C.; et al. Maternal, fetal and perinatal alterations associated with obesity, overweight and gestational diabetes: An observational cohort study (PREOBE). *BMC Public Health* **2016**, *16*, 207. [CrossRef] [PubMed]

35. Chisaguano, A.M.; Lozano, B.; Moltó-Puigmartí, C.; Castellote, A.I.; Rafecas, M.; López-Sabater, M.C. Elaidic acid, vaccenic acid and rumenic acid (c9,t11-CLA) determination in human plasma phospholipids and human milk by fast gas chromatography. *Anal. Methods* **2013**, *5*, 1264. [CrossRef]

36. Borghi, E.; de Onis, M.; Garza, C.; Van den Broeck, J.; Frongillo, E.A.; Grummer-Strawn, L.; Van Buuren, S.; Pan, H.; Molinari, L.; Martorell, R.; et al. Construction of the World Health Organization child growth standards: Selection of methods for attained growth curves. *Stat. Med.* **2006**, *25*, 247–265. [CrossRef] [PubMed]

37. Duggan, M.B. Anthropometry as a tool for measuring malnutrition: Impact of the new WHO growth standards and reference. *Ann. Trop. Paediatr.* **2010**, *30*, 1–17. [CrossRef]

38. Goldstein, S.; Naglieri, J.A. *Encyclopedia of Child Behavior and Development*; Springer: Berlin/Heidelberg, Germany, 2011; ISBN 9780387775791.

39. Faucher, M.A.; Barger, M.K. Gestational weight gain in obese women by class of obesity and select maternal/newborn outcomes: A systematic review. *Women Birth* **2015**, *28*, e70–e79. [CrossRef] [PubMed]

40. Durie, D.E.; Thornburg, L.L.; Glantz, J.C. Effect of second-trimester and third-trimester rate of gestational weight gain on maternal and neonatal outcomes. *Obstet. Gynecol.* **2011**, *118*, 569–575. [CrossRef]

41. Hinkle, S.N.; Sharma, A.J.; Dietz, P.M. Gestational weight gain in obese mothers and associations with fetal growth. *Am. J. Clin. Nutr.* **2010**, *92*, 644–651. [CrossRef]

42. Patterson, E.; Wall, R.; Fitzgerald, G.F.; Ross, R.P.; Stanton, C. Health implications of high dietary omega-6 polyunsaturated Fatty acids. *J. Nutr. Metab.* **2012**, *2012*, 539426. [CrossRef]

43. Alberico, S.; Montico, M.; Barresi, V.; Monasta, L.; Businelli, C.; Soini, V.; Erenbourg, A.; Ronfani, L.; Maso, G.; Multicentre Study Group on Mode of Delivery in Friuli Venezia Giulia. The role of gestational diabetes, pre-pregnancy body mass index and gestational weight gain on the risk of newborn macrosomia: Results from a prospective multicentre study. *BMC Pregnancy Childbirth* **2014**, *14*, 23. [CrossRef] [PubMed]

44. Olsen, I.E.; Groveman, S.A.; Lawson, M.L.; Clark, R.H.; Zemel, B.S. New intrauterine growth curves based on United States data. *Pediatrics* **2010**, *125*, e214–e224. [CrossRef] [PubMed]

45. Moltó-Puigmartí, C.; Castellote, A.I.; Carbonell-Estrany, X.; López-Sabater, M.C. Differences in fat content and fatty acid proportions among colostrum, transitional, and mature milk from women delivering very preterm, preterm, and term infants. *Clin. Nutr.* **2011**, *30*, 116–123. [CrossRef] [PubMed]

46. Haddad, I.; Mozzon, M.; Frega, N.G. Trends in fatty acids positional distribution in human colostrum, transitional, and mature milk. *Eur. Food Res. Technol.* **2012**, *235*, 325–332. [CrossRef]

47. Lawn, J.E.; Blencowe, H.; Oza, S.; You, D.; Lee, A.C.C.; Waiswa, P.; Lalli, M.; Bhutta, Z.; Barros, A.J.D.; Christian, P.; et al. Every Newborn: Progress, priorities, and potential beyond survival. *Lancet* **2014**, *384*, 189–205. [CrossRef]

48. Morales, E.; Bustamante, M.; Gonzalez, J.R.; Guxens, M.; Torrent, M.; Mendez, M.; Garcia-Esteban, R.; Julvez, J.; Forns, J.; Vrijheid, M.; et al. Genetic Variants of the FADS Gene Cluster and ELOVL Gene Family, Colostrums LC-PUFA Levels, Breastfeeding, and Child Cognition. *PLoS ONE* **2011**, *6*, e17181. [CrossRef] [PubMed]

49. Shivappa, N.; Steck, S.E.; Hurley, T.G.; Hussey, J.R.; Hébert, J.R. Designing and developing a literature-derived, population-based dietary inflammatory index. *Public Health Nutr.* **2014**, *17*, 1689–1696. [CrossRef] [PubMed]

50. Yang, L.G.; Song, Z.X.; Yin, H.; Wang, Y.Y.; Shu, G.F.; Lu, H.X.; Wang, S.K.; Sun, G.J. Low n-6/n-3 PUFA Ratio Improves Lipid Metabolism, Inflammation, Oxidative Stress and Endothelial Function in Rats Using Plant Oils as n-3 Fatty Acid Source. *Lipids* **2016**, *51*, 49–59. [CrossRef]

51. Nishimura, R.Y.; Barbieri, P.; de Castro, G.S.F.; Jordão, A.A.; da Silva Castro Perdoná, G.; Sartorelli, D.S. Dietary polyunsaturated fatty acid intake during late pregnancy affects fatty acid composition of mature breast milk. *Nutrition* **2014**, *30*, 685–689. [CrossRef]

52. Hadley, K.B.; Ryan, A.S.; Forsyth, S.; Gautier, S.; Salem, N., Jr. The Essentiality of Arachidonic Acid in Infant Development. *Nutrients* **2016**, *8*, 216. [CrossRef]

53. Lassek, W.D.; Gaulin, S.J.C. Maternal milk DHA content predicts cognitive performance in a sample of 28 nations. *Matern. Child Nutr.* **2015**, *11*, 773–779. [CrossRef] [PubMed]

54. Isanaka, S.; Villamor, E.; Shepherd, S.; Grais, R.F. Assessing the impact of the introduction of the World Health Organization growth standards and weight-for-height z-score criterion on the response to treatment of severe acute malnutrition in children: Secondary data analysis. *Pediatrics* **2009**, *123*, e54–e59. [CrossRef] [PubMed]

55. Rombaldi Bernardi, J.; de Souza Escobar, R.; Ferreira, C.F.; Pelufo Silveira, P. Fetal and neonatal levels of omega-3: Effects on neurodevelopment, nutrition, and growth. *Sci. World J.* **2012**, *2012*, 202473.

56. Pedersen, L.; Lauritzen, L.; Brasholt, M.; Buhl, T.; Bisgaard, H. Polyunsaturated fatty acid content of mother's milk is associated with childhood body composition. *Pediatr. Res.* **2012**, *72*, 631–636. [CrossRef] [PubMed]

57. Buss, J. Limitations of Body Mass Index to Assess Body Fat. *Workplace Health Saf.* **2014**, *62*, 264. [CrossRef] [PubMed]

58. Vanderwall, C.; Randall Clark, R.; Eickhoff, J.; Carrel, A.L. BMI is a poor predictor of adiposity in young overweight and obese children. *BMC Pediatr.* **2017**, *17*, 135. [CrossRef]

59. Macé, K.; Shahkhalili, Y.; Aprikian, O.; Stan, S. Dietary fat and fat types as early determinants of childhood obesity: A reappraisal. *Int. J. Obes.* **2006**, *30*, S50–S57. [CrossRef]

60. Innis, S.M. Impact of maternal diet on human milk composition and neurological development of infants. *Am. J. Clin. Nutr.* **2014**, *99*, 734S–741S. [CrossRef]

61. Bernard, J.Y.; Armand, M.; Garcia, C.; Forhan, A.; De Agostini, M.; Charles, M.A.; Heude, B. The association between linoleic acid levels in colostrum and child cognition at 2 and 3 y in the EDEN cohort. *Pediatr. Res.* **2015**, *77*, 829–835. [CrossRef]

62. Guxens, M.; Mendez, M.A.; Molto-Puigmarti, C.; Julvez, J.; Garcia-Esteban, R.; Forns, J.; Ferrer, M.; Vrijheid, M.; Lopez-Sabater, M.C.; Sunyer, J. Breastfeeding, Long-Chain Polyunsaturated Fatty Acids in Colostrum, and Infant Mental Development. *Pediatrics* **2011**, *128*, e880–e889. [CrossRef]

63. Bernard, J.Y.; Armand, M.; Peyre, H.; Garcia, C.; Forhan, A.; De Agostini, M.; Charles, M.A.; Heude, B.; EDEN Mother-Child Cohort Study Group (Etude des Déterminants pré- et postnatals précoces du développement et de la santé de l'Enfant). Breastfeeding, Polyunsaturated Fatty Acid Levels in Colostrum and Child Intelligence Quotient at Age 5–6 Years. *J. Pediatr.* **2017**, *183*, 43–50. [CrossRef] [PubMed]

64. Horta, B.L.; Loret de Mola, C.; Victora, C.G. Breastfeeding and intelligence: A systematic review and meta-analysis. *Acta Paediatr.* **2015**, *104*, 14–19. [CrossRef] [PubMed]

65. Jedrychowski, W.; Perera, F.; Jankowski, J.; Butscher, M.; Mroz, E.; Flak, E.; Kaim, I.; Lisowska-Miszczyk, I.; Skarupa, A.; Sowa, A. Effect of exclusive breastfeeding on the development of children's cognitive function in the Krakow prospective birth cohort study. *Eur. J. Pediatr.* **2012**, *171*, 151–158. [CrossRef] [PubMed]

66. WHO. *Breastfeeding*; WHO: Geneva, Switzerland, 2018.

67. Nyaradi, A.; Li, J.; Hickling, S.; Foster, J.; Oddy, W.H. The role of nutrition in children's neurocognitive development, from pregnancy through childhood. *Front. Hum. Neurosci.* **2013**, *7*, 97. [CrossRef] [PubMed]

68. Carson, V.; Hunter, S.; Kuzik, N.; Wiebe, S.A.; Spence, J.C.; Friedman, A.; Tremblay, M.S.; Slater, L.; Hinkley, T. Systematic review of physical activity and cognitive development in early childhood. *J. Sci. Med. Sport* **2016**, *19*, 573–578. [CrossRef] [PubMed]

69. Worobey, J. Physical activity in infancy: Developmental aspects, measurement, and importance. *Am. J. Clin. Nutr.* **2014**, *99*, 729S–733S. [CrossRef]

Article

Fruit and Vegetable Consumption and Potential Moderators Associated with All-Cause Mortality in a Representative Sample of Spanish Older Adults

Beatriz Olaya [1,2,*], **Cecilia A. Essau** [3], **Maria Victoria Moneta** [1,2], **Elvira Lara** [2,4,5], **Marta Miret** [2,4,5], **Natalia Martín-María** [2,4,5], **Darío Moreno-Agostino** [2,4,5], **José Luis Ayuso-Mateos** [2,4,5], **Adel S. Abduljabbar** [6] **and Josep Maria Haro** [1,2,6]

[1] Research, Innovation and Teaching Unit, Parc Sanitari Sant Joan de Déu, Carrer Doctor Pujadas 42, 08830 Sant Boi de Llobregat, Spain

[2] Instituto de Salud Carlos III, Centro de Investigación Biomédica en Red de Salud Mental (CIBERSAM), Calle Monforte de Lemos 3–5, 28029 Madrid, Spain

[3] Department of Psychology, University of Roehampton, Whitelands College, London SW15 4JD, UK

[4] Department of Psychiatry, Instituto de Investigación Sanitaria, Hospital Universitario de La Princesa (IIS-Princesa), Calle de Diego de León 62, 28006 Madrid, Spain

[5] Department of Psychiatry, Universidad Autónoma de Madrid, Madrid, Calle Arzobispo Morcillo 4, 28029 Madrid, Spain

[6] Psychology Deparment, King Saud University, Riyadh 11451, Saudi Arabia

* Correspondence: beatriz.olaya@pssjd.org; Tel.: +34-93-640-63-50 (ext: 1-2540); Fax: +34-93-652-00-51

Received: 17 June 2019; Accepted: 31 July 2019; Published: 2 August 2019

Abstract: This study sought to determine the association between levels of fruit and vegetable consumption and time to death, and to explore potential moderators. We analyzed a nationally-representative sample of 1699 older adults aged 65+ who were followed up for a period of 6 years. Participants were classified into low (≤3 servings day), medium (4), or high (≥5) consumption using tertiles. Unadjusted and adjusted cox proportional hazard regression models (by age, gender, cohabiting, education, multimorbidity, smoking, physical activity, alcohol consumption, and obesity) were calculated. The majority of participants (65.7%) did not meet the recommendation of five servings per day. High fruit and vegetable intake increased by 27% the probability of surviving among older adults with two chronic conditions, compared to those who consumed ≤3 servings per day (HR = 0.38, 95%CI = 0.21–0.69). However, this beneficial effect was not found for people with none, one chronic condition or three or more, indicating that this protective effect might not be sufficient for more severe cases of multimorbidity. Given a common co-occurrence of two non-communicable diseases in the elderly and the low frequency of fruit and vegetable consumption in this population, interventions to promote consuming five or more servings per day could have a significant positive impact on reducing mortality.

Keywords: survival; fruit and vegetable consumption; interaction; older adults; multimorbidity

1. Introduction

Fruit and vegetable consumption has consistently been associated with beneficial effects on health [1,2]. Several meta-analyses have indicated a reduced risk of non-communicable diseases such as cancer [3], stroke [4], diabetes [5], hypertension [6], and heart diseases [3]. According to the WHO, 16 million disability-adjusted life-years (DALYs) and 1.7 million deaths worldwide are attributable to low fruit and vegetable consumption [7]. High fruit and vegetable intake has been associated with a reduced risk of all-cause mortality [8,9] due to cardiovascular diseases (CVD) [10], but also to other non-cardiovascular diseases [11], such as cancer [12], although findings are inconsistent [13].

Many public health guidelines recommend a daily intake of a minimum of five servings per day of fruit and vegetable, although these recommendations vary across regions. For example, the Eurodiet core report [14], the World Cancer Research Fund [15], and the WHO/FAO [16] recommend at least 400 g/day, 600 g/day in Denmark [17], and 640/800 g/day in the USA [3].

Older adults have unique nutritional needs that might require special adaptations of the nutritional clinical guidelines and public health policies addressed to this population [18]. Chronic diseases, multimorbidity, and geriatric conditions such as polypharmacy, mobility-difficulties, and oral problems, are common in older adults and might be associated with malnutrition [19]. The majority of studies of the health benefits of fruit and vegetable consumption have traditionally focused on children, adolescents, or young adults, but few have included older adults [18]. These few studies seem to suggest that fruit and vegetable intake can prevent the onset of depression [20], cognitive decline [21], disability [22], and frailty [23], and can decrease the risk of disease-specific and all-cause mortality in this population [18]. Moreover, the potential benefit of fruit and vegetable intake in reducing the risk of mortality in older adults might depend on various circumstances. For example, a study conducted in several population-based cohorts from Eastern Europe [24] reported that fruit and vegetable consumption was strongly and inversely associated with risk of total and CVD mortality among smokers, compared with non-smokers.

The present study sought to determine the association between different levels of fruit and vegetable consumption and risk of all-cause mortality in a representative sample of Spanish community-dwelling older adults and to determine potential moderators of this association, including multimorbidity and other lifestyle factors such as smoking, alcohol consumption, and physical activity.

2. Methods

Study Sample

This study used data from the "Edad con Salud", a longitudinal household survey of the non-institutionalized adult population in Spain. The first wave took place between 2011 and 2012 and was part of the Collaborative Research on Ageing in Europe (COURAGE) study [25]. The participants were re-evaluated twice, between 2014 and 2015, and in 2018. A stratified multistage clustered design was used, in which strata included Autonomous Communities except Ceuta and Melilla. People aged 50+ were oversampled. The Spanish Statistical Office provided a list of households, and individuals were randomly selected from within the household by the interviewer. A total of 4753 persons participated at baseline with a final response rate of 69.9%.

Interviews were conducted face to face by trained lay interviewers using Computer-Assisted Personal Interviewing (CAPI) at respondents' homes. Participants answered a questionnaire adapted from the Study on global AGEing and adult health [26] (SAGE) which was translated from English into Spanish using World Health Organization translation guidelines for assessment instruments [27]. Quality control procedures were implemented during the fieldwork [28]. At the beginning of the interview, the interviewer judged whether the selected person had cognitive limitations that would prevent correct understanding of the survey questions. In such cases, a proxy respondent answered a short version of the interview. Proxy interviews were excluded from the present analysis ($n = 170$). We focused on people aged 65 or older with complete data in all the covariates at baseline, resulting in a final n of 1699. Ethical approval was obtained from the Clinical Investigation Ethics Committee, Parc Sanitari Sant Joan de Déu, Barcelona (PIC-12-11; PIC-71-12) and from the Clinical Investigation Ethics Committee, Hospital Universitario la Princesa, Madrid (PI-364; 2399). Informed consent was obtained from each participant.

3. Measures

3.1. Fruit and Vegetable Consumption

Participants were asked the following questions: *"How many servings of fruit do you eat on a typical day?"* and *"How many servings of vegetables do you eat on a typical day?"*. Respondents were shown a card indicating with pictures and in written explanation what was considered a serving of fruit and vegetables, according to the WHO recommendations [7]. One standard serving (portion) included 80 g, translated into different units of cups depending on the type of fruit and vegetable and standard cup measures available in the country. For example, a piece of banana or apple was considered as one serving. Tubers (such as potatoes) were not included. The total number of fruit and vegetables was added up (ranging from 0 to 18), and a categorical variable was created to indicate the level of consumption using tertiles: low ($\leq$3 servings day), medium (4 servings day), and high ($\geq$5 servings day).

3.2. Other Covariates

Socio-demographic information at baseline included age, gender, educational level (no education/primary school, secondary school, and high school/university studies), and current marital status (never married, widowed, separated or divorced, recorded as "not cohabiting", and married or cohabiting with someone, recorded as "cohabiting"). A binary variable (*ever smoked*) was created including those who had never smoked and those who were current smokers or ex-smokers (including smoking tobacco or using smokeless tobacco). Level of physical activity was evaluated with the Global Physical Activity Questionnaire [29], and participants were classified into high, medium, and low levels [30]. Respondents were asked if they had ever consumed alcohol, the number of days, and standard drinks on average. They were classified as lifetime abstainers (never consumed alcohol), occasional drinkers (did not consume alcohol in the last 30 days or in the last 7 days), non-heavy drinkers (did consume alcohol in the last 30 days and in the last 7 days), infrequent heavy drinkers (did consume alcohol 1–2 days per week, with five or more standard drinks in the last 7 days for men and four or more for women), and frequent heavy drinkers (did consume alcohol 3 or more days per week with five or more standard drinks in the last 7 days for men and four or more for women). Due to the low frequency of heavy drinkers in our sample, non-heavy drinkers, infrequent, and frequent heavy-drinkers were merged into the single category of "frequent drinker".

A combined method, consisting of self-reported physician's diagnosis and/or symptom-based algorithms [31], was used to assess the following medical conditions: arthritis, asthma, chronic obstructive pulmonary disease (COPD), angina pectoris, stroke, hypertension, and diabetes. For diabetes, only a self-reported diagnosis was considered. The presence of hypertension was based on self-reported diagnosis or presence of systolic blood pressure $\geq$ 140 mmHg or diastolic blood pressure $\geq$ 90 mmHg measured at the time of the interview [32,33]. The number of chronic conditions (CC) was calculated (none or one, 2, and 3 or more CC). Interviewers measured participants' height and weight using a stadiometer and a routinely calibrated electronic weighing scale, respectively. Body Mass Index (BMI) was calculated as weight (in kilograms) divided by the square of height (in meters). A BMI of 30 or higher was used as cut-off point for obesity [34].

3.3. Mortality

The National Death Index, a civil registry for all Spanish residents, was used to ascertain the vital status and date of death for all participants from 25 July 2011 to March 2018. Vital status was also updated during household visits in the follow-up assessment by asking respondents' relatives. A final update was conducted on the 31 October 2018 by once again consulting the National Death Index. Twelve participants that appeared as deceased had no information about their date of death. Thus, we estimated their date of death as occurring at the mid-point between the date of interview at baseline and 31 October 2018.

3.4. Statistical Analysis

Unweighted frequencies, weighted proportions, and means were used for descriptive analyses. Distinct levels of fruit and vegetable consumption were compared using the Rao-Scott chi-squared test statistic (which adjusts for complex sample design) [35] for categorical variables and one-way ANOVA test for continuous variables.

Mortality was the outcome for the analyses. Kaplan–Meier survival curves and log-rank test statistics were used to estimate the time to death (from the first interview) stratified by levels of fruit and vegetable consumption. Participants who were alive at the end of the observational period (31 October 2018) were censored.

We conducted unadjusted and adjusted Cox proportional hazards regression models to explore the association between fruit and vegetable consumption and risk of all-cause mortality. The adjusted model included levels of fruit and vegetable consumption (with "low" level as the reference category) plus other potential confounders at baseline (gender, age, educational level, cohabiting, smoking status, level of physical activity, obesity, and number of chronic conditions). Interactions between levels of fruit and vegetable consumption and covariates were explored in the adjusted models. Hazard ratios (HRs) with their 95% confidence intervals (CI) were calculated.

The assumption of proportionality was explored by calculating plots of cumulative hazard functions across the independent variables. Violation in the assumption was not found. All analyses were performed using Stata version 13 for Windows (SE version 13, StataCorp: College Station, TX, USA) taking into account complex sampling design. Weights were used to adjust for differential probabilities of selection within households, and post-stratification corrections to the weights were made to match the samples to the socio-demographic distributions of the Spanish population. Statistical significance was set at $p < 0.05$.

3.5. Results

The mean age of the total sample was 74.8 years (95% CI = 74.49–75.12) ranging from 65 to 104, with 54.8% females, 54.7% cohabiting with someone, and 46.3% reporting low levels of education (Table 1). Some 37.2% reported being smokers or ex-smokers, 33.4% presented obesity, and 33.2% had low levels of physical activity. A total of 56.2% reported having two or more chronic conditions. Fruit and vegetable servings per day ranged from 0 to 18, with a mean of four servings (95% CI = 3.82–4.19). Participants with lower consumption (equal to or less than three servings per day) were more likely to be men, smokers, not frequently engaging in physical activity, frequent drinkers, and have a low educational level.

The minimum number of days of survival was 19 and the maximum 2688, with a mean of 2323.94 days (SD = 553.7). We observed 322 confirmed deceased cases (132 women and 190 men). The Kaplan–Meier estimated curves (Figure 1) showed that the level of fruit and vegetable consumption had a significant negative effect on survival. In the adjusted Cox proportional hazards regression model, only the interaction term *fruit and vegetable consumption* number of chronic conditions* was found to be significant ($p = 0.035$). Table 2 presents the unadjusted and adjusted HRs and 95%CI. In the unadjusted model, both medium (HR = 1.68, 95%CI = 1.3–2.18) and high fruit (HR = 1.12, 95%CI = 1.1–1.14) consumption were significantly associated with higher risk of death. Other significant predictors of higher risk of mortality were being male, low levels of physical activity (compared with high), lower levels of education, being a smoker (current or past), and having three or more chronic conditions. The adjusted model including the interaction term between fruit and vegetable consumption and number of chronic conditions is presented in Table 2. In order to interpret the effect of fruit and vegetable consumption on time to death in the presence of interaction, HRs were calculated according to the number of chronic conditions (Table 3). The proportion of deceased people was significantly higher ($p < 0.001$) among those who had three or more CCs (n = 112, 25.8%) compared to those with none or one CC (n = 127, 16.3%) or two CCs (n = 83, 15.5%). Subjects who consumed five or more servings of fruit and vegetable per day and had two chronic conditions were at 27% less risk of mortality (HR = 0.38, 95%CI = 0.21–0.69, *p* = 0.002) compared with participants consuming three or fewer servings, while other covariates held constant. However, fruit and vegetable consumption had no impact on time to death among subjects who reported none or one CC, or three or more.

Table 1. Baseline characteristics of the sample and comparison between levels of fruit and vegetable consumption.

	Fruit & Vegetable Consumption				
	Total Sample (*n* = 1699)	**Low** (*n* = 669, 39.6%)	**Medium** (*n* = 448, 26.1%)	**High** (*n* = 582, 34.3%)	*p* **Value**
Death, *n (%)*	322 (18.6)	150 (21.9)	78 (18.7)	94 (14.7)	0.019
Age, *mean (95%CI)*	74.80 (74.49–75.12)	74.98 (74.46–75.5)	75.06 (74.43–75.68)	74.4 (73.57–75.24)	0.476
Females, *n (%)*	956 (54.8)	341 (49.7)	266 (58.8)	349 (57.6)	0.013
Cohabiting, *n (%)*	935 (54.7)	376 (54.4)	254 (56.9)	305 (53.4)	0.653
Educational level, *n (%)*					0.023
No education/Primary school	804 (46.3)	343 (51.7)	216 (47.5)	245 (39)	-
Secondary school	492 (30.1)	181 (27)	125 (30.6)	186 (33.4)	-
High school/University	403 (23.6)	145 (21.3)	107 (22)	151 (27.6)	0.791
Ever smoked, *n (%)*	620 (37.2)	278 (41.8)	150 (32.5)	192 (35.5)	0.008
Obesity, *n (%)*	606 (33.4)	237 (33.8)	156 (34.5)	213 (32.1)	0.803
Alcohol consumption, *n (%)*					<0.001
Lifetime abstainer	618 (36.5)	218 (31.9)	167 (38.7)	233 (40.1)	-
Occasional drinker	524 (31)	197 (28.6)	138 (30.4)	189 (34.2)	-
Frequent drinker	557 (32.5)	254 (39.5)	143 (30.9)	160 (25.7)	-
Number CC, *n (%)*					0.369
None or one	780 (43.8)	305 (43.9)	197 (42.2)	278 (44.9)	-
Two	482 (29.7)	188 (28.5)	130 (28.6)	164 (31.8)	-
Three or more	437 (26.5)	176 (27.6)	121 (29.2)	140 (23.3)	-
Level PA, *n (%)*					0.011
High	427 (25.2)	151 (21.6)	108 (21)	168 (32.7)	-
Medium	700 (41.6)	268 (40.6)	185 (45.1)	247 (40.1)	-
Low	572 (33.2)	250 (37.8)	155 (33.9)	167 (27.2)	-

Note: CC = Chronic conditions; PA = Physical activity; 95% CI = 95% Confidence interval; Unweighted frequencies, weighted proportions and means.

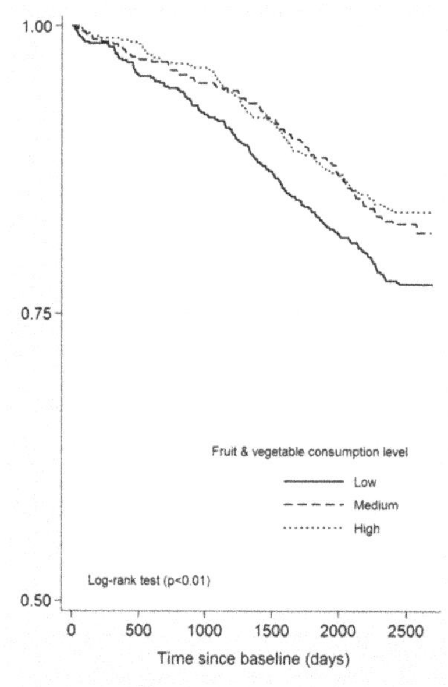

Figure 1. Kaplan–Meier estimated curves for cumulative survival by levels of fruit and vegetable consumption.

Table 2. Hazard ratios, Confidence intervals and *p* values for the unadjusted and adjusted Cox proportional hazards models (*n* = 1699).

	Unadjusted		Adjusted [a]	
	HR (95% CI)	*p* Value	HR (95% CI)	*p* Value
Fruit & veg consumption				
Low (ref.)	-	-	-	-
Medium	0.83 (0.61–1.13)	0.802	1.0 (0.62–1.60)	0.989
High	**1.68 (1.3–2.18)**	**<0.001**	1.11 (0.70–1.74)	0.659
Age	**1.12 (1.1–1.14)**	**<0.001**	**1.13 (1.1–15)**	**<0.001**
Gender				
Female (ref.)	-	-	-	-
Male	**1.96 (1.49–2.57)**	**<0.001**	**3.31 (2.21–4.99)**	**<0.001**
Marital status				
Not cohabiting (ref.)	-	-	-	-
Cohabiting	0.84 (0.66–1.1)	0.169	0.9 (0.66–1.24)	0.523
Educational level				
No education/Primary school	-	-	-	-
Secondary school	**0.57 (0.39–0.83)**	**0.004**	0.77 (0.54–1.11)	0.168
High school/University	**0.67 (0.49–0.92)**	**0.013**	1.0 (0.74–1.35)	0.997
Level PA				
High (ref.)	-	-	-	-
Medium	1.45 (0.99–2.13)	0.056	1.28 (0.84–1.93)	0.246
Low	**2.3 (1.59–3.32)**	**<0.001**	**1.61 (1.07–2.43)**	**0.021**
Smoking status	-	-	-	-
Never smoked	-	-	-	-
Ever smoked	**1.38 (1.07–1.78)**	**0.013**	1.01 (0.73–1.39)	0.973
Alcohol consumption				
Lifetime abstainer (ref.)	-	-	-	-
Occasional drinker	1.13 (0.84–1.5)	0.414	0.97 (0.69–1.36)	0.87
Frequent drinker	0.93 (0.69–1.24)	0.625	**0.64 (0.46–0.91)**	**0.012**
Obesity				
Non-obese (ref.)	-	-	-	-
Obese	1.14 (0.82–1.57)	0.43	1.11 (0.8–1.55)	0.515
Number CC				
None/one (ref.)	-	-	-	-
Two	0.95 (0.67–1.37)	0.802	1.56 (0.91–2.65)	0.102
Three or more	**1.68 (1.3–2.18)**	**<0.001**	1.47 (0.99–2.18)	0.056
Fruit & veg × number CC				
Medium/two CC	-	-	0.63 (0.31–1.28)	0.199
Medium/three+ CC	-	-	0.97 (0.47–2.0)	0.930
High/two CC	-	-	**0.34 (0.18–0.65)**	**0.001**
High/three+ CC	-	-	0.74 (0.39–1.42)	0.360

Note: HR = Hazard ratio; 95% CI = 95% Confidence interval; CC = Chronic conditions; PA = Physical activity. In bold, significant effect. [a] The adjusted model included all the variables and the interaction term simultaneously.

Table 3. Adjusted hazard ratios of the effect of fruit and vegetable consumption by number of chronic conditions on all-cause mortality. Cox regression model (*n* = 1,699).

	None or One CC				Two CC				Three or More CC			
Level of Fruit & Vegetable Consumption	HR	95%CI		*p* Value	HR	95%CI		*p* Value	HR	95%CI		*p* Value
Low (ref.)	-	-	-	-	-	-	-	-	-	-	-	-
Medium	1.0	0.62	1.6	0.989	0.63	0.33	0.19	0.151	0.96	0.54	0.73	0.905
High	1.11	0.7	1.74	0.659	**0.38**	**0.21**	**0.69**	**0.002**	0.82	0.5	1.35	0.428

Note: HR = Hazard ratio; 95%CI = 95% Confidence interval; CC = Chronic conditions. In bold, significant HR. Model included the *fruit & vegetable consumption*number CC* interaction term and the following variables: age, gender, educational level, cohabiting, ever smoked, obesity, alcohol consumption, and level of physical activity (see Table 2).

Comparison between "medium" and "high" levels of fruit and vegetable consumption; (1) in the none or one CC group: $p = 0.66$; (2) in the two CC: $p = 0.13$; (3) and in the three or more CC: $p = 0.596$

Survival curves as a function of levels of fruit and vegetable consumption and number of chronic conditions are displayed in Figure 2 for a specific pattern of covariates (i.e., females, not cohabiting, never smoked, no education/primary school studies, non-obese, lifetime abstainers, high level of physical activity, and age 74.8 years old). For people with two chronic conditions, the probability of surviving until the end of the study was significantly greater if they consumed five or more servings per day of fruit and vegetables, compared with those who ate three or fewer.

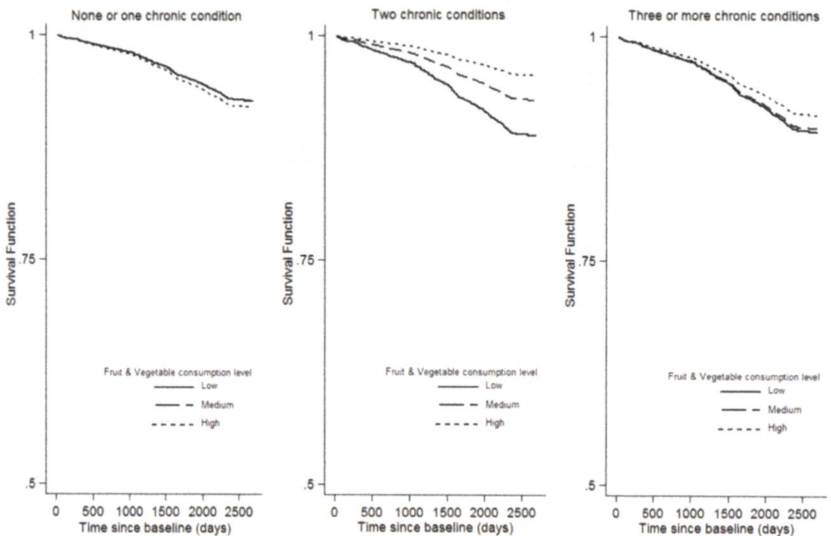

Figure 2. Survival function according to the level of fruit and vegetable consumption and number of chronic conditions ($n = 1699$). The first subfigure (left) shows the survival curves according to the three levels of fruit and vegetable intake for people with none or one chronic condition, the second (middle) for those with two chronic conditions and the third (right) shows survival curves associated with fruit and vegetable consumption for respondents with three or more chronic conditions. Note: Survival functions calculated from the adjusted Cox proportional hazards model presented in Table 2. All covariates were set equal to zero.

4. Discussion

This study sought to determine the effect of fruit and vegetable consumption on all-cause mortality in a representative sample of Spanish community-dwelling older adults who were followed up for a period of approximately 6 years. Our results show that consuming five or more servings per day increases the probability of surviving in the general older population with two chronic conditions by 27%, compared to those who consume three or fewer servings per day. However, this beneficial effect of fruit and vegetable consumption is not found among participants with none or one chronic condition, or three or more.

Most participants (65.7%) in the study did not adhere to the WHO recommendation of consuming a minimum of five servings of fruit and vegetables per day, with the median of servings per day being four. In a very large Spanish sample of university graduates, the mean consumption of fruits and vegetables was 343 g/day and 525 g/day [36], respectively, with an equivalence of approximately four and six portions of 80 g. However, few epidemiological studies have described the patterns of fruit and vegetable consumption among the older Spanish population. For example, in the Seniors-ENRICA study, a population-based cohort of Spanish older adults aged 60+, a total of 22.5% participants reported having five or more portions of fruit and vegetables a day [23], slightly inferior to the 34.3% found in our study. These discrepancies could be explained by the different tools used to assess fruit and vegetable consumption.

Previous population-based studies have repeatedly reported an inverse association between fruit and vegetable consumption, and all-cause and disease-specific mortality in the older population [37–39], although some studies have reported inconsistent findings about whether this greater risk of all-cause mortality might be mainly due to CVD-related or non-cardiovascular-related deaths, such as cancer [40]. Our findings contribute to this evidence by showing that this effect is exerted through a protective effect in the presence of multimorbidity. Additionally, the consumption of both fruit and vegetables rather than the consumption of only fruit or vegetables seems to be especially beneficial for reducing the risk of CVD [39] and non-cardiovascular diseases [3]. Our findings also support the recommendation of a minimum of five servings per day of fruit and vegetables whereas there were no differences in terms of increased risk of mortality among older adults with low (equal to or less than three servings/d) or medium (four servings/d) consumption. This is in line with previous studies that investigated the risk of all-cause mortality associated with a dose-response of fruit and vegetable consumption in a large population-based cohort aged 45–83 [8]. The authors found that consuming fewer than five servings a day was associated with progressively shorter survival and higher mortality rate, whereas consuming more than five servings did not add any benefits with respect to survival.

A systematic review conducted by Nunes et al [41] showed an overall positive association between multimorbidity (defined as the presence of two or more chronic diseases) and mortality (HR = 1.44, 95%CI = 134–1.55). In the unadjusted model, we found that only three or more chronic conditions were related to a 62% higher probability of having a shorter survival and dying, compared to those with none or one chronic condition. We found that the beneficial effect of consuming five or more fruit and vegetable servings per day is exerted in those having two CCs, but not three or more. Additionally, this protective effect seems to be beyond the confounding effects of other risk factors, such as obesity, physical activity, smoking, gender, or educational level. Participants suffering from three or more chronic conditions might represent complex patients, who might be in need of intensive care. The presence of multiple diseases is related to interactions between morbidities, inadequate use of medication, polypharmacy [42], and frailty [43]. Thus, the protective effect of high intake of fruit and vegetables might not be sufficient to reduce the risk of death in people with three or more CCs. It is also possible that older adults with three or more CCs have been given a prescription of a balanced diet, or have been advised to quit or reduce smoking and alcohol intake [44], which might in turn explain the lack of association between fruit and vegetable intake and time to death in this particular subgroup. Despite this, the beneficial effect of consuming five or more servings per day of fruit and vegetables could be huge. Taking into account that an important proportion of Spanish older adults do not reach the recommended five servings per day of fruit and vegetables along with the high prevalence of multimorbidity in this population, interventions promoting fruit and vegetable consumption among older adults might have a positive impact on reducing the risk of death and increasing their quality of life. Future research is needed to learn whether fruit and vegetable intake is particularly beneficial in reducing the risk of death for a particular pair of diseases.

There are several mechanisms by which fruit and vegetable consumption can reduce the risk of mortality in older adults. Fruit and vegetables contain a variety of nutrients and phytochemicals (i.e., fibre, vitamin C, carotenoids, antioxidants, potassium, and flavonoids) that act through several biological mechanisms to reduce the risk of chronic conditions and premature mortality [3]. Greater intake of fruit and vegetables has also been linked to a greater adherence to the Mediterranean diet in older adults (characterized by abundant consumption of olive oil, minimally processed, locally grown vegetables, fruits, nuts, legumes, and cereals, and proteins coming mainly from fish and shellfish) [45] and to reduced consumption of sweet foods [46] which in turn might also prevent

CVD [47], several types of cancer [48,49], cognitive decline, and dementia [50,51], while increasing longevity [52]. Our study did not include data on adherence to the Mediterranean diet or other potential dietary risk factors for non-communicable diseases and risk of mortality, such as consumption of red and processed meat [53] or ultraprocessed food [54]. More studies are needed to determine whether the beneficial effect of fruit and vegetable intake on the probability of survival among people with multimorbidity is maintained or attenuated by the presence of these diet-related risk factors. Additionally, the way in which fruits and vegetables are consumed (e.g., raw or cooked) might also play an important role in the potential protective factor among older adults with chronic conditions. Another mechanism by which fruit and vegetable consumption might impact the risk of mortality among older adults is the presence of unhealthy lifestyles among those who consume less fruit and vegetables. Previous research has indicated an inverse association between fruit and vegetable intake and smoking [55], alcohol consumption [56], obesity [57], and sedentarism [58]. The beneficial effects of consuming fruit and vegetables, such as lower systemic inflammation [59], reduced oxidative stress [60], and decreased platelet aggregation [61], may partially reduce the effects of smoking and alcohol intake [55,56]. However, we did not find significant interactions between fruit and vegetable consumption and smoking status, alcohol consumption, obesity, or low levels of physical activity. Future research is needed to replicate these results.

Our study had some limitations. First, health variables, such as fruit and vegetable consumption, tobacco and physical activity, were self-reported, thus potentially leading to measurement errors or misclassification. Additionally, recall bias might also be present. Second, it was assumed that the fruit and vegetable intake pattern was unchanged during the follow-up period. Third, it is possible that the beneficial effect of fruit and vegetable consumption is not observed among participants with none or one chronic condition because they are more likely to survive during the follow-up period. Thus, longer periods of follow-up might be needed. Fourth, measuring fruit and vegetable consumption might be problematic. For example, the study did not extensively measure the dietary habits of the sample through a 24-hour dietary recall or a frequency questionnaire; thus, some measurement bias might have been introduced. Questions concerning the number of fruit and vegetable servings were asked once, yet they may be prone to seasonable bias as well. Additionally, these questions were aggregated, and the effect of this variable could be due to specific sorts of fruits and vegetables. Fourth, residual confounding might explain our findings. For example, consuming vitamin supplements or specific diet patterns such as the Mediterranean diet could be related to both fruit and vegetable consumption and mortality. However, findings were adjusted for several potential confounders, such as smoking status, alcohol consumption, physical activity, and obesity.

In sum, the finding that a high level of fruit and vegetable consumption (reaching the threshold of five or more servings per day) significantly reduces the risk of mortality among older adults with two chronic conditions has several implications. As has been shown in the present study, fruit and vegetable intake in the general population of older adults does not approach recommended levels. Interventions to increase fruit and vegetable intake in older adults should take into account their unique nutritional needs and barriers, as well as several characteristics that might influence their fruit and vegetable intake, such as appetite loss, tooth loss and oral problems, changes in perception of hunger, taste acuity and sense of smell (sometimes associated with drugs' side effects), and mobility difficulties in shopping [18]. These factors should be taken into account when designing interventions to promote fruit and vegetable consumption geared to the older population.

Author Contributions: Conceptualization, B.O., C.A.E., J.L.A.-M., and J.M.H.; Methodology, B.O., M.V.M., E.L., M.M., N.M.M., D.M.-A., J.L.A.-M., and J.M.H.; Formal analysis, B.O. and M.V.M.; Investigation, B.O., M.V.M., E.L., M.M., N.M.M., D.M.-A., J.L.A.-M., and J.M.H.; Resources, B.O., M.V.M., E.L., M.M., N.M.M., D.M.-A., J.L.A.-M., and J.M.H.; Data curation, B.O., M.V.M., E.L., N.M.M., D.M.-A.; Writing–Original draft preparation, B.O., C.A.E. and A.A.; Writing–Review & Editing, all authors; Supervision, C.A.E., J.L.A.-M., and J.M.H.; Project administration, B.O., E.L., M.M., N.M.M., D.M.-A., J.L.A.-M., and J.M.H., Funding acquisition, M.M., J.L.A.-M., and J.M.H.

Funding: This work was supported by the Seventh Framework Programme [grant number 223071-COURAGE Study]; the Instituto de Salud Carlos III-FIS [grant numbers PS09/00295, PS09/01845, PI12/01490, PI13/00059, PI16/00218, and PI16/01073]; the European Regional Development Fund (ERDF) "A Way to Build Europe" [grant numbers PI12/01490 and PI13/00059]; Horizon 2020 Research and Innovation Programme [grant number 635316-ATHLOS]; and the Centro de Investigación Biomédica en Red de Salud Mental. B.O.'s work is supported by the PERIS program 2016-2020 "Ajuts per a la Incorporació de Científics i Tecnòlegs" [grant number SLT006/17/00066], with the support of the Health Department of the Generalitat de Catalunya. E.L.'s work is supported by the Sara Borrell postdoctoral programme (CD18/00099) of the Instituto de Salud Carlos III (Spain) and co-funded by European Union (ERDF/ESF, "Investing in your future"). Funding sources did not have any role in study design, data collection, analysis and interpretation of data, writing of the article, or the decision to submit it for publication.

Conflicts of Interest: The authors declare no conflict of interest.

References

1. Fulton, S.L.; McKinley, M.C.; Young, I.S.; Cardwell, C.R.; Woodside, J.V. The effect of increasing fruit and vegetable consumption on overall diet: A systematic review and meta-analysis. *Crit. Rev. Food Sci. Nutr.* **2016**, *56*, 802–816. [CrossRef] [PubMed]

2. Mytton, O.T.; Nnoaham, K.; Eyles, H.; Scarborough, P.; Ni Mhurchu, C. Systematic review and meta-analysis of the effect of increased vegetable and fruit consumption on body weight and energy intake. *BMC Public Health* **2014**, *14*, 886.

3. Aune, D.; Giovannucci, E.; Boffetta, P.; Fadnes, L.T.; Keum, N.N.; Norat, T.; Greenwood, D.C.; Riboli, E.; Vatten, L.J.; Tonstad, S. Fruit and vegetable intake and the risk of cardiovascular disease, total cancer and all-cause mortality—A systematic review and dose-response meta-analysis of prospective studies. *Int. J. Epidemiol.* **2017**, *46*, 1029–1056. [CrossRef] [PubMed]

4. Hu, D.; Huang, J.; Wang, Y.; Zhang, D.; Qu, Y. Fruits and vegetables consumption and risk of stroke. *Stroke* **2014**, *45*, 1613–1619. [CrossRef] [PubMed]

5. Li, M.; Fan, Y.; Zhang, X.; Hou, W.; Tang, Z. Fruit and vegetable intake and risk of type 2 diabetes mellitus: Meta-analysis of prospective cohort studies. *BMJ Open* **2014**, *4*, e005497. [PubMed]

6. Li, B.; Li, F.; Wang, L.; Zhang, D. Fruit and vegetables consumption and risk of hypertension: A meta-analysis. *J. Clin. Hypertens.* **2016**, *18*, 468–476.

7. WHO, Promoting Fruit and Vegetable Consumption Around the World. Available online: https://www.who.int/dietphysicalactivity/fruit/en/index2.html (accessed on 5 June 2019).

8. Bellavia, A.; Larsson, S.C.; Bottai, M.; Wolk, A.; Orsini, N. Fruit and vegetable consumption and all-cause mortality: A dose-response analysis. *Am. J. Clin. Nutr.* **2013**, *98*, 454–459. [PubMed]

9. Buil-Cosiales, P.; Zazpe, I.; Toledo, E.; Corella, D.; Salas-Salvadó, J.; Diez-Espino, J.; Ros, E.; Navajas, J.F.C.; Santos-Lozano, J.M.; Arós, F.; et al. Fiber intake and all-cause mortality in the Prevención con Dieta Mediterránea (PREDIMED) study. *Am. J. Clin. Nutr.* **2014**, *100*, 1498–1507. [CrossRef] [PubMed]

10. Wang, X.; Ouyang, Y.; Liu, J.; Zhu, M.; Zhao, G.; Bao, W.; Hu, F.B. Fruit and vegetable consumption and mortality from all causes, cardiovascular disease, and cancer: Systematic review and dose-response meta-analysis of prospective cohort studies. *BMJ* **2014**, *349*, g4490. [CrossRef]

11. Miller, V.; Mente, A.; Dehghan, M.; Rangarajan, S.; Zhang, X.; Swaminathan, S.; Dagenais, G.; Gupta, R.; Mohan, V.; Lear, S.; et al. Fruit, vegetable, and legume intake, and cardiovascular disease and deaths in 18 countries (PURE): A prospective cohort study. *Lancet* **2017**, *390*, 2037–2049. [CrossRef]

12. Oyebode, O.; Gordon-Dseagu, V.; Walker, A.; Mindell, J.S. Fruit and vegetable consumption and all-cause, cancer and CVD mortality: Analysis of health survey for England data. *J. Epidemiol. Community Health* **2014**, *68*, 856–862. [CrossRef] [PubMed]

13. Liu, Y.; Sobue, T.; Otani, T.; Tsugane, S. Vegetables, fruit consumption and risk of lung cancer among middle-aged Japanese men and women: JPHC study. *Cancer Causes Control* **2004**, *15*, 349–357. [CrossRef] [PubMed]

14. Rajala, M. Nutrition and diet for healthy lifestyles in Europe: Science and policy implications. *Public Health Nutr.* **2001**, *4*, 339–340. [PubMed]

15. World Cancer Research Fund; American Institute for Cancer Research. *Diet Nutrition Physical Activity and Cancer: A Global Perspective*; Continuous Update Project Expert Report 2018; World Cancer Research Fund/American Institute for Cancer Research: Washington, DC, USA, 2018. Available online: http://dietandcancerreport.org (accessed on 27 April 2019).

16. WHO/FAO. *Expert Report on Diet, Nutrition and the Prevention of Chronic Diseases*; Technical Report Series 916; World Health Organisation: Geneva, Switzerland, 2003.

17. Yngve, A.; Wolf, A.; Poortvliet, E.; Elmadfa, I.; Brug, J.; Ehrenblad, B.; Franchini, B.; Haraldsdóttir, J.; Krølner, R.; Maes, L.; et al. Fruit and vegetable intake in a sample of 11-year-old children in 9 European countries: The pro children cross-sectional survey. *Ann. Nutr. Metab.* **2005**, *49*, 236–245. [CrossRef] [PubMed]

18. Nicklett, E.J.; Kadell, A.R. Fruit and vegetable intake among older adults: A scoping review. *Maturitas* **2013**, *75*, 305–312. [PubMed]

19. Amarya, S.; Singh, K.; Sabharwal, M. Changes during aging and their association with malnutrition. *J. Clin. Gerontol. Geriatr.* **2015**, *6*, 78–84.

20. Tsai, A.C.; Chang, T.-L.; Chi, S.-H. Frequent consumption of vegetables predicts lower risk of depression in older Taiwanese—Results of a prospective population-based study. *Public Health Nutr.* **2012**, *15*, 1087–1092. [CrossRef]

21. Loef, M.; Walach, H. Fruit, vegetables and prevention of cognitive decline or dementia: A systematic review of cohort studies. *J. Nutr. Health Aging* **2012**, *16*, 626–630. [CrossRef]

22. Gopinath, B.; Russell, J.; Flood, V.M.; Burlutsky, G.; Mitchell, P. Adherence to dietary guidelines positively affects quality of life and functional status of older adults. *J. Acad. Nutr. Diet.* **2014**, *114*, 220–229. [CrossRef]

23. García-Esquinas, E.; Rahi, B.; Peres, K.; Colpo, M.; Dartigues, J.-F.; Bandinelli, S.; Feart, C.; Rodríguez-Artalejo, F. Consumption of fruit and vegetables and risk of frailty: A dose-response analysis of 3 prospective cohorts of community-dwelling older adults. *Am. J. Clin. Nutr.* **2016**, *104*, 132–142. [CrossRef]

24. Stefler, D.; Pikhart, H.; Kubinova, R.; Pajak, A.; Stepaniak, U.; Malyutina, S.; Simonova, G.; Peasey, A.; Marmot, M.G.; Bobak, M. Fruit and vegetable consumption and mortality in Eastern Europe: Longitudinal results from the health, alcohol and psychosocial factors in Eastern Europe study. *Eur. J. Prev. Cardiol.* **2016**, *23*, 493–501. [CrossRef] [PubMed]

25. Leonardi, M.; Chatterji, S.; Koskinen, S.; Ayuso-Mateos, J.L.; Haro, J.M.; Frisoni, G.; Frattura, L.; Martinuzzi, A.; Tobiasz-Adamczyk, B.; Gmurek, M.; et al. Determinants of health and disability in ageing population: The courage in Europe project (collaborative research on ageing in Europe). *Clin. Psychol. Psychother.* **2014**, *21*, 193–198. [PubMed]

26. Kowal, P.; Chatterji, S.; Naidoo, N.; Biritwum, R.; Fan, W.; Lopez Ridaura, R.; Maximova, T.; Arokiasamy, P.; Phaswana-Mafuya, N.; Williams, S.; et al. Data resource profile: The World Health Organization study on global AGEing and adult health (SAGE). *Int. J. Epidemiol.* **2012**, *41*, 1639–1649. [CrossRef] [PubMed]

27. WHO. Process of Translation and Adaptation of Instruments. Available online: http://www.who.int/substance_abuse/research_tools/translation/en/ (accessed on 31 March 2019).

28. Üstün, T.; Chatterji, S.; Mechbal, A.; Murray, C.; Groups, W.C. Quality assurance in surveys: Standards, guidelines and procedures. In *Household Sample Surveys in Developing and Transtion Countries*; United Nations: New York, NY, USA, 2005.

29. Bull, F.C.; Maslin, T.S.; Armstrong, T. Global physical activity questionnaire (GPAQ): Nine country reliability and validity study. *J. Phys. Act. Health* **2009**, *6*, 790–804. [CrossRef]

30. Olaya, B.; Moneta, M.V.; Doménech-Abella, J.; Miret, M.; Bayes, I.; Ayuso-Mateos, J.L.; Haro, J.M. Mobility difficulties, physical activity, and all-cause mortality risk in a nationally representative sample of older adults. *J. Gerontol. Ser. A Biol. Sci. Med. Sci.* **2018**, *73*, 1272–1279. [CrossRef]

31. Garin, N.; Koyanagi, A.; Chatterji, S.; Tyrovolas, S.; Olaya, B.; Leonardi, M.; Lara, E.; Koskinen, S.; Tobiasz-Adamczyk, B.; Ayuso-Mateos, J.L.; et al. Global multimorbidity patterns: A cross-sectional, population-based, multi-country study. *J. Gerontol. A Biol. Sci. Med. Sci.* **2016**, *71*, 205–214. [PubMed]

32. Basu, S.; Millett, C. Social epidemiology of hypertension in middle-income countries: Determinants of prevalence, diagnosis, treatment, and control in the WHO SAGE study. *Hypertension* **2013**, *62*, 18–26. [CrossRef]

33. Mancia, G.; Fagard, R.; Narkiewicz, K.; Redón, J.; Zanchetti, A.; Böhm, M.; Christiaens, T.; Cifkova, R.; De Backer, G.; Dominiczak, A.; et al. 2013 ESH/ESC Guidelines for the management of arterial hypertension. *J. Hypertens.* **2013**, *31*, 1281–1357. [CrossRef]

34. World Health Organization WHO. Global Database on Body Mass Index. Available online: http://www.euro.who.int/en/health-topics/disease-prevention/nutrition/a-healthy-lifestyle/body-mass-index-bmi (accessed on 2 May 2019).

35. Rao, J.N.K.; Scott, A.J. On chi-squared tests for multiway contingency tables with cell proportions estimated from survey data. *Ann. Stat.* **1984**, *12*, 46–60. [CrossRef]
36. Buil-Cosiales, P.; Martinez-Gonzalez, M.A.; Ruiz-Canela, M.; Díez-Espino, J.; García-Arellano, A.; Toledo, E. Consumption of fruit or fiber-fruit decreases the risk of cardiovascular disease in a Mediterranean young cohort. *Nutrients* **2017**, *9*, 295. [CrossRef]
37. Hodgson, J.M.; Prince, R.L.; Woodman, R.J.; Bondonno, C.P.; Ivey, K.L.; Bondonno, N.; Rimm, E.B.; Ward, N.C.; Croft, K.D.; Lewis, J.R. Apple intake is inversely associated with all-cause and disease-specific mortality in elderly women. *Br. J. Nutr.* **2016**, *115*, 860–867. [CrossRef]
38. Iimuro, S.; Yoshimura, Y.; Umegaki, H.; Sakurai, T.; Araki, A.; Ohashi, Y.; Iijima, K.; Ito, H. Japanese elderly diabetes intervention trial study group dietary pattern and mortality in Japanese elderly patients with type 2 diabetes mellitus: Does a vegetable- and fish-rich diet improve mortality? An explanatory study. *Geriatr. Gerontol. Int.* **2012**, *12*, 59–67. [CrossRef]
39. Buil-Cosiales, P.; Toledo, E.; Salas-Salvadó, J.; Zazpe, I.; Farràs, M.; Basterra-Gortari, F.J.; Diez-Espino, J.; Estruch, R.; Corella, D.; Ros, E.; et al. Association between dietary fibre intake and fruit, vegetable or whole-grain consumption and the risk of CVD: Results from the PREvención con DIeta MEDiterránea (PREDIMED) trial. *Br. J. Nutr.* **2016**, *116*, 534–546. [CrossRef]
40. Hung, H.-C.; Joshipura, K.J.; Jiang, R.; Hu, F.B.; Hunter, D.; Smith-Warner, S.A.; Colditz, G.A.; Rosner, B.; Spiegelman, D.; Willett, W.C. Fruit and vegetable intake and risk of major chronic disease. *J. Natl. Cancer Inst.* **2004**, *96*, 1577–1584. [CrossRef]
41. Nunes, B.P.; Flores, T.R.; Mielke, G.I.; Thumé, E.; Facchini, L.A. Multimorbidity and mortality in older adults: A systematic review and meta-analysis. *Arch. Gerontol. Geriatr.* **2016**, *67*, 130–138. [CrossRef]
42. Calderón-Larrañaga, A.; Poblador-Plou, B.; González-Rubio, F.; Gimeno-Feliu, L.A.; Abad-Díez, J.M.; Prados-Torres, A. Multimorbidity, polypharmacy, referrals, and adverse drug events: Are we doing things well? *Br. J. Gen. Pract.* **2012**, *62*, e821–e826. [CrossRef]
43. De Mello, C.A.; Engstrom, E.M.; Alves, L.C. Health-related and socio-demographic factors associated with frailty in the elderly: A systematic literature review. *Cad. Saude Publica* **2014**, *30*, 1143–1168. [CrossRef]
44. Hurst, J.R.; Dickhaus, J.; Maulik, P.K.; Miranda, J.J.; Pastakia, S.D.; Soriano, J.B.; Siddharthan, T.; Vedanthan, R.; GACD Multi-Morbidity Working Group. Global alliance for chronic disease researchers' statement on multimorbidity. *Lancet Glob. Health* **2018**, *6*, e1270–e1271. [CrossRef]
45. Trichopoulou, A.; Lagiou, P. Healthy traditional Mediterranean diet: An expression of culture, history, and lifestyle. *Nutr. Rev.* **2009**, *55*, 383–389. [CrossRef]
46. Bermejo, L.M.; Aparicio, A.; Andrés, P.; López-Sobaler, A.M.; Ortega, R.M. The influence of fruit and vegetable intake on the nutritional status and plasma homocysteine levels of institutionalised elderly people. *Public Health Nutr.* **2007**, *10*, 266–272. [CrossRef]
47. Martínez-González, M.A.; Gea, A.; Ruiz-Canela, M. The Mediterranean diet and cardiovascular health: A critical review. *Circ. Res.* **2019**, *124*, 779–798. [CrossRef]
48. Barak, Y.; Fridman, D. Impact of Mediterranean diet on cancer: Focused literature review. *Cancer Genom. Proteom.* **2017**, *14*, 403–408.
49. Schwingshackl, L.; Hoffmann, G. Adherence to Mediterranean diet and risk of cancer: An updated systematic review and meta-analysis of observational studies. *Cancer Med.* **2015**, *4*, 1933–1947. [CrossRef]
50. Samieri, C.; Grodstein, F.; Rosner, B.A.; Kang, J.H.; Cook, N.R.; Manson, J.E.; Buring, J.E.; Willett, W.C.; Okereke, O.I. Mediterranean diet and cognitive function in older age. *Epidemiology* **2013**, *24*, 490–499. [CrossRef]
51. Anastasiou, C.A.; Yannakoulia, M.; Kosmidis, M.H.; Dardiotis, E.; Hadjigeorgiou, G.M.; Sakka, P.; Arampatzi, X.; Bougea, A.; Labropoulos, I.; Scarmeas, N. Mediterranean diet and cognitive health: Initial results from the Hellenic longitudinal investigation of ageing and diet. *PLoS ONE* **2017**, *12*, e0182048. [CrossRef]
52. Trichopoulou, A.; Critselis, E. Mediterranean diet and longevity. *Eur. J. Cancer Prev.* **2004**, *13*, 453–456. [CrossRef]
53. Schwingshackl, L.; Schwedhelm, C.; Hoffmann, G.; Lampousi, A.M.; Knüppel, S.; Iqbal, K.; Bechthold, A.; Schlesinger, S.; Boeing, H. Food groups and risk of all-cause mortality: A systematic review and meta-analysis of prospective studies. *Am. J. Clin. Nutr.* **2017**, *105*, 1462–1473. [CrossRef]

54. Schnabel, L.; Kesse-Guyot, E.; Allès, B.; Touvier, M.; Srour, B.; Hercberg, S.; Buscail, C.; Julia, C. Association between ultraprocessed food consumption and risk of mortality among middle-aged adults in France. *JAMA Intern. Med.* **2019**, *179*, 490–498. [CrossRef]

55. Dauchet, L.; Montaye, M.; Ruidavets, J.-B.; Arveiler, D.; Kee, F.; Bingham, A.; Ferrières, J.; Haas, B.; Evans, A.; Ducimetière, P.; et al. Association between the frequency of fruit and vegetable consumption and cardiovascular disease in male smokers and non-smokers. *Eur. J. Clin. Nutr.* **2010**, *64*, 578–586. [CrossRef]

56. Kesse, E.; Clavel-Chapelon, F.; Slimani, N.; van Liere, M. Do eating habits differ according to alcohol consumption? Results of a study of the French cohort of the European prospective investigation into cancer and nutrition (E3N-EPIC). *Am. J. Clin. Nutr.* **2001**, *74*, 322–327. [CrossRef]

57. Sharma, S.P.; Chung, H.J.; Kim, H.J.; Hong, S.T. Paradoxical effects of fruit on obesity. *Nutrients* **2016**, *8*, 633. [CrossRef]

58. Jezewska-Zychowicz, M.; Gębski, J.; Guzek, D.; Świątkowska, M.; Stangierska, D.; Plichta, M.; Wasilewska, M. The associations between dietary patterns and sedentary behaviors in Polish adults (LifeStyle study). *Nutrients* **2018**, *10*, 1004. [CrossRef]

59. Esmaillzadeh, A.; Kimiagar, M.; Mehrabi, Y.; Azadbakht, L.; Hu, F.B.; Willett, W.C. Fruit and vegetable intakes, C-reactive protein, and the metabolic syndrome. *Am. J. Clin. Nutr.* **2006**, *84*, 1489–1497. [CrossRef]

60. Kris-Etherton, P.M.; Hecker, K.D.; Bonanome, A.; Coval, S.M.; Binkoski, A.E.; Hilpert, K.F.; Griel, A.E.; Etherton, T.D. Bioactive compounds in foods: Their role in the prevention of cardiovascular disease and cancer. *Am. J. Med.* **2002**, *113* (Suppl. 9B), 71S–88S. [CrossRef]

61. Vita, J.A. Polyphenols and cardiovascular disease: Effects on endothelial and platelet function. *Am. J. Clin. Nutr.* **2005**, *81*, 292S–297S. [CrossRef]

 *nutrients*

Article

Changes in Nutritional Status and Musculoskeletal Health in a Geriatric Post-Fall Care Plan Setting

Romy Conzade [1,2,3], Steven Phu [1,2], Sara Vogrin [1,2], Ebrahim Bani Hassan [1,2], Walter Sepúlveda-Loyola [1,2,4], Barbara Thorand [3] and Gustavo Duque [1,2,*]

[1] Australian Institute for Musculoskeletal Science (AIMSS), The University of Melbourne and Western Health, St. Albans, Victoria 3021, Australia

[2] Department of Medicine—Western Health, Melbourne Medical School, The University of Melbourne, St. Albans, Victoria 3021, Australia

[3] Institute of Epidemiology, Helmholtz Zentrum München, German Research Center for Environmental Health (GmbH), 85764 Neuherberg, Germany

[4] Department of Physiotherapy, Londrina State University (UEL) and University North of Paraná (UNOPAR), Londrina, Paraná 86041-120, Brazil

[*] Correspondence: gustavo.duque@unimelb.edu.au; Tel.: +61-3-8395-8121

Received: 25 May 2019; Accepted: 5 July 2019; Published: 9 July 2019

Abstract: Understanding how changes in nutritional status influence musculoskeletal recovery after falling remains unclear. We explored associations between changes in nutritional status and musculoskeletal health in 106 community-dwelling older adults aged ≥65 years, who attended the Falls and Fractures Clinic at Sunshine Hospital in St Albans, Australia after falling. At baseline and after 6 months, individuals were assessed for Mini Nutritional Assessment (MNA®), grip strength, gait speed, Timed Up and Go (TUG) test, Short Physical Performance Battery (SPPB), and bone turnover marker levels. Associations were examined using multiple linear regression, adjusted for baseline covariates and post-fall care plans. Over 6 months, the prevalence of malnutrition or risk thereof decreased from 29% to 15% using MNA <24/30. Specifically, 20 individuals (19%) improved, 7 (7%) deteriorated, and 73 (69%) maintained nutritional status, including 65 (61%) who remained well-nourished and 8 (8%) who remained malnourished/at risk. A 1-point increase in MNA score over 6 months was associated with an increase of 0.20 points (95% confidence interval 0.10, 0.31, $p < 0.001$) in SPPB score. Improvement in nutritional status was associated with improvement in physical performance, providing a basis for interventional studies to ascertain causality and evaluate nutritional models of care for post-fall functional recovery in older adults.

Keywords: malnutrition; MNA; nutrition; physical performance; bone turnover; sarcopenia; osteosarcopenia; falls; elderly; prospective

1. Introduction

About a third of community-dwelling older adults aged ≥65 years in Western countries fall each year and the frequency of falls and fall-related injuries (fractures or head trauma) increase with age [1]. Falls increase the risk of hospitalization and nursing home admission, as well as morbidity and mortality [2]. Investigating modifiable risk factors of falls is a key priority area for healthcare systems, which strive to identify conditions that prevent falls.

Poor nutritional status has been considered an important modifiable risk factor for falls [3]. Compared to well-nourished older adults, risk of experiencing falls has been shown in a meta-analysis of prospective studies to be 45% higher in malnourished individuals or those at risk of malnutrition ($n = 9510$) [4], based on the validated Mini Nutritional Assessment (MNA®) tool [5]. In an interventional study by Swanenburg et al., the combination of a three-month calcium/vitamin D supplementation plus protein/exercise was associated with a 89% reduction in the rate of falls over 12 months compared to only calcium/vitamin D supplementation in older women aged ≥65 years with low mineral density (*n* = 20) [6]. Altogether, these findings suggest that nutritional status should be properly considered when assessing the risk of falls in community-dwelling older adults; yet it is currently not included in most fall risk screening tools [7].

Poor nutritional status can be a consequence of underlying comorbid conditions [8], which may increase the risk of falls due to clinical and adverse effects on cognitive, functional, and physical performance [9]. Poor nutritional status due to inadequate nutritional intake, especially of proteins, can also be detrimental for maintaining the integrity and function of skeletal muscle and bone [10–12], possibly increasing the risk for sarcopenia, osteoporosis or both [13–16]. Sarcopenia-associated risk of falling and increased bone vulnerability have a synergistic impact on falls and fractures occurrence [17,18]. The impact of nutritional status on the risk of falls can thus be explored through the pathway of musculoskeletal health as an important contributor to falls risk [19,20].

Evidence suggests that malnutrition based on MNA is able to predict musculoskeletal decline in various healthcare settings [21–23], but the relationship between nutritional changes and musculoskeletal outcomes remains under-researched [24], in particular among those who fall. Improving knowledge about how nutritional changes may influence relevant musculoskeletal outcomes might be important to effective targeting of multidisciplinary post-fall interventions for older adults living in the community. This study aimed to investigate changes in nutritional status in older adults with a history of falling using the validated MNA®, and to determine associations between changes in nutritional status and relevant musculoskeletal outcomes. We hypothesized that improvement in nutritional status is associated with greater musculoskeletal recovery.

2. Materials and Methods

2.1. Study Design and Individuals

This retrospective observational study examined associations between changes in nutritional status and musculoskeletal outcomes among community-dwelling older adults who attended the Falls and Fractures Clinic at the Australian Institute for Musculoskeletal Science (Western Health-Sunshine Hospital) in St Albans, VIC, Australia. A multidisciplinary team at the clinic, including a geriatrician, a fracture liaison nurse, an accredited exercise physiologist, and a bone densitometrist, provides comprehensive care for older adults with a history of more than two falls in the previous year, or a single fall with established gait and/or balance problem, and/or clinical or radiological risk of falls and/or fractures. We analyzed information from baseline attendance between October 2016 and December 2018 and from follow-up attendance after a median time of 6 months (interquartile range (Q1–Q3) 6–8 months). All measurements obtained were part of standard care practices at this health service. The Western Health Low Risk Ethics Panel approved the registration of the Falls and Fractures Clinic Databank (DB2017.13, date of approval 23 October 2018) and the research protocol of the present study (QA2018.90_48118, date of approval 5 December 2018). Participant consent was waived due to use of de-identified data collected as part of standard care at the clinic and due to the low risk nature of the study beyond the initial consent to attend the clinic.

2.2. Demographic and Clinical Measures

Demographic data was obtained from the patient medical record including age, gender, and residential location. Comprehensive clinical assessment was performed by the geriatrician and the nurse as part of routine care practices on clinic attendance including comorbidities, family history, fracture history, osteoporosis risk assessment (e.g., hormone replacement therapy, menopause age, smoking, alcohol), falls risk (e.g., hearing and visual deficit, altered elimination, impaired mobility), assessment for postural drop, and list of current medications. For the purpose of this study, a Charlson age-comorbidity index (CACI) was generated, with an index of ≥5 being suggestive of severe comorbidity [25]. The CACI calculation is explained in Supplementary Table S1. Polypharmacy was defined as use of ≥5 prescribed or regularly taken medications, including drugs and dietary supplements. Depression was screened using the Short Form Geriatric Depression Scale (GDS), with a score of ≥6/15 points considered as "suggestive of depression" [26].

2.3. Nutritional Status

Nutritional status was evaluated by the nurse using the Mini Nutritional Assessment (MNA®), which is a validated screening and assessment tool for older adults in community and hospital settings. The full MNA consists of 18 items (6 questions in the screening part, also called the MNA Short-Form (MNA-SF), and 12 questions in the assessment part) capturing anthropometric measures, dietary intake, appetite, general health, and mobility [5]. The screening part first identifies older adults as "well-nourished" (MNA-SF ≥ 12/14) or "at nutritional risk" (MNA-SF < 12/14), so that the full MNA is performed only if an individual is "at nutritional risk". The full categorized MNA then classifies individuals into "malnourished" (MNA < 17/30), "at risk of malnutrition" (17/30 ≤ MNA < 24/30) or "well nourished" (MNA ≥ 24/30) [5]. To calculate a full continuous MNA score, individuals with MNA-SF ≥ 12/14 in the screening part were adjusted into a full MNA score (MNA-SF + 16 points) to obtain a full score ranging 0–30 points. Weight and height were measured using standardized scales to the nearest 0.1 kg and 0.01 m, respectively.

2.4. Biochemical Measures

Fasting venous blood was collected for the measurement of serum albumin, 25-hydroxyvitamin D (25OHD), parathyroid hormone (PTH), hemoglobin, and C-terminal telopeptide of type 1 collagen (CTx). Serum albumin, and hemoglobin levels were determined using automated standard laboratory methods. Serum 25OHD levels were measured by chemiluminescence immunoassay on a LIAISON® XL analyzer (DiaSorin S.p.A., Saluggia, Italy). Circulating intact PTH was measured by immunochemoluminometric assay performed on ADVIA Centaur® (Siemens Healthcare Diagnostics, Deerfield, MA, USA). Serum CTx levels were measured by electrochemiluminescence immunoassay on a Cobas® 6000 analyzer (Roche Diagnostics International Ltd, Rotkreuz Switzerland). Cut-off values for subnormal levels were 25OHD < 75 nmol/L [27], PTH > 6.9 pmol/L, and hemoglobin <130 g/L (men), <120 g/L (women). Estimated-glomerular filtration rate (eGFR) was calculated from serum creatinine as an indicator of renal function (MDRD formula [28]), with subnormal cut-off values of <60 mL/min/1.73 m^2. All measurements were performed at the pathology networks affiliated with the Western Health-Sunshine Hospital in St. Albans, Australia.

2.5. Post-Fall Care Plan

After review of the results of the complete assessment of the individuals' risk of falls and fractures by the multi-disciplinary team, individuals were provided with individualized care plans that included pharmacological (e.g., osteoporosis treatment, vitamin D supplements, protein supplements), and non-pharmacological recommendations (e.g., nutrition advice, physical exercise), with a focus on preventing new or recurrent episodes of falls and/or osteoporotic fractures. The care plan was patient-centered through consideration of the risk assessment, individual patient circumstances, and preferences. In consultation with their local general practitioners, individuals were involved in the management of their respective care plan. The current study looks into changes in nutritional status and musculoskeletal components over a period of 6 months.

2.6. Musculoskeletal Outcome Measures

Musculoskeletal outcome measures were evaluated by the exercise physiologist as part of standard patient assessment on attendance at baseline and 6-month follow-up. Grip strength (kg) was measured with a handheld JAMAR hydraulic dynamometer (Sammons Preston Inc., Bolingbrook, IL, USA). Individuals had to squeeze the device as hard as possible 3 times in each hand; the highest value was recorded. Gait speed (m/sec) was evaluated using a sensitive walkway (GAITRite system, 16′ model, CIR Systems Inc., Havertown, PA, USA), which recorded spatiotemporal gait speed over 4.8 m with individuals walking at usual speed. The best result of two trials was considered. The Timed Up and Go (TUG) test (sec) measured the time taken to stand up from a standard chair, walk a distance of 3 m, turn, walk back to the chair, and sit down again [29]. The Short Physical Performance Battery (SPPB) is a group of measures that combines the results of the gait speed, chair stand, and balance tests [30]. We included serum CTx levels (assessment described under biochemical measures) as a measure of bone turnover.

2.7. Osteopenia/Osteoporosis and Sarcopenia

Body composition and areal bone mineral density (BMD) at three sites (lumbar spine, total hip, and femoral neck) were assessed by the bone densitometrist using a Horizon dual energy X-ray absorptiometry (DXA) machine (Hologic Inc., Bedford, MA, USA). DXA scans were only performed at baseline, and osteopenia/osteoporosis was defined as a BMD T-score <-1.0 SD on at least one of the three regions. As recommended by the Australian and New Zealand Society for Sarcopenia and Frailty Research (ANZSSFR) [31], sarcopenia was defined according to the EWGSOP 2010 definition by fulfillment of low height-adjusted appendicular lean mass (ALM/height^2) combined with low grip strength or slow gait speed [32]. (ALM/height^2) was calculated automatically by the DXA machine. We applied the EWGSOP cut-offs for low ALM/height^2: $\leq$7.26 kg/m^2 ($\male$), $\leq$5.5 kg/m^2 ($\female$); for low grip strength: $<$30 kg ($\male$), $<$20 kg ($\female$) and for slow gait speed: $\leq$0.8 m/sec [32]. Osteosarcopenia was defined as the simultaneous presence of osteopenia/osteoporosis and sarcopenia.

2.8. Statistical Analysis

Statistical analysis was performed using SAS, version 9.4 (SAS Institute Inc., USA). Statistical significance was based on a two-sided p-value <0.05. Normality was assessed using the Shapiro-Wilk test. Mean (standard deviation (SD)) or median (25th percentile (Q1), and 75th percentile (Q3)) were reported for continuous data, and number (percentage (%)) for categorical data. Change (Δ) in continuous variables was calculated as the difference between follow-up and baseline, e.g., ΔMNA = [MNA score (follow-up)—MNA score (baseline)]. To compare baseline and follow-up results, differences were tested using paired t-test or Wilcoxon signed-rank test for normally-distributed or not normally-distributed paired samples, respectively.

As proof of concept, multiple linear regression analyses were first performed to test the cross-sectional associations between baseline MNA score and baseline musculoskeletal outcomes. Analyses were adjusted for baseline variables (age, sex, GDS, CACI, and number of medications—all continuous except sex).

For main analysis, individuals were divided into four subgroups based on change in MNA category from baseline to follow-up: (1) Improved nutritional status from baseline to follow-up; (2) Deteriorated; (3) Maintained but remained malnourished or at risk of malnutrition; (4) Maintained and remained well-nourished (reference group). Comparison of clinical, biochemical and musculoskeletal outcome measures between the subgroups vs. the reference group was analyzed using t-test or Wilcoxon rank-sum test for normally-distributed or not normally-distributed variables, respectively.

To explore the longitudinal associations between changes in nutritional status and musculoskeletal outcomes, multiple linear regression analyses were performed for each of the musculoskeletal outcome. Change in nutritional status was considered as both continuous (ΔMNA) and categorized exposure. Analyses were adjusted for the baseline outcome, baseline variables (age, sex, GDS, CACI, and number of medications—all continuous except sex), and care plan variables (osteoporosis treatment, vitamin D supplements use, protein supplements use, and physical activity—all categorical). When change in nutritional status was used as continuous exposure (ΔMNA), analyses were additionally adjusted for baseline MNA score. To avoid deletion of information-rich participants, missing values for four binary variables were coded as a separate category. Scatter and residual plots were examined to determine if ΔMNA was related to musculoskeletal changes in a linear manner and if the errors components were independent, homogenous with respect to the variance, and had a mean of zero. If these assumptions were violated, the outcome and/or independent variables were log-transformed to ensure good model fit.

To control for multiple testing, we ranked our hypotheses. Our primary hypothesis is that improvement in nutritional status is associated with greater musculoskeletal recovery. Our secondary hypotheses are that individuals who deteriorated and remained malnourished or at risk of malnutrition are associated with poorer musculoskeletal recovery. As such, p-values from multiple linear regression analyses were interpreted in the view of multiple comparisons. If the p-value was fairly large ($0.01 \leq p < 0.05$), we did not interpret them as definitely true, but considered that they may be likely false positive, while very small p-values ($p < 0.01$ and $p < 0.001$) were interpreted as likely real findings.

3. Results

3.1. Descriptive Characteristics, Including Change in Nutritional Status

Out of 254 patients screened at the Falls and Fractures Clinic between October 2016 and December 2018, 106 (76% female) consecutive individuals with median age of 79 (Q1, Q3 72, 82) years were re-assessed at 6-month follow-up. Table 1 presents descriptive characteristics of this study sample. On attendance at baseline, polypharmacy and severe comorbidity were quite prevalent (67% taking ≥5 medications and 45% with a CACI ≥5). The median number of reported falls in the past year was 2 falls. Most individuals (92%) were osteopenic/osteoporotic, and 22% were sarcopenic. On attendance at follow-up, the median number of reported falls in the past 6 months decreased to 0 falls. Moreover, 91 (86%) individuals reported using vitamin D supplements and 5 (5%) protein supplements, 70 (66%) reported having an osteoporosis treatment, and 51 (48%) reported being physically active, as part of the post-fall care plans recommended.

Table 1. Descriptive characteristics of the study sample, including change in nutritional status.

Characteristic	Descriptive Statistics	All ($n = 106$)
Baseline		
Age, year	Median (Q1, Q3)	79 (72, 82)
Female	n (%)	80 (75.5)
BMI at baseline, kg/m^2	Median (Q1, Q3)	27.8 (23.8, 31.9)
Weight at baseline, kg	Median (Q1, Q3)	69.6 (58.5, 85.0)
Height, m	Mean (SD)	1.59 (0.09)
Current smoker [g]	n (%)	11 (10.4)
Number of falls (past 12 months)	Median (Q1, Q3)	2 (1, 2)
Number of fractures (past 5 years)	Median (Q1, Q3)	1 (1, 1)
Severe comorbidity (CACI ≥5)	n (%)	48 (45.3)
Polypharmacy (≥5 medications)	n (%)	71 (67.0)
Suggestive of depression [a] (GDS ≥6/15)	n (%)	26 (24.5)
Osteopenia/osteoporosis [c]	n (%)	97 (91.5)
Sarcopenia [b]	n (%)	22 (20.8)
Osteosarcopenia [b]	n (%)	22 (20.8)
Nutritional status		
MNA at baseline, score	Median (Q1, Q3)	29 (23, 30)
Malnourished or at risk at baseline (MNA <24/30)	n (%)	31 (29.3)

Table 1. *Cont.*

Characteristic	Descriptive Statistics	All ($n = 106$)
Follow-up		
ΔBMI [h], kg/m^2	Median (Q1, Q3)	0.6 (−0.2, 1.5) *
Δweight [f], kg	Median (Q1, Q3)	0.6 (−1.0, 3.0) *
Number of falls [d] (past 6 months)	Median (Q1, Q3)	0 (0, 1)
Number of fractures [d] (past 6 months)	Median (Q1, Q3)	0 (0, 0)
Osteoporosis treatment [e]	n (%)	70 (66.0)
Vitamin D supplement use [e]	n (%)	91 (85.9)
Protein supplement use [e]	n (%)	5 (4.7)
Physically active [j]	n (%)	51 (48.1)
Nutritional status		
ΔMNA [d], score	Median (Q1, Q3)	0 (0, 3.3) *
Malnourished or at risk at follow-up [d] (MNA <24/30)	n (%)	16 (15.1)

Q1, Q3 = 25th, 75th percentile; CACI = Charlson age-comorbidity index; GDS = Geriatric Depression Scale; MNA= Mini Nutritional Assessment; BMI = body mass index; number of missing values: [a] 1, [b] 3, [c] 5, [d] 6, [e] 10, [f] 14 [g] 15, [h] 16, [j] 27; *: $p < 0.05$ for paired *t*-test or Wilcoxon signed-rank test for the difference between baseline and follow-up values.

The prevalence of malnutrition or risk of malnutrition based on an MNA score <24/30 was 29% at baseline and 15% at 6-month follow-up. Most individuals maintained or improved nutritional status with at least 75% not having a decrease in MNA score (median change of 0.0 (0.0, 3.3) points, $p = 0.001$) (Table 1). Specifically, 73 individuals (69%) maintained nutritional status, including 65 (61%) who remained well-nourished and 8 (8%) who remained malnourished or at risk of malnutrition. Moreover, 20 individuals (19%) improved and 6 individuals (7%) deteriorated nutritional status, as illustrated in Figure 1.

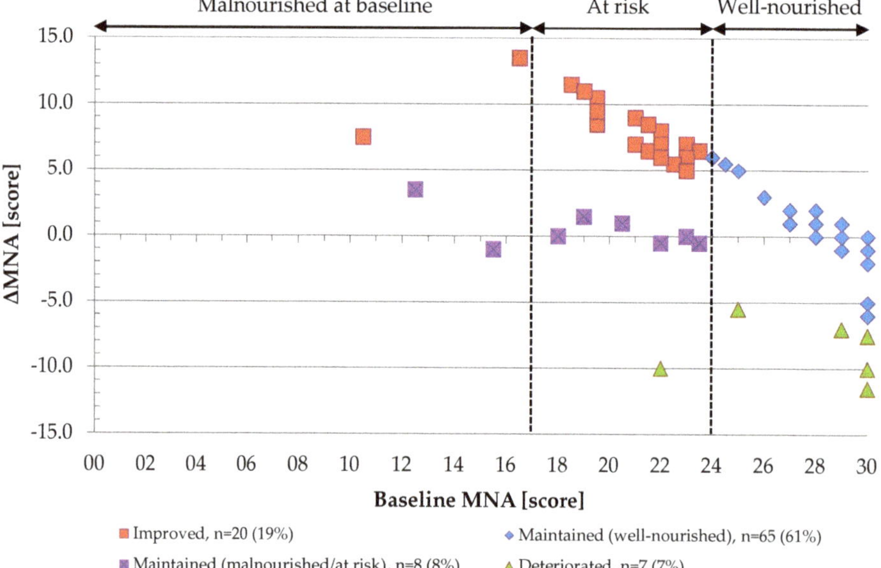

Figure 1. Change in MNA score (ΔMNA) by baseline MNA score across subgroups of nutritional status change. One dot represents one individual. Six participants could not be represented due to missing value in MNA score at follow-up.

3.2. Nutritional Status and Musculoskeletal Health at Baseline

In age-sex adjusted analyses, a 1-point increase in baseline MNA score was associated with an increase of 0.02 m/sec (95% CI 0.00, 0.03, $p = 0.022$) in gait speed and of 0.14 points (95% CI 0.05, 0.23, $p = 0.003$) in SPPB score. Additional adjustment for baseline variables weakened the associations, which became non-significant (Supplementary Table S2).

3.3. Changes in Nutritional Status and Musculoskeletal Health

Table 2 compares clinical, biochemical and musculoskeletal measures between subgroups of nutritional status change using the maintained (well-nourished) nutritional status group as reference group. The reference group was associated with significant increase in BMI ($p = 0.002$), weight ($p = 0.031$), 25OHD levels ($p = 0.013$), gait speed ($p < 0.001$), SPPB score ($p = 0.008$) and decrease in CTx levels ($p < 0.001$). Individuals who improved nutritional status were associated with greater increase in BMI ($p = 0.002$) and weight ($p = 0.006$) compared to the reference group, and similar increase in SPPB score ($p = 0.177$). They were not associated with significant improvement in gait speed ($p = 0.193$), but their TUG time significantly decreased ($p = 0.001$). Those who maintained (malnourished/at risk) or deteriorated nutritional status were not linked to significant improvement in any of the variables.

In multiple linear regression analyses (Table 3), change in nutritional status over 6 months showed the strongest associations with SPPB. After adjusting for baseline and care plan covariates, a 1-point increase in MNA score over 6 months was associated with an increase of 0.20 points (95% CI 0.10, 0.31, $p < 0.001$) in SPPB score. In subgroup analyses, individuals who improved nutritional status had for 3.30 sec (95% CI −6.34, −0.26, $p = 0.033$) a larger decrease in time for the TUG test compared to the reference group. Conversely, those who deteriorated in nutritional status had a larger decrease in SPPB score by 1.74 points (95% CI −3.29, −0.20, $p = 0.028$).

Table 2. Comparison of clinical, biochemical and musculoskeletal measures, stratified by subgroup of nutritional status change.

Characteristic	Descriptive Statistics	Maintained (Well-Nourished) n = 65 (*ref*)	Improved n = 20	Maintained (Malnourished/at Risk) n = 8	Deteriorated n = 7
Clinical and biochemical					
Age, year	Median (Q1, Q3)	77 (71, 81)	79 (73, 83)	80 (72, 84)	81 (79, 82)
BMI at baseline, kg/m^2	Median (Q1, Q3)	29.4 (24.8, 34.2)	25.6 (20.8, 29.6) [†]	23.8 (18.9, 29.8) [†]	24.1 (21.4, 27.9) [†]
ΔBMI [l], kg/m^2	Median (Q1, Q3)	0.4 (−0.3, 1.3) *	1.2 (0.7, 2.5) *, [†]	−0.8 (−2.5, 0.6)	0.0 (−0.3, 1.5)
Weight at baseline, kg	Median (Q1, Q3)	74.0 (64.0, 87.7)	64.1 (51.0, 76.6) [†]	64.8 (42.3, 70.4) [†]	64.7 (50.4, 74.0)
Δweight [k], kg	Median (Q1, Q3)	0.5 (−0.9, 2.4) *	3.0 (1.0, 5.9) *, [†]	−2.0 (−9.3, 0.0) [†]	−2.1 (−4.0, 2.8)
Number of falls at follow-up [f]	Median (Q1, Q3)	0 (0, 0)	0 (0, 1)	1 (0, 2)	2 (0, 2) [†]
Albumin at baseline, g/L	Median (Q1, Q3)	38.0 (36.0, 40.0)	38.0 (36.5, 40.0)	34.0 (29.5, 39.5)	38.0 (34.0, 41.0)
Δalbumin [c], g/L	Median (Q1, Q3)	0.0 (−2.0, 2.0)	0.0 (−2.0, 2.0)	0.5 (−1.5, 4.5)	1.0 (−2.0, 2.0)
25OHD at baseline, nmol/L	Mean (SD)	65.5 (22.5)	74.7 (20.2)	66.0 (23.7)	77.1 (28.3)
Δ25OHD [a], nmol/L	Median (Q1, Q3)	4.0 (−7.0, 25.0) *	9.5 (−11.0, 17.5)	8.5 (−5.0, 24.5)	12.0 (−1, 23.0)
PTH at baseline [f], pmol/L	Median (Q1, Q3)	6.9 (5.3, 10.3)	7.5 (5.8, 11.2)	6.0 (4.7, 10.5)	5.5 (3.9, 5.7) [†]
ΔPTH [g], pmol/L	Median (Q1, Q3)	−0.1 (−2.0, 2.3)	0.8 (−1.4, 3.8)	1.4 (1.1, 4.9) [†]	−0.1 (−1.4, 2.6)
Calcium, mmol/L	Mean (SD)	2.4 (0.1)	2.4 (0.1)	2.5 (0.1)	2.5 (0.1)
Phosphate, mmol/L	Mean (SD)	1.2 (0.2)	1.2 (0.2)	1.2 (0.2)	1.2 (0.1)
Hemoglobin at baseline [a], g/L	Mean (SD)	130.1 (13.7)	131.5 (13.2)	135.4 (18.9)	130.6 (13.5)
Δhemoglobin [c], g/L	Median (Q1, Q3)	1.0 (−4.0, 5.0)	1.0 (−5.0, 5.0)	1.5 (−6.5, 10.0)	5.0 (−11.0, 15.0)
eGFR at baseline, mL/min/1.73 m^2	Median (Q1, Q3)	75.0 (59.0, 86.0)	67.0 (52.5, 84.5)	85.0 (79.5, 87.5)	59.0 (55.0, 68.0)
ΔeGFR [d], mL/min/1.73 m^2	Median (Q1, Q3)	0.0 (−4.0, 4.0)	−3.0 (−8.0, 1.0)	−3.0 (−13.0, 0.0)	8.0 (0.0, 14.0) [†]
Musculoskeletal					
ALM/height^2 at baseline, kg/m^2	Median (Q1, Q3)	6.8 (6.0, 8.1)	6.0 (5.2, 6.8) [†]	6.2 (5.5, 6.3) [†]	5.7 (5.3, 6.6) [†]
Grip strength at baseline [a], kg	Median (Q1, Q3)	22.0 (17.0, 28.0)	20.5 (18.0, 26.0)	19.0 (15.0, 24.5)	16.0 (10.0, 24.0)
Δgrip strength [d], kg	Median (Q1, Q3)	0.0 (−3.0, 2.0)	0.0 (−2.0, 0.5)	0.5 (−1.5, 1.5)	−3.0 (−4.0, 2.0)
Gait speed at baseline [e], m/sec	Median (Q1, Q3)	0.7 (0.5, 1.0)	0.6 (0.5, 0.7)	0.7 (0.5, 0.9)	0.8 (0.6, 0.9)
Δgait speed [h], m/sec	Median (Q1, Q3)	0.1 (−0.1, 0.2) *	0.1 (−0.1, 0.2)	−0.0 (−0.1, −0.0) [†]	−0.1 (−0.2, 0) [†]
TUG at baseline [f], sec	Median (Q1, Q3)	15.2 (10.2, 21.3)	19.4 (16.5, 24.0) [†]	15.5 (10.9, 19.9)	18.3 (12.6, 22.0)
ΔTUG [k], sec	Median (Q1, Q3)	−0.5 (−2.2, 1.3)	−3.2 (−7.4, −0.6) *, [†]	−1.2 (−2.9, 3.4)	−2.6 (−2.9, 1.1)
SPPB at baseline [b], score (/12)	Median (Q1, Q3)	7.0 (5.0, 10.0)	6.0 (5.0, 7.0)	6.0 (4.0, 9.0)	7.0 (4.0, 8.0)
ΔSPPB [g], score	Median (Q1, Q3)	1.0 (0.0, 2.0) *	1.0 (0.0, 2.0) *	0.0 (−1.5, 1.0)	−1.5 (−2.0, 0.0) [†]
CTx at baseline [j], ng/L	Median (Q1, Q3)	330 (245, 447)	284 (204, 604)	278 (180, 361)	290 (162, 308)
ΔCTx [m], ng/L	Median (Q1, Q3)	−127 (−224, −15) *	−116 (−268, −9)	75.0 (−151.0, 102.0)	130 (8, 149) [†]

SD = standard deviation; Q1, Q3 = 25th, 75th percentile; ALM/height^2 = height-adjusted appendicular lean mass; BMI = body mass index; 25OHD = 25-hydroxyvitamin D; PTH = parathyroid hormone; EGFR = estimated glomerular filtration rate; TUG test = Timed Up and Go test; SPPB = Short Physical Performance Battery; CTX = C-terminal telopeptide of type 1 collagen; number of missing values: [a] 1, [b] 2, [c] 3, [d] 4, [e] 5, [f] 6, [g] 9, [h] 11, [j] 12, [k] 14, [l] 16, [m] 19; *: $p < 0.05$ for *t*-test or Wilcoxon signed-rank test for the difference between baseline and follow-up values within each group; [†]: $p < 0.05$ for *t*-test or Wilcoxon rank-sum test for the difference between the improved, maintained (malnourished/at risk) or deteriorated group vs. the maintained (well-nourished) nutritional status group (*ref*), respectively.

Table 3. Results of multiple linear regression analyses testing the association between changes in nutritional status and musculoskeletal outcomes.

Change in Musculoskeletal Outcome	Change in Nutritional Status		n	Age-Sex Adjusted β (95% CI)	Multivariable Adjusted β (95% CI)
Δgrip strength, kg	Categorized [a]	Improved vs. *ref*	20	−0.17 (−1.98, 1.63)	0.34 (−1.70, 2.73)
		Maintained (malnourished/at risk) vs. *ref*	8	−1.13 (−3.78, 1.52)	−1.22 (−4.07, 1.63)
		Deteriorated vs. *ref*	7	−2.12 (5.00, 0.75)	−1.48 (−4.41, 1.44)
	Continuous [b]	1-point higher in ΔMNA	96	0.10 (−0.10, 0.30)	0.09 (−0.11, 0.30)
Δgait speed, m/sec	Categorized [a]	Improved vs. *ref*	16	−0.03 (−0.12, 0.07)	0.04 (−0.07, 0.15)
		Maintained (malnourished/at risk) vs. *ref*	8	−0.14 (−0.27, −0.01) [i]	−0.07 (−0.21, 0.07)
		Deteriorated vs. *ref*	4	−0.15 (−0.33, 0.02)	−0.14 (−0.31, 0.04)
	Continuous [b]	1-point higher in ΔMNA	89	0.01 (0.00, 0.03) [i]	0.01 (0.00, 0.02) [i]
ΔTUG, sec	Categorized [a]	Improved vs. *ref*	17	−3.41 (−5.82, −0.99) [ii]	−3.30 (−6.34, −0.27) [i]
		Maintained (malnourished/at risk) vs. *ref*	8	0.04 (−3.25, 3.33)	−0.28 (−4.13, 3.56)
		Deteriorated vs. *ref*	3	−1.87 (−7.09, 3.34)	−1.74 (−7.35, 3.87)
	Continuous [b]	1-point higher in ΔMNA	86	−0.18 (−0.46, 0.10)	−0.13 (−0.41, 0.15)
ΔSPPB, score	Categorized [a]	Improved vs. *ref*	18	0.40 (−0.54, 1.33)	1.05 (−0.06, 2.15)
		Maintained (malnourished/at risk) vs. *ref*	8	−1.18 (−2.47, 0.11)	−0.72 (−2.13, 0.69)
		Deteriorated vs. *ref*	6	−2.21 (−3.69, −0.72) [ii]	−1.74 (−3.29, −0.20) [i]
	Continuous [b]	1-point higher in ΔMNA	91	0.21 (0.11, 0.31) [iii]	0.20 (0.10, 0.31) [iii]
Δlog (CTx), ng/L	Categorized [a]	Improved vs. *ref*	15	0.17 (−0.32, 0.66)	0.00 (−0.56, 0.56)
		Maintained (malnourished/at risk) vs. *ref*	5	0.10 (−0.69, 0.89)	0.10 (−0.79, 1.00)
		Deteriorated vs. *ref*	5	0.60 (−0.19, 1.40)	0.69 (−0.09, 1.48)
	Continuous [b]	1-point higher in ΔMNA	81	−0.02 (−0.08, 0.04)	−0.04 (−0.09, 0.02)

MNA = Mini Nutritional Assessment; TUG test = Timed Up and Go test; SPPB = Short Physical Performance Battery; CTX = C-terminal telopeptide of type 1 collagen; [iii] $p < 0.001$, [ii] $p < 0.01$, [i] $p < 0.05$: p-value of multiple linear regression models testing the associations between changes in nutritional status and musculoskeletal outcomes with the maintained (well-nourished) nutritional status group as reference group (*ref*); [a] β coefficient is the change in musculoskeletal outcome associated with change in MNA category from baseline to follow-up. Age-sex model adjusted for the baseline outcome, sex, and age. Multivariable model adjusted for the baseline outcome, baseline variables (sex, age, GDS, CACI, number of medications—all continuous except sex), and care plan variables (osteoporosis treatment, vitamin D supplement use, protein supplement use, physical activity—all categorical); [b] β coefficient is the change in musculoskeletal outcome associated with a unit increase in nutritional status over 6 months (ΔMNA). Models also adjusted for baseline MNA score.

4. Discussion

Poor nutritional status is subject to intense discussion in geriatric research mainly due to its high prevalence in older adults with falls, associations with higher morbidity and mortality risk, and effects on increased healthcare spending [33]. We assessed change in nutritional status in older adults with a history of falling and how it is related to relevant musculoskeletal changes during post-fall recovery. We found that improvement in nutritional status, based on increase in the MNA score over 6 months, was associated with improvement in physical performance, based on increase in the SPPB score over time.

4.1. Changes in Nutritional Status

Approximately one-third (29%) of the studied 106 older adults were malnourished or at risk of malnutrition at baseline. This prevalence is comparable to other studies of community-dwelling older adults using the MNA® (6%–32%) [34]. The prevalence of malnutrition or risk thereof decreased to 15% at follow-up. Comparable observational studies in community-dwelling older adults are lacking, but studies within inpatient settings reported a similar reduction in malnutrition prevalence (10%–13%) based on MNA category change between admission and discharge. However, older adults receiving inpatient services differ significantly in nutritional status and health recovery goals post-discharge to the community, so that results cannot be compared to community-dwelling older adults with confidence [24,35,36].

Most individuals (89%) maintained or improved nutritional status. It is worth noting that 8% of them remained malnourished or at risk of malnutrition at follow-up. A smaller number (7%) deteriorated to an extent sufficient to downgrade MNA category. This implies that while improvement or stabilization of nutritional status is possible during post-fall recovery, a number of individuals may not reach a well-nourished state, despite provision of individualized care plans, which included education and prescription of protein supplements, when indicated. This may have long-term implications for musculoskeletal recovery and quality of life and highlights the need for adequate follow-up of nutritional assessment. Moreover, while it is possible that subtle improvement or deterioration occurred within the stable group, the degree of change may not have been sufficient to alter MNA category.

Our study further demonstrated that MNA change was consistent with significant anthropometric (weight and BMI) changes. This is an important finding, as it is valuable to have a validated nutrition assessment tool to monitor nutrition progress over time, rather than relying only on anthropometric or biochemistry measures such as albumin, which may be confounded by clinical factors such as inflammation [37,38].

4.2. Changes in Nutritional Status and Musculoskeletal Health

Over 6 months, there were significant improvements in physical performance (based on gait speed, SPPB score, and TUG test performance) and in CTx levels. For gait speed and SPPB score, improvements were within a range indicating clinically meaningful changes [39], supporting that performance measures may offer a powerful mechanism to act on healthcare needs of older adults at risk for falls.

The observational design of this study prevents us from attributing changes of nutritional status and musculoskeletal outcomes to specific post-fall recommendations or other causes. Care plans were individualized through consideration of patient circumstances and treatment preferences. Incorporation of patients' decisions about treatment choices and their active involvement in managing their own care plan forms an integrative part of patient-centered medicine [40].

Our research investigated whether changes in nutritional status were reflected by changes in relevant musculoskeletal outcomes post-fall recovery. Improvement in nutritional status, based on a 1-point increase in MNA score over 6 months, was strongly associated with improvement in physical performance, based on an increase of 0.20 points (95% CI 0.10, 0.31, $p < 0.001$) in SPPB score over time. In subgroup analyses, the improved group was significantly associated with decrease in time to perform the TUG test and the deteriorated group with decrease in the SPPB score over time, compared to the reference group. These tools are interrelated and provide valid and reliable measurements of physical performance in community-dwelling older adults, incorporating elements of mobility and balance, with the addition of strength in both the SPPB and TUG test [41]. Nevertheless, caution is warranted when interpreting findings from subgroup analyses, because *p*-values were fairly large ($0.01 \leq p < 0.05$).

The impact of change in nutritional status on physical performance may be explained by direct or indirect mechanisms. First, increased adequacy of nutritional intake (in terms of quantity and quality) may contribute to recovery of muscle mass and function [42]. This affects physical performance, leading to functional and mobility improvements [10,11]. Improved nutritional status may also be an indicator of decreased comorbidity, which has positive effects on cognitive, functional, and physical performance [8]. Detailed data on changes in disease-related and medical factors could not be considered in this study and may have influenced changes between nutritional status groups and the time taken to recover musculoskeletal health. Finally, there was no association between changes in nutritional status and CTx levels. Physical performance may be more likely than bone turnover to improve alongside nutritional status due to recovery of muscle mass and function.

Our findings support the hypothesis that adequate nutritional follow-up support might increase relevant functional abilities during recovery from a fall. Two recent intervention studies, involving over 200 older adults aged ≥65 years each, showed that nutrition interventions (including enriched diets and/or oral nutritional supplements, home visits and/or telephone follow-ups) yielded significant improvements in weight and functional status over 3 months [43]. Another randomized control study involving over 150 geriatric patients aged >65 years at nutritional risk demonstrated the positive effect of individualized dietician counseling at home after discharge from hospital [44]. Perhaps this study design [44] can be used to conduct larger randomized controlled trials evaluating the effectiveness of specific nutritional interventions and models of care to improve nutritional and musculoskeletal measures in older adults at risk for falls.

4.3. Strengths and Limitations

Strengths of the study include the use of validated nutritional and musculoskeletal assessment tools, and the repeated measurements at two time points. The follow-up period of 6 months makes the study appropriate for documenting changes in nutritional status and musculoskeletal outcomes. There are also a number of limitations. The MNA lacks sensitivity to detect subtle changes in nutritional status [36]. As a result, the four nutritional status change subgroups are not evenly represented and are dominated by those who maintained well-nourished nutritional status. Another limitation arises from the selection bias associated with the follow-up design of this study, whereby only those willing to attend a follow-up session were assessed.

A control group was not feasible as routine geriatric care needed to be provided, which included individualized care plans for all patients. This limits the ability to attribute changes observed in musculoskeletal outcomes to specific recommendations. Importantly, it remains unclear whether change in nutritional status has a causal role in change in physical performance, or whether it is a case of reverse causation. The possibility of the temporal association between nutritional status and physical performance being due to residual confounding by unmeasured genetic, lifestyle or environmental factors cannot be ruled out. Finally, our findings may not be generalized because of the heterogeneous and convenience nature of the database.

5. Conclusions

Approximately one third of community-dwelling older adults with a history of falling were malnourished or at risk of malnutrition at baseline, and nearly one fifth improved nutritional status at 6-month follow-up. Improvement in nutritional status, based on increase in the MNA score over 6 months, was associated with improvement in physical performance, based on increase in the SPPB score over time. Larger intervention studies are required to ascertain causality and to evaluate specific nutritional interventions and models of care to improve nutritional status and functional recovery in older adults at risk for falls.

Supplementary Materials: The following are available online at http://www.mdpi.com/2072-6643/11/7/1551/s1, Table S1: The Charlson age-comorbidity index (CACI), Table S2: Cross-sectional associations of baseline nutritional status with baseline musculoskeletal outcomes.

Author Contributions: R.C. and G.D. conceived and designed the study; S.P., E.G., and G.D. were involved in data collection and quality control; R.C. and G.D. coordinated the ethics application, R.C. analyzed the data, under the supervision of S.V., and wrote the first draft of the paper; W.S.-L. and B.T. helped interpreting the data; all authors contributed to the critical revision of the paper and approved the submitted version.

Funding: This study was funded by the Australian Institute for Musculoskeletal Science (AIMSS). Romy Conzade received a PhD scholarship from the 'Studienstiftung des deutschen Volkes'. The funding sponsors had no role in the design of the study, in the collection, analyses, or interpretation of data, in the writing of the manuscript, and in the decision to publish the results.

Acknowledgments: We thank all patients attending the Falls and Fractures Clinic for their extraordinary commitment and good will. We also thank Liz Liberts for her guidance on the ethics application process, Solange Bernardo for her assistance with the assessments of the participants, and Florian Schederecker for sharing his expertise on longitudinal data analysis.

Conflicts of Interest: The authors declare no conflict of interest.

Abbreviations

25OHD	25-hydroxyvitamin D
ALM	Appendicular lean mass
ANZSSFR	Australian and New Zealand Society for Sarcopenia and Frailty Research
BMD	Bone mineral density
BMI	Body mass index
CACI	Charlson age-comorbidity index
CI	Confidence interval
CTx	C-terminal telopeptide of type 1 collagen
DXA	Dual energy X-ray absorptiometry
EWGSOP	European Working Group on Sarcopenia in Older People
eGFR	Estimated glomerular filtration rate
GDS	Geriatric Depression Scale
MNA	Mini Nutritional Assessment
PTH	Parathyroid hormone
Q1	25th percentile
Q3	75th percentile
SD	Standard deviation
SPPB	Short Physical Performance Battery
TUG test	Timed Up and Go test

References

1. Gillespie, L.D.; Robertson, M.C.; Gillespie, W.J.; Lamb, S.E.; Gates, S.; Cumming, R.G.; Rowe, B.H. Interventions for preventing falls in older people living in the community. *Cochrane Database Syst. Rev.* **2009**. [CrossRef]

2. WHO (World Health Organization). WHO Global Report on Falls Prevention in Older Age. Available online: https://www.who.int/ageing/publications/Falls_prevention7March.pdf (accessed on 10 May 2018).

3. Torres, M.J.; Feart, C.; Samieri, C.; Dorigny, B.; Luiking, Y.; Berr, C.; Barberger-Gateau, P.; Letenneur, L. Poor nutritional status is associated with a higher risk of falling and fracture in elderly people living at home in France: the Three-City cohort study. *Osteoporos. Int.* **2015**, *26*, 2157–2164. [CrossRef]

4. Trevisan, C.; Crippa, A.; Ek, S.; Welmer, A.K.; Sergi, G.; Maggi, S.; Manzato, E.; Bea, J.W.; Cauley, J.A.; Decullier, E.; et al. Nutritional Status, Body Mass Index, and the Risk of Falls in Community-Dwelling Older Adults: A Systematic Review and Meta-Analysis. *J. Am. Med. Dir. Assoc.* **2018**. [CrossRef]

5. Vellas, B.; Villars, H.; Abellan, G.; Soto, M.E.; Rolland, Y.; Guigoz, Y.; Morley, J.E.; Chumlea, W.; Salva, A.; Rubenstein, L.Z.; et al. Overview of the MNA—Its history and challenges. *J. Nutr. Health Aging* **2006**, *10*, 456–463.

6. Swanenburg, J.; de Bruin, E.D.; Stauffacher, M.; Mulder, T.; Uebelhart, D. Effects of exercise and nutrition on postural balance and risk of falling in elderly people with decreased bone mineral density: randomized controlled trial pilot study. *Clin. Rehabil.* **2007**, *21*, 523–534. [CrossRef]

7. Park, S.H. Tools for assessing fall risk in the elderly: a systematic review and meta-analysis. *Aging Clin. Exp. Res.* **2018**, *30*, 1–16. [CrossRef]

8. Vellas, B.J.; Hunt, W.C.; Romero, L.J.; Koehler, K.M.; Baumgartner, R.N.; Garry, P.J. Changes in nutritional status and patterns of morbidity among free-living elderly persons: A 10-year longitudinal study. *Nutrition* **1997**, *13*, 515–519. [CrossRef]

9. Afrin, N.; Honkanen, R.; Koivumaa-Honkanen, H.; Lukkala, P.; Rikkonen, T.; Sirola, J.; Williams, L.J.; Kroger, H. Multimorbidity predicts falls differentially according to the type of fall in postmenopausal women. *Maturitas* **2016**, *91*, 19–24. [CrossRef]

10. Mithal, A.; Bonjour, J.P.; Boonen, S.; Burckhardt, P.; Degens, H.; El Hajj Fuleihan, G.; Josse, R.; Lips, P.; Morales Torres, J.; Rizzoli, R.; et al. Impact of nutrition on muscle mass, strength, and performance in older adults. *Osteoporos. Int.* **2013**, *24*, 1555–1566. [CrossRef]

11. Misu, S.; Asai, T.; Doi, T.; Sawa, R.; Ueda, Y.; Saito, T.; Nakamura, R.; Murata, S.; Sugimoto, T.; Yamada, M.; et al. Association between gait abnormality and malnutrition in a community-dwelling elderly population. *Geriatr. Gerontol. Int.* **2017**, *17*, 1155–1160. [CrossRef]

12. Bonjour, J.P. Dietary protein: an essential nutrient for bone health. *J. Am. Coll. Nutr.* **2005**, *24*, 526S–536S. [CrossRef]

13. Sanchez-Rodriguez, D.; Marco, E.; Ronquillo-Moreno, N.; Miralles, R.; Vazquez-Ibar, O.; Escalada, F.; Muniesa, J.M. Prevalence of malnutrition and sarcopenia in a post-acute care geriatric unit: Applying the new ESPEN definition and EWGSOP criteria. *Clin. Nutr.* **2017**, *36*, 1339–1344. [CrossRef]

14. Salminen, H.; Saaf, M.; Johansson, S.E.; Ringertz, H.; Strender, L.E. Nutritional status, as determined by the Mini-Nutritional Assessment, and osteoporosis: a cross-sectional study of an elderly female population. *Eur. J. Clin. Nutr.* **2006**, *60*, 486–493. [CrossRef]

15. Huo, Y.R.; Suriyaarachchi, P.; Gomez, F.; Curcio, C.L.; Boersma, D.; Gunawardene, P.; Demontiero, O.; Duque, G. Comprehensive nutritional status in sarco-osteoporotic older fallers. *J. Nutr. Health Aging* **2015**, *19*, 474–480. [CrossRef]

16. Reiss, J.; Iglseder, B.; Alzner, R.; Mayr-Pirker, B.; Pirich, C.; Kassmann, H.; Kreutzer, M.; Dovjak, P.; Reiter, R. Sarcopenia and osteoporosis are interrelated in geriatric inpatients. *Z. Gerontol. Geriatr.* **2019**. [CrossRef]

17. Zhang, X.; Huang, P.; Dou, Q.; Wang, C.; Zhang, W.; Yang, Y.; Wang, J.; Xie, X.; Zhou, J.; Zeng, Y. Falls among older adults with sarcopenia dwelling in nursing home or community: A meta-analysis. *Clin. Nutr.* **2019**. [CrossRef]

18. Hirschfeld, H.P.; Kinsella, R.; Duque, G. Osteosarcopenia: where bone, muscle, and fat collide. *Osteoporos. Int.* **2017**, *28*, 2781–2790. [CrossRef]

19. Tinetti, M.E.; Kumar, C. The patient who falls: "It's always a trade-off". *Jama* **2010**, *303*, 258–266. [CrossRef]

20. Deandrea, S.; Lucenteforte, E.; Bravi, F.; Foschi, R.; La Vecchia, C.; Negri, E. Risk factors for falls in community-dwelling older people: a systematic review and meta-analysis. *Epidemiology* **2010**, *21*, 658–668. [CrossRef]

21. Nuotio, M.; Tuominen, P.; Luukkaala, T. Association of nutritional status as measured by the Mini-Nutritional Assessment Short Form with changes in mobility, institutionalization and death after hip fracture. *Eur. J. Clin. Nutr.* **2016**, *70*, 393–398. [CrossRef]

22. Inoue, T.; Misu, S.; Tanaka, T.; Kakehi, T.; Ono, R. Acute phase nutritional screening tool associated with functional outcomes of hip fracture patients: A longitudinal study to compare MNA-SF, MUST, NRS-2002 and GNRI. *Clin. Nutr.* **2019**, *38*, 220–226. [CrossRef]

23. Vivanti, A.; Ward, N.; Haines, T. Nutritional status and associations with falls, balance, mobility and functionality during hospital admission. *J. Nutr. Health Aging* **2011**, *15*, 388–391. [CrossRef]

24. Whitley, A.; Skliros, E.; Graven, C.; McIntosh, R.; Lasry, C.; Newsome, C.; Bowie, A. Changes in Nutritional and Functional Status in Longer Stay Patients Admitted to a Geriatric Evaluation and Management Unit. *J. Nutr. Health Aging* **2017**, *21*, 686–691. [CrossRef]

25. Charlson, M.E.; Pompei, P.; Ales, K.L.; MacKenzie, C.R. A new method of classifying prognostic comorbidity in longitudinal studies: Development and validation. *J. Chronic Dis.* **1987**, *40*, 373–383. [CrossRef]

26. Sheikh, J.I.; Yesavage, J.A. Geriatric Depression Scale (GDS): Recent evidence and development of a shorter version. *Clin. Gerontol. J. Aging Ment. Health* **1986**, *5*, 165–173.

27. Vitamin, D. Supplementation Guidelines-Osteoporosis Australia. Available online: https://www.osteoporosis. org.au/sites/default/files/files/oa_medical_vitd_2nd_ed.pdf (accessed on 4 July 2019).

28. Levey, A.S.; Bosch, J.P.; Lewis, J.B.; Greene, T.; Rogers, N.; Roth, D. A more accurate method to estimate glomerular filtration rate from serum creatinine: a new prediction equation. Modification of Diet in Renal Disease Study Group. *Ann. Intern. Med.* **1999**, *130*, 461–470. [CrossRef]

29. Podsiadlo, D.; Richardson, S. The timed "Up & Go": A test of basic functional mobility for frail elderly persons. *J. Am. Geriatr. Soc.* **1991**, *39*, 142–148.

30. Guralnik, J.M.; Simonsick, E.M.; Ferrucci, L.; Glynn, R.J.; Berkman, L.F.; Blazer, D.G.; Scherr, P.A.; Wallace, R.B. A short physical performance battery assessing lower extremity function: Association with self-reported disability and prediction of mortality and nursing home admission. *J. Gerontol.* **1994**, *49*, M85–M94. [CrossRef]

31. Zanker, J.; Scott, D.; Reijnierse, E.M.; Brennan-Olsen, S.L.; Daly, R.M.; Girgis, C.M.; Grossmann, M.; Hayes, A.; Henwood, T.; Hirani, V.; et al. Establishing an Operational Definition of Sarcopenia in Australia and New Zealand: Delphi Method Based Consensus Statement. *J. Nutr. Health Aging* **2019**, *23*, 105–110. [CrossRef]

32. Cruz-Jentoft, A.J.; Baeyens, J.P.; Bauer, J.M.; Boirie, Y.; Cederholm, T.; Landi, F.; Martin, F.C.; Michel, J.P.; Rolland, Y.; Schneider, S.M.; et al. Sarcopenia: European consensus on definition and diagnosis: Report of the European Working Group on Sarcopenia in Older People. *Age Ageing* **2010**, *39*, 412–423. [CrossRef]

33. Agarwal, E.; Miller, M.; Yaxley, A.; Isenring, E. Malnutrition in the elderly: A narrative review. *Maturitas* **2013**, *76*, 296–302. [CrossRef]

34. Kaiser, M.J.; Bauer, J.M.; Ramsch, C.; Uter, W.; Guigoz, Y.; Cederholm, T.; Thomas, D.R.; Anthony, P.S.; Charlton, K.E.; Maggio, M.; et al. Frequency of malnutrition in older adults: a multinational perspective using the mini nutritional assessment. *J. Am. Geriatr. Soc.* **2010**, *58*, 1734–1738. [CrossRef]

35. Collins, J.; Porter, J.; Truby, H.; Huggins, C.E. How does nutritional state change during a subacute admission? Findings and implications for practice. *Eur. J. Clin. Nutr.* **2016**, *70*, 607–612. [CrossRef]

36. McDougall, K.E.; Cooper, P.L.; Stewart, A.J.; Huggins, C.E. Can the Mini Nutritional Assessment (MNA) Be Used as a Nutrition Evaluation Tool for Subacute Inpatients over an Average Length of Stay? *J. Nutr. Health Aging* **2015**, *19*, 1032–1036. [CrossRef]

37. Kruizenga, H.M.; Wierdsma, N.J.; van Bokhorst-de van der Schueren, M.A.; Haollander, H.J.; Jonkers-Schuitema, C.F.; van der Heijden, E.; Melis, G.C.; van Staveren, W.A. Screening of nutritional status in The Netherlands. *Clin. Nutr.* **2003**, *22*, 147–152. [CrossRef]

38. Cabrerizo, S.; Cuadras, D.; Gomez-Busto, F.; Artaza-Artabe, I.; Marin-Ciancas, F.; Malafarina, V. Serum albumin and health in older people: Review and meta analysis. *Maturitas* **2015**, *81*, 17–27. [CrossRef]

39. Perera, S.; Mody, S.H.; Woodman, R.C.; Studenski, S.A. Meaningful change and responsiveness in common physical performance measures in older adults. *J. Am. Geriatr. Soc.* **2006**, *54*, 743–749. [CrossRef]

40. Morley, J.E.; Vellas, B. Patient-Centered (P4) Medicine and the Older Person. *J. Am. Med. Dir. Assoc.* **2017**, *18*, 455–459. [CrossRef]

41. Mijnarends, D.M.; Meijers, J.M.; Halfens, R.J.; ter Borg, S.; Luiking, Y.C.; Verlaan, S.; Schoberer, D.; Cruz Jentoft, A.J.; van Loon, L.J.; Schols, J.M. Validity and reliability of tools to measure muscle mass, strength, and physical performance in community-dwelling older people: A systematic review. *J. Am. Med. Dir. Assoc.* **2013**, *14*, 170–178. [CrossRef]

42. Robinson, S.M.; Reginster, J.Y.; Rizzoli, R.; Shaw, S.C.; Kanis, J.A.; Bautmans, I.; Bischoff-Ferrari, H.; Bruyere, O.; Cesari, M.; Dawson-Hughes, B.; et al. Does nutrition play a role in the prevention and management of sarcopenia? *Clin. Nutr.* **2018**, *37*, 1121–1132. [CrossRef]

43. Neelemaat, F.; Bosmans, J.E.; Thijs, A.; Seidell, J.C.; van Bokhorst-de van der Schueren, M.A. Post-discharge nutritional support in malnourished elderly individuals improves functional limitations. *J. Am. Med. Dir. Assoc.* **2011**, *12*, 295–301. [CrossRef] [PubMed]
44. Beck, A.M.; Kjaer, S.; Hansen, B.S.; Storm, R.L.; Thal-Jantzen, K.; Bitz, C. Follow-up home visits with registered dietitians have a positive effect on the functional and nutritional status of geriatric medical patients after discharge: A randomized controlled trial. *Clin. Rehabil.* **2013**, *27*, 483–493. [CrossRef] [PubMed]

Article

High Prevalence of Hypovitaminosis D in Institutionalized Elderly Individuals is Associated with Summer in a Region with High Ultraviolet Radiation Levels

Sara Estéfani S. Sousa [1], Márcia Cristina Sales [2], José Rodolfo T. Araújo [3], Karine C.M. Sena-Evangelista [1,4], Kenio C. Lima [3,5] and Lucia F.C. Pedrosa [1,3,4,*]

[1] Postgraduate Program in Nutrition, Federal University of Rio Grande do Norte, Av. Senador Salgado Filho, 3000, Lagoa Nova, 59078970 Natal, RN, Brazil

[2] School of Medicine, State University of Roraima, Rua Sete de Setembro, 231, Canarinho, 69306530 Boa Vista, RR, Brazil

[3] Postgraduate Program of Health Sciences, Federal University of Rio Grande do Norte, Rua General Gustavo Cordeiro de Farias, Petrópolis, 59010180 Natal, RN, Brazil

[4] Department of Nutrition, Federal University of Rio Grande do Norte, Av. Senador Salgado Filho, 3000, Lagoa Nova, 59078970 Natal, Brazil

[5] Department of Dentistry, Federal University of Rio Grande do Norte, Av. Senador Salgado Filho, 1787, Lagoa Nova, 59056000 Natal, RN, Brazil

* Correspondence: lfcpedrosa@gmail.com; Tel.: +55-84-3342-2291

Received: 13 May 2019; Accepted: 1 July 2019; Published: 4 July 2019

Abstract: Vitamin D may play a significant role in regulating the rate of aging. The objective of the study was to assess vitamin D status and its associated factors in institutionalized elderly individuals. A total of 153 elderly individuals living in Nursing Homes (NH) were recruited into the study. Serum 25-hydroxyvitamin D [25(OH)D] concentration was used as the biomarker of vitamin D status, and it was considered as the dependent variable in the model. The independent variables were the type of NH, age-adjusted time of institutionalization, age, sex, skin color, body mass index, waist and calf circumference, physical activity practice, mobility, dietary intake of vitamin D and calcium, vitamin D supplementation, use of antiepileptics, and season of the year. Serum 25(OH)D concentrations less than or equal to 29 ng/mL were classified as insufficient vitamin D status. The prevalences of inadequate dietary intake of vitamin D and calcium were 95.4% and 79.7%, respectively. The prevalence of hypovitaminosis D was 71.2%, and the mean serum concentration of 25(OH)D was 23.9 ng/mL (95% confidence interval [CI]: 22.8–26.1). Serum 25(OH)D concentration was associated with the season of summer ($p = 0.046$). There were no associations with other independent variables (all $p > 0.05$). The present results showed that a high prevalence of hypovitaminosis D was significantly associated with summer in institutionalized elderly individuals.

Keywords: vitamin D; elderly individuals; nursing homes

---

1. Introduction

The population aged 60 or over is growing at a faster rate than the total population in almost all regions worldwide. Most notably, the current challenge in demographics is the rapidly increasing population of older adults. In 2012, people aged 60 or over represented almost 11.5 per cent of our total global population of 7 billion. By 2050, the proportion is projected to nearly double to 22 per cent [1].

The number of elderly individuals in Brazil was estimated to be more than 30.2 million in 2017. Thus, they represented approximately 7 per cent of the Brazilian population of 209.3 million inhabitants in this period. In the Northeast of Brazil, the proportion of elderly individuals increased from 5.8% in 2000 to 7.2% in 2010 [2].

A widespread health problem experienced among elderly individuals is multiple micronutrient inadequacy, including vitamin D, which may lead mainly to frailty, sarcopenia, and cognitive decline [3,4].

Vitamin D regulates aging by controlling the activity of autophagy, which slows down aging processes by removing dysfunctional mitochondria. Vitamin D also reduces mitochondrial dysfunction, oxidative stress, inflammation, calcium signaling, epigenetic changes and DNA disorders, including telomere shortening, which act upon aging processes [5].

The diagnosis of vitamin D deficiency is a controversial issue. Serum levels of 25-hydroxyvitamin D [25(OH)D] above 30 ng/mL are associated with increased efficacy of intestinal calcium absorption and stabilization of serum parathyroid hormone (PTH) values, which are suggested as being the threshold for fracture prevention [6]. The Endocrine Society suggests that the 25(OH)D concentrations that are considered deficient, insufficient, and sufficient are ≤20, ≥21 and ≤29, and ≥30 ng/mL, respectively [7]. However, the Brazilian Society of Clinical Pathology/Laboratory Medicine and the Brazilian Society of Endocrinology and Metabolism recently proposed concentrations of 25(OH)D above 20 ng/mL as being adequate for the health of the population up to 60 years old, while values between 30 and 60 ng/mL are adopted for groups at risk of vitamin D deficiency, including elderly individuals [8].

Vitamin D deficiency among elderly individuals can result in cardiovascular risk, mortality, low quality-of-life scores, decreased physical functionality, secondary hyperparathyroidism, and increased risk of fractures [3,9–11]. Age-related changes in body composition, such as decreased muscle mass and increased adipose tissue, can lead to a decrease in serum vitamin D concentrations. Furthermore, the muscular weakness presented by the elderly population can be potentiated by the deficiency of vitamin D [12]. Moreover, Pilz et al. (2012) observed a significantly increased mortality risk in the female Nursing Home (NH) residents with the lowest 25(OH)D levels [13].

Several factors are involved in changes in vitamin D status in aging, such as the seasonality, which is currently observed as being an important predictor because it has an impact on the behavior and lifestyle of individuals [6,14,15]. Even under favorable environmental conditions, vitamin D bioavailability in elderly individuals can be negatively affected by the reduced capacity of its cutaneous synthesis. Furthermore, this population can be vulnerable to vitamin D deficiency as a result of low exposure to sunlight, their particular type of skin pigmentation, adiposity, use of antiepileptic medications, reduced kidney function, and low dietary vitamin D intake [9,15–17].

In light of these considerations, hypovitaminosis D has been frequently found among institutionalized elderly individuals living in countries with low ultraviolet radiation (UV) indexes, where elderly people spend more time indoors [13,18], as well as in high-latitude regions where there is a high UV index [16,19,20].

Although several studies carried out on elderly people have revealed a poorer vitamin D status during the winter seasons because of the lower sunlight exposure [19–22], contradictory findings have been found in the literature. One study developed with elderly people from a Mediterranean area suggested significant differences in serum 25(OH)D concentrations between seasons, with the lowest concentrations occurring in the summer and the highest during the autumn, although their vitamin D intake was significantly lower in the autumn and winter [23].

Based on these findings, our study was developed to answer the question about the status of vitamin D in institutionalized elderly individuals living in a region with high levels of UV radiation, while considering various particularities of aging. For this reason, we have drawn up a theoretical model that encompasses diverse variables and contemplates these conditions.

In addition to this, institutionalization impacts the interactions of elderly individuals with the environment as well as their nutritional status, which, in turn, are influenced by their social status, psychological disorders, and general health conditions [24,25]. In this scenario, the objective of the present study is to assess serum 25(OH)D concentrations and associated factors in institutionalized elderly individuals living in a region of northeastern Brazil with high ultraviolet radiation levels.

2. Materials and Methods

2.1. Study Population

The study was carried out with 304 elderly individuals living in nine Nursing Homes (NH) in the city of Natal (Rio Grande do Norte state), which is in the northeast of Brazil. All the individuals included in the study were at least 60 years old and living in these NHs. In Brazil, the Statute of the Elderly [26] considers elderly individuals to be aged 60 years or older. Elderly individuals who had marked physical and mental impairments, as well as difficult venous access, were excluded from the study. After exclusion, the initial sample of 304 elderly individuals was narrowed down to 153 participants (Figure 1).

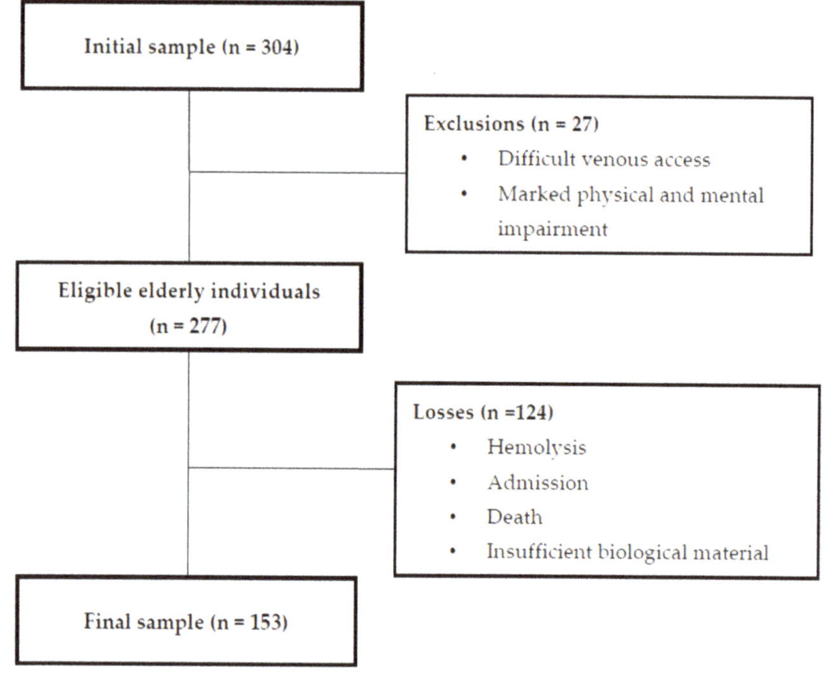

Figure 1. Flowchart of the study participants.

The study was conducted in accordance with the Declaration of Helsinki, and the protocol was approved by the Research Ethics Committee of the Federal University of Rio Grande do Norte (#263/11; CAAE 0290.0.051.000-11). All participants or their guardians provided written informed consent before they participated in the study.

The sample size calculation for the cross-sectional studies was defined on the basis of the mean and standard deviation of the concentration of 24.0 (7.6) ng/mL of 25(OH)D, corresponding to 10 random subjects in the sample of this study, and to the power of the test of 95%, resulting in 154 elderly individuals. Thus, this study obtained a response rate of 99.3%. For association studies, the sample size would be 156 elderly individuals, considering the population size of 304 elderly individuals, a hypothetical frequency of vitamin D deficiency of 71.2% [19], a confidence limit of 5%, and a design effect equal to 1 [27].

2.2. Theoretical Model of the Study

The theoretical model of this study was developed according to determinant factors of vitamin D status. Serum 25(OH)D concentration was defined as the dependent variable, and the independent variables were those likely to influence vitamin D status: Age, sex, skin color, body mass index (BMI), physical activity practice, mobility, diet, use of vitamin D supplements, and use of antiepileptic medications. The sociodemographic, nutritional, lifestyle, and biological variables were included in the "proximal" layers. Type of NH, age-adjusted time of institutionalization, and level of education were components of the "distal" layer. The season of the year was considered as a cross-sectional influence because it has a direct relationship with 25(OH)D concentrations (Figure 2). This study also addressed the relationship between the dependent variable and possible clinical implications in cases of vitamin D-deficient elderly individuals—multimorbidity, falls, sarcopenia, depression, cognition, and functional status.

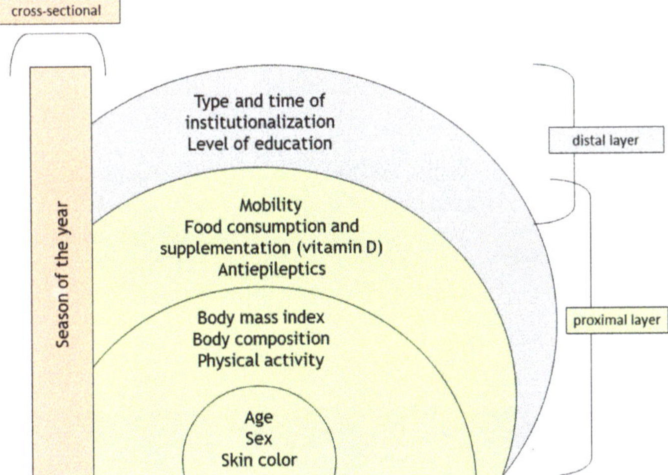

Figure 2. Theoretical model of the determinants of vitamin D status in institutionalized elderly individuals.

Age-adjusted time of institutionalization was calculated to determine how much time after the advanced age period (as a percentage, defined as the time after turning 60 years of age) elderly individuals remained institutionalized, according to a previous study [28].

Weight, height, and waist and calf circumference were measured to determine the anthropometric variables. Skin color was obtained by self-classification among the five categories adopted by the Brazilian Institute of Geography and Statistics (IBGE): Black, mixed race ("pardo" in official Portuguese), Caucasian, Asian, and indigenous (Native Brazilian) [29]. Body weight was assessed with a Balmak® electronic scale (Santa Barbara do Oeste, Brazil) with a capacity of 300 kg and an accuracy of 50 g. In elderly individuals with disabilities or permanent mobility restrictions, weight was measured with a Seca® 985 bed scale (Bolton, England). Height was defined using the average of the two measurements and a portable Caumaq® stadiometer (Cachoeira do Sul, Brazil). For physically impaired elderly individuals, height was estimated using the equation of Chumlea et al. (1987) [30]. BMI was classified while considering the cutoff points proposed by Lipschitz [31].

Calf perimeter values less than 31 cm were indicative of loss of muscle mass in elderly individuals. Results for waist perimeter greater than 102 cm in males and 88 cm in females were considered to indicate a higher risk of diseases [32].

Physical activity practice was determined through information provided by caregivers, categorized into "yes" or "no" answers. Information on diseases, multimorbidity, polypharmacy, use of antiepileptic drugs, and vitamin D supplementation was collected from the medical records of the NHs. Multimorbidity was considered as the presence of two or more diseases with clinical diagnoses included in medical records. Polypharmacy was characterized for those individuals who took five or more medications daily [33].

Data on the mobility variable were collected from the caregivers of the elderly individuals using a Barthel scale [34]. Sarcopenia was measured according to previously defined criteria [35], based on discrimination of reduced muscle mass concentrations (calf circumference assessment) associated with a reduction in strength (grip test) and/or functional status. Functional status was assessed according to the scale set by Katz [36]. Cognitive function was assessed according to Pfeiffer's Short Portable Mental Status Questionnaire, which evaluates short- and long-term memory, orientation, information on everyday facts, and mathematical reasoning. Individuals were classified into four categories: intact mental function (0–2 errors), mild cognitive impairment (3–4 errors), moderate (5–7 errors), and severe (8–10 errors) [37,38].

The elderly subjects were categorized by seasons. The season used was the date corresponding to 30 days prior to blood collection. This was to take into account the half-life of 25(OH)D [39]. The Laboratory of Environmental and Tropical Variables of the Federal University of Rio Grande do Norte provided UV values for each season of the year [39].

2.3. Analysis of 25(OH)D Serum Concentration

Blood samples after overnight fasting (8–12 h) were collected by standard venipuncture. Serum concentrations of 25(OH)D were measured using the chemiluminescent Liaison® test from kit DiaSorin® (Saluggia, Italy). Serum 25(OH)D values equal to or below 29 ng/mL (72 nmol/L) were considered as "insufficient", while values between 30–60 ng/mL (75–150 nmol/L) were classified as "sufficient" [8]. In the present study, the name "hypovitaminosis D" was used for elderly individuals with a 25(OH)D value equal to or below 29 ng/mL.

2.4. Dietary Intake

Food consumption data were collected according to the method of weighed food records (WFRs), as described by Sales et al. (2018) [28]. Habitual intake of each nutrient was estimated by removing the effect of intrapersonal variance, based on the method developed by Nusser et al. (1996) [40]. Subsequently, the results were adjusted for energy intake [41]. The prevalences of inadequate intake of calcium and vitamin D were estimated by using the Estimated Average Requirement (EAR) as the cut-off point [42].

2.5. Statistical Analysis

The data were analyzed using the statistical software IBM SPSS Statistics 21 (IBM Corp, Armonk, NY, USA). Initially, a descriptive analysis was performed with variables at the individual level, which allowed the sample to be characterized for socioeconomic, health, anthropometric, nutritional, and vitamin D statuses. Data on serum levels of 25(OH)D and dietary intake were checked for missing data and outliers that could hinder a multivariate analysis. In addition, variables were excluded from further analysis when they had more than 25% of missing data and outliers, identified under a multivariate perspective through measurement of the Mahalanobis distance (D^2)—distance greater than 3.0. There was no record of sample losses and multivariate outliers after adjustment of intrapersonal variability, interpersonal variability, and energy from dietary intake. The normality was checked by using the Kolmogorov–Smirnov test. Subsequently, the variables color, body mass index, cognition, functional status, and seasons of the year were dichotomized according to the predominance of elderly individuals in groups and/or clinical characteristics. Subsequently, to check for statistical associations, Pearson's chi-square test was run, considering the outcome variable (25(OH)D) and other variables relative to the theoretical model. The magnitude of the association was assessed using prevalence ratios (PR) with the corresponding confidence intervals (CIs). The significance level α was set at $p < 0.05$ (two-tailed test).

3. Results

The power of the study was 99.7% (with continuity correction), considering a 95% confidence interval. The calculation was obtained from 109 exposed participants and 44 unexposed participants, resulting in a prevalence of vitamin D insufficiency of 71% and 28%, respectively. Thus, the prevalence ratio (PR) was equal to 2.5, with a difference of 43% among the prevalences.

The mean age of the elderly individuals was 81.7 (9.2) years, ranging from 60–103. Thus, they were a long-living population, and there were more females (78.4%) than males. It was found that 64.5% of them had some level of education, 70.6% lived in nonprofit institutions, and that the average age-adjusted time of institutionalization was 29.4%. The BMI values indicated that 46.4% of the elderly individuals were thin.

It was found that 60.3% showed signs of loss of muscle mass, according to calf circumference. Most of the elderly individuals did not practice physical activity, had moderate-to-severe cognitive impairment, and showed multimorbidity (Table 1).

Table 1. General, anthropometric, lifestyle, and health characteristics of institutionalized elderly individuals ($n = 153$).

Variables	n (%)	95% CI
Sex		
Male	33 (21.6)	3.30–15.00
Female	120 (78.4)	3.30–71.90
Skin color [a]		
White	70 (54.7)	46.10–54.10
Black	25 (19.5)	12.50–26.60
Brown	27 (21.1)	14.10–28.90
Other	6 (4.7)	1.60–8.60
Marital status [b]		
Single	66 (52.0)	42.50–60.60
Widowed	32 (25.2)	18.10–33.10
Married	17 (13.4)	7.90–19.70
Separated/Divorced	12 (9.4)	3.70–17.30
Level of education [c]		
Illiterate	43 (35.5)	27.30–43.80
Literate	32 (26.4)	18.20–34.70
Elementary I (years 1–5)	16 (13.2)	6.60–19.00
Elementary II (years 6–9)	6 (5.0)	1.70–9.10
High School	18 (14.9)	8.30–21.50
Higher Education	6 (5.0)	1.70–9.10
Type of institution		
Profit	45 (29.4)	22.20–37.20
Nonprofit	108 (70.6)	62.80–77.80
Time of institutionalization (mean) [d]	29.38%	24.72–34.04
BMI (kg/m^2) [e]		
Thin	55 (46.4)	39.10–54.00
Normal weight	37 (31.1)	23.90–39.50
Excess weight	27 (22.7)	15.10–29.40
Abdominal circumference [f]		
Obese	51 (40.5)	31.80–50.00
Non-obese	75 (59.5)	50.00–68.20
Calf circumference [g]		
Sign of lean mass loss	88 (60.3)	51.10–68.50
No sign of lean mass loss	58 (39.7)	31.50–47.90

Table 1. *Cont.*

Variables	*n* (%)	95% CI
Physical activity [h]		
Inactive	87 (70.2)	67.80–81.50
Active	37 (29.8)	18.50–32.20
Mobility		
Without mobility	20 (13.1)	7.80–19.00
With mobility	133 (86.9)	81.00–92.20
Sarcopenia [g]	66 (45.2)	46.60–63.00
Cognition [i]		
Preserved mental function	13 (9.0)	4.90–13.90
Mild cognitive impairment	9 (6.3)	2.80–10.40
Moderate cognitive impairment	26 (18.1)	11.80–25.00
Severe cognitive impairment	96 (66.7)	58.40–74.30
Multimorbidity [b]	84 (66.1)	57.50–74.00
Polypharmacy [j]	59 (38.6)	34.80–50.70
Use of antiepileptic medications	27 (18.8)	12.50–25.70

[a] n = 128; [b] n = 127; [c] n = 121; [d] n = 133; [e] n = 151; [f] n = 126; [g] n = 146; [h] n = 124; [i] n = 144; [j] n = 139.

Concerning functional status, 31.8% of the elderly individuals were classified as "A" (independent for eating, continence, locomotion, toilet use, dressing, and bathing), followed by 16.2% in the item "G" (dependent on performing the above six functions). Dietary intake of vitamin D was 2.8 (3.2) µg/day, resulting in a high prevalence of inadequacy (95.4%), in addition to low frequency in the use of vitamin D supplementation (5.9%). The prevalence of hypovitaminosis D in the study population was 71.2%, with a mean value of 23.9 (16.6) ng/mL for 25(OH)D/mL. No participant was categorized in the winter season according to the criteria applied to the distribution of elderly individuals by seasons (Table 2).

Table 2. Dietary and environmental characteristics of institutionalized elderly individuals (*n* = 153).

Variables	Mean (SD)	95% CI
Vitamin D intake (µg/day)	2.8 (3.2)	2.32–3.33
Calcium intake (mg/day)	997.9 (291.8)	921.18–1074.64
Serum 25(OH)D (ng/mL) [a]	23.9 (16.6–31.0)	-
Local UV Index		
Summer	6.6 (0.5)	6.45–6.66
Spring	6.6 (0.5)	6.51–6.80
Fall	6.3 (0.4)	6.48–6.99
Vitamin D intake		
Possibly inadequate	146 (95.4)	92.20–98.70
Possibly adequate	7 (4.6)	1.30–7.80
Calcium intake		
Possibly inadequate	122 (79.7)	73.20–85.60
Possibly adequate	31 (20.3)	14.40–26.80
Vitamin D status [b]		
Insufficient	109 (71.2)	64.70–78.40
Sufficient	44 (28.8)	21.60–35.30
Distribution by seasons		
Summer	92 (60.1)	51.60–68.00
Spring	46 (30.1)	22.90–37.90
Fall	15 (9.8)	5.20–15.00

[a] median (inter-quartile range between 25 and 75, respectively); [b] according to the diagnosis of vitamin D status.

There was an association between 25(OH)D and summer ($p = 0.046$), and there were no associations with any other independent variables (all $p > 0.05$) (Table 3).

Table 3. Association between serum 25(OH)D concentration and variables of the theoretical model.

| | Serum 25(OH)D (ng/mL) | | | | | |
| | Hypovitaminosis D (≤29) | | Sufficient (30–60) | | | |
	n	%	*n*	%	PR [a] (95% CI)	*p*
Sex						
Male	25	75.8	8	24.2	1.08 (0.86–1.36)	0.518
Female	84	70.0	36	30.0		
Age						
75 years or older	84	73.0	31	27.0	1.11 (0.86–1.43)	0.392
60–74 years	25	65.8	13	34.2		
Skin color						
Non-white	36	62.1	22	37.9	0.80 (0.63–1.02)	0.063
White	54	77.1	16	22.9		
Type of institution						
Nonprofit	72	66.7	36	33.3	0.43 (0.18–1.03)	0.053
Profit	37	82.2	8	17.8		
Body Mass Index						
Non-normal weight	83	72.3	31	27.2	0.63 (0.25–1.57)	0.314
Normal weight	30	81.1	7	18.9		
Abdominal circumference						
With abdominal fat	39	76.5	12	23.5	0.85 (0.68–1.06)	0.181
Without abdominal fat	49	65.3	26	34.7		
Calf circumference						
Sign of muscle mass loss	62	70.5	26	29.5	1.05 (0.86–1.29)	0.628
No sign of muscle mass loss	43	74.1	15	25.9		
Physical activity						
Inactive	75	68.8	34	31.2	0.43 (0.16–1.12)	0.078
Active	31	83.8	6	16.2		
Mobility						
Yes	17	85.0	3	15.0	1.23 (0.99–1.53)	0.145
No	92	69.2	41	30.8		
Multimorbidity						
Yes	59	70.2	25	29.8	0.91 (0.77–1.23)	0.828
No	31	72.1	12	27.9		
Falls						
Yes	4	80.0	1	20.0	1.12 (0.72–1.76)	0.669
None	99	71.2	40	28.8		
Sarcopenia						
Yes	57	71.3	23	28.8	1.00 (0.81–1.23)	0.996
No	47	71.2	19	28.8		
Cognition						
Moderate to severe impairment	89	73.0	33	27.0	1.33 (0.90–1.99)	0.083
Preserved to mild impairment	12	54.5	10	45.5		
Depression						
Depressed	6	54.5	5	45.5	0.75 (0.13–1.30)	0.198
Non-depressed	86	72.9	32	27.1		
Functional status						
Reduced independence	68	67.3	33	32.7	0.88 (0.71–1.08)	0.251
Independent	36	76.6	11	23.4		
Seasons of the year						
Other seasons	38	62.3	23	37.7	0.49 (0.24–1.00)	0.046
Summer	71	77.2	21	22.8		

Table 3. *Cont.*

| | Serum 25(OH)D (ng/mL) | | | | | |
| | Hypovitaminosis D (≤29) | | Sufficient (30–60) | | | |
	n	%	*n*	%	PR [a] (95% CI)	*p*
Vitamin D intake						
Possibly inadequate	104	71.2	42	28.8	1.00 (0.62–1.61)	0.991
Possibly adequate	5	71.4	2	28.6		
Dietary Calcium						
Possibly inadequate	117	95.9	5	4.1	1.02 (0.93–1.13)	0.576
Possibly adequate	29	93.5	2	6.5		
Vitamin D supplementation						
Yes	87	70.2	37	29.8	1.05 (0.65–1.69)	0.825
No	6	66.7	3	33.3		
Antiepileptic medications						
Yes	19	70.4	8	29.6	0.98 (0.75–1.28)	0.882
No	84	71.8	33	28.2		

[a] Prevalence Ratio.

4. Discussion

The high prevalence of hypovitaminosis D found in the institutionalized elderly individuals of the present study was only significantly associated with the summer season, and this was found in the theoretical model involving several proximal and distal factors that determine vitamin D status. Moreover, the seasons of the year were considered as cross-sectional variables. The evaluation of the seasons of the year together with the assessment of the dietary intake of vitamin D (whether it was adequate or inadequate) was a relevant point of the present study, in that it was necessary to consider the two external sources of vitamin D obtained by the population.

Vitamin D is the only nutrient acquired from external sources, e.g., natural sunlight. It is estimated that an individual's highest daily vitamin D intake can be obtained by exposure to sunlight. However, in elderly individuals, the capacity of the cutaneous synthesis of vitamin D is undermined by the gradual loss of this physiological mechanism [9,17].

In epidemiological studies on vitamin D status, physical activity is often used as a proxy for time spent outdoors and, indirectly, for exposure to sunlight. Consequently, the elderly individuals who spend their lives at home indoors and those who spend less time doing outdoor physical activities have decreased levels of 25(OH)D [43].

Many of the elderly individuals were distributed by the summer season (60.1%), but although the region where the study was conducted has an annual average of UV indexes ranging from high to very high [44], it can be inferred that the studied population did not benefit from this environmental factor, as shown by the contradictory significant association between hypovitaminosis D and summer. We know that in tropical regions, the summer heat is extremely uncomfortable, causing most elderly individuals to avoid sunlight exposure. Similar reasons for this paradoxical finding were previously discussed, emphasizing different weather conditions between areas or countries [23]. In addition, their failure to benefit from the summer season can also be explained because activities in the NHs follow a pattern—they start early in the day with personal hygiene and provision of food and medicines, and end with little or no sunlight exposure, because the times are inappropriate. Furthermore, factors such as health problems, logistics of the NH, and indoor leisure activities affect the access of the elderly individuals to sunlight exposure more often when compared to their non-institutionalized peers [19,45].

Other findings from our study, such as the high inadequacies of dietary vitamin D and calcium, physical inactivity, and multimorbidities may additionally have nullified the response to environmental conditions that would otherwise be favorable for vitamin D status. In regard to elderly Brazilian subjects, even in the summer, Caucasian males in the youngest group of individuals presented a significant increase in 25(OH)D levels, demonstrating that a seasonal variation in 25(OH)D concentrations was influenced by factors such as gender, ethnicity, and age [20].

Okan et al. noted that 25(OH)D levels and the UV index of elderly people living in NHs were significantly lower than those living in their own homes, though there was a positive correlation between UV index and 25(OH)D observed in both groups [46].

These data reinforce the need to discuss vitamin D status in the context of the institutionalization of elderly individuals, and demands a model that encompasses diversified variables, such as those used in our study.

Although aging has been reported as a factor that interferes with vitamin D metabolism [9,17], the 25(OH)D concentrations in elderly Brazilian individuals with regular sunlight exposure and physical activity practice were similar to the values found in young and healthy individuals. These results confirm that vitamin D metabolism is activated by sunlight exposure, despite the physiological limitations of elderly individuals [20]. Physical activity is considered to be one of the factors associated with higher 25(OH)D levels [14]. In light of this evidence, some benefits can be achieved with changes in the lifestyles of elderly individuals.

Most of the elderly individuals lived in nonprofit institutions and had lived 29% of their life as an elderly individual in an institutional setting, which was considered a distal factor of the study model and was associated with an increased risk of vitamin D insufficiency and hypovitaminosis D when compared to elderly individuals living in the community [19]. In the same study, the prevalences of hypovitaminosis D in institutionalized elderly individuals and outpatients from a city in southeastern Brazil were 71.2% and 43.8%, respectively. This result is similar to that found in the present study. In this regard, confounding factors arising from disease burden in elderly individuals may also account for the high prevalence of vitamin D insufficiency. In a previous study, frail elderly individuals who had been recently institutionalized showed a 41% prevalence of vitamin D deficiency and a 45% prevalence of vitamin D insufficiency [47].

In our study, the population profiles and the infrastructures of the NHs were similar. Therefore, the studied elderly individuals did not show significant differences, whether they were in nonprofit public or for-profit private institutions. As shown in Table 3, the high prevalence of hypovitaminosis D in for-profit private institutions demonstrates that the institutionalization, by itself, can be addressed as a determinant of health problems in NHs. Furthermore, the study population lives in the Northeast of Brazil, a region with high social inequalities that may affect genetic distinctions in comparison to other areas of the country. A study developed in the same region found evidence that longevity in this region was characterized by the presence of longer telomeres in women associated with a low level of education [48].

All these factors may partly explain the findings reported in this study. However, the high prevalence of vitamin D insufficiency, per se, in the context of low dietary intake of vitamin D and calcium as well as physical inactivity, represents signs of deterioration in vitamin D status, thus affecting the health of institutionalized elderly individuals.

Other points should be addressed in the field of aging. This study focuses on this particular generation of long-living individuals, and according to the theoretical model proposed, age is a proximal factor of interference in 25(OH)D concentrations. In this regard, there is increasing evidence that aging can proceed at variable rates and be regulated by vitamin D. Normal concentrations of vitamin D are capable of maintaining these processes at their regular low rates, and this slows down the aging process and also helps to prevent the onset of some age-related diseases. When vitamin D is deficient, there is an increase in the activities of these aging processes, which not only accelerates the rate of aging, but also creates conditions that initiate the onset of age-related diseases [5].

The present study showed limitations regarding its temporal approach (cross-sectional), hindering the establishment of a causal relationship, and also due to lack of evaluation of sun exposure. Future studies should focus on reproducibility, involving a protocol with longitudinal observation and comparison among institutionalized and non-institutionalized elderly individuals. Furthermore, as shown by the benefits found in institutionalized elderly individuals in Finland [49] who received vitamin D food fortification, the literature reinforces the need for special attention in health care for elderly people, with a focus on modifiable risk factors.

5. Conclusions

A high prevalence of hypovitaminosis D was found in institutionalized elderly individuals living in a region with high ultraviolet radiation levels, and this was significantly associated with summer. These findings suggest the probability of the presence of interfering vitamin D factors mediated by aging. In this scenario, aging conditions which increase the risk of vitamin D deficiency have to be taken into consideration. Furthermore, the institutionalization of elderly individuals needs to be carefully handled to create alternatives that mitigate aging-related health problems.

Author Contributions: Conceptualization, S.E.S.S., K.C.L., and L.F.C.P.; data curation, S.E.S.S., M.C.S., and J.R.T.A.; formal analysis, S.E.S.S. and K.C.L.; funding acquisition, K.C.L. and L.F.C.P.; investigation, S.E.S.S. and L.F.C.P.; methodology, S.E.S.S., M.C.S., J.R.T.A., K.C.M.S.-E., K.C.L., and L.F.C.P.; project administration, K.C.L. and L.F.C.P.; supervision, K.C.M.S.-E., K.C.L., and L.F.C.P.; validation, S.E.S.S., M.C.S., J.R.T.A., K.C.L., and L.F.C.P.; visualization, S.E.S.S.; writing—original draft, S.E.S.S. and L.F.C.P.; writing—review & editing, S.E.S.S., M.C.S., J.R.T.A., K.C.M.S.-E., K.C.L., and L.F.C.P.

Funding: This research was supported by the Research Support Foundation of Rio Grande do Norte (Fundação de Apoio à Pesquisa do Rio Grande do Norte—FAPERN) and the National Council for Scientific and Technological Development (Conselho Nacional de Desenvolvimento Científico e Tecnológico—CNPq, Brazil; grant no 77228/2013). This study was also financed by the Coordenação de Aperfeiçoamento de Pessoal de Nível Superior—Brasil (CAPES)—Finance Code 001.

Conflicts of Interest: The authors declare no conflict of interest.

References

1. UNFPA. *Ageing in the Twenty-First Century: A Celebration and A Challenge*; UNFPA: New York, NY, USA, 2012; ISBN 9780897149815.
2. BRASIL Sinopse Do Censo Demográfico. Available online: http://www.ibge.gov.br/home/estatistica/populacao/censo2010/sinopse/default_sinopse.shtm (accessed on 22 August 2017).
3. Veldurthy, V.; Wei, R.; Oz, L.; Dhawan, P.; Jeon, Y.H.; Christakos, S. Vitamin D, calcium homeostasis and aging. *Bone Res.* **2016**, *4*, 16041. [CrossRef] [PubMed]
4. Ter Borg, S.; Verlaan, S.; Hemsworth, J.; Mijnarends, D.M.; Schols, J.M.G.A.; Luiking, Y.C.; De Groot, L.C. Micronutrient intakes and potential inadequacies of community-dwelling older adults: A systematic review. *Br. J. Nutr.* **2015**, *113*, 1195–1206. [CrossRef] [PubMed]
5. Berridge, M.J. Vitamin D deficiency accelerates ageing and age-related diseases: A novel hypothesis. *J. Physiol.* **2017**, *595*, 6825–6836. [CrossRef] [PubMed]
6. Holick, M.F. Vitamin D Deficiency. *N. Engl. J. Med.* **2007**, *357*, 266–281. [CrossRef] [PubMed]
7. Holick, M.F.; Binkley, N.C.; Bischoff-Ferrari, H.A.; Gordon, C.M.; Hanley, D.A.; Heaney, R.P.; Murad, M.H.; Weaver, C.M. Endocrine Society Evaluation, treatment, and prevention of vitamin D deficiency: An Endocrine Society clinical practice guideline. *J. Clin. Endocrinol. Metab.* **2011**, *96*, 1911–1930. [CrossRef] [PubMed]
8. Ferreira, C.E.S.; Maeda, S.S.; Batista, M.C.; Lazaretti-Castro, M.; Vasconcellos, L.S.; Madeira, M.; Soares, L.M.; Borba, V.Z.C.; Moreira, C.A. Posicionamento oficial da Sociedade Brasileira de Patologia Clínica/Medicina Laboratorial (SBPC/ML) e da Sociedade Brasileira de Endocrinologia e Metabologia (SBEM) sobre intervalos de referência da vitamina D [25(OH)D]. *J. Bras. Patol. Med. Lab.* **2017**, *53*, 377–381.
9. Flicker, L. Vitamin D and the endocrinology of ageing. *Curr. Opin. Endocr. Metab. Res.* **2019**, *5*, 7–10. [CrossRef]
10. Feng, X.; Guo, T.; Wang, Y.; Kang, D.; Che, X.; Zhang, H.; Cao, W.; Wang, P. The vitamin D status and its effects on life quality among the elderly in Jinan, China. *Arch. Gerontol. Geriatr.* **2016**, *62*, 26–29. [CrossRef]
11. Salminen, M.; Saaristo, P.; Salonoja, M.; Vaapio, S.; Vahlberg, T.; Lamberg-Allardt, C.; Aarnio, P.; Kivelä, S.-L. Vitamin D status and physical function in older Finnish people: A one-year follow-up study. *Arch. Gerontol. Geriatr.* **2015**, *61*, 419–424. [CrossRef]

12. Dawson-Hughes, B. Vitamin D and muscle function. *J. Steroid Biochem. Mol. Biol.* **2017**, *173*, 313–316. [CrossRef]
13. Pilz, S.; Dobnig, H.; Tomaschitz, A.; Kienreich, K.; Meinitzer, A.; Friedl, C.; Wagner, D.; Piswanger-Sölkner, C.; März, W.; Fahrleitner-Pammer, A. Low 25-hydroxyvitamin D is associated with increased mortality in female nursing home residents. *J. Clin. Endocrinol. Metab.* **2012**, *97*, 653–657. [CrossRef] [PubMed]
14. Skaaby, T.; Husemoen, L.L.N.; Thuesen, B.H.; Pisinger, C.; Hannemann, A.; Jørgensen, T.; Linneberg, A. Longitudinal associations between lifestyle and vitamin D: A general population study with repeated vitamin D measurements. *Endocrine* **2016**, *51*, 342–350. [CrossRef] [PubMed]
15. Liu, B.A.; Gordon, M.; Labranche, J.M.; Murray, T.M.; Vieth, R.; Shear, N.H. Seasonal prevalence of vitamin D deficiency in institutionalized older adults. *J. Am. Geriatr. Soc.* **1997**, *45*, 598–603. [CrossRef] [PubMed]
16. Mosekilde, L. Vitamin D and the elderly. *Clin. Endocrinol.* **2005**, *62*, 265–281. [CrossRef] [PubMed]
17. De Jongh, R.T.; Van Schoor, N.M.; Lips, P. Changes in vitamin D endocrinology during aging in adults. *Mol. Cell. Endocrinol.* **2017**, *453*, 144–150. [CrossRef] [PubMed]
18. Samefors, M.; Östgren, C.J.; Mölstad, S.; Lannering, C.; Midlöv, P.; Tengblad, A. Vitamin D deficiency in elderly people in Swedish nursing homes is associated with increased mortality. *Eur. J. Endocrinol.* **2014**, *170*, 667–675. [CrossRef]
19. Saraiva, G.L.; Cendoroglo, M.S.; Ramos, L.R.; Araújo, L.M.Q.; Vieira, J.G.H.; Maeda, S.S.; Borba, V.Z.C.; Kunii, I.; Hayashi, L.F.; Lazaretti-Castro, M. Prevalência da deficiência, insuficiência de vitamina D e hiperparatiroidismo secundário em idosos institucionalizados e moradores na comunidade da cidade de São Paulo, Brasil. *Arq. Bras. Endocrinol. Metabol.* **2007**, *51*, 437–442. [CrossRef]
20. Maeda, S.S.; Kunii, I.S.; Hayashi, L.F.; Lazaretti-Castro, M. Increases in summer serum 25-hydroxyvitamin D (25OHD) concentrations in elderly subjects in São Paulo, Brazil vary with age, gender and ethnicity. *BMC Endocr. Disord.* **2010**, *10*, 12. [CrossRef]
21. Andersen, R.; Mølgaard, C.; Skovgaard, L.T.; Brot, C.; Cashman, K.D.; Chabros, E.; Charzewska, J.; Flynn, A.; Jakobsen, J.; Karkkainen, M.; et al. Teenage girls and elderly women living in northern Europe have low winter vitamin D status. *Eur. J. Clin. Nutr.* **2005**, *59*, 533–541. [CrossRef]
22. Mowé, M.; Bohmer, T.; Haug, E. Serum calcidiol and calcitriol concentrations in elderly people: Variations with age, sex, season and disease. *Clin. Nutr.* **1996**, *15*, 201–206. [CrossRef]
23. Pérez-Llamas, F.; López-Contreras, M.J.; Blanco, M.J.; López-Azorín, F.; Zamora, S.; Moreiras, O. Seemingly paradoxical seasonal influences on vitamin D status in nursing-home elderly people from a Mediterranean area. *Nutrition* **2008**, *24*, 414–420. [CrossRef] [PubMed]
24. Pereira Machado, R.S.; Santa Cruz Coelho, M.A. Risk of malnutrition among Brazilian institutionalized elderly: A study with the Mini Nutritional Assessment (MNA) questionnaire. *J. Nutr. Health Aging* **2011**, *15*, 532–535. [CrossRef] [PubMed]
25. Hoffman, R. Micronutrient deficiencies in the elderly—Could ready meals be part of the solution? *J. Nutr. Sci.* **2017**, *6*, 4. [CrossRef]
26. Brazil. The Statute of the Elderly, Law 10741. 2003. Available online: http://www.planalto.gov.br/ccivil_03/leis/2003/l10.741.htm (accessed on 15 April 2017).
27. Open Source Epidemiologic Statistics for Public Health. Available online: https://www.openepi.com (accessed on 22 January 2018).
28. Sales, M.C.M.C.; de Oliveira, L.P.; de Araújo Cabral, N.L.; de Sousa, S.E.S.; das Graças Almeida, M.; Lemos, T.M.A.M.; de Oliveira Lyra, C.; de Lima, K.C.; Sena-Evangelista, K.C.M.; de Fatima Campos Pedrosa, L.; et al. Plasma zinc in institutionalized elderly individuals: Relation with immune and cardiometabolic biomarkers. *J. Trace Elem. Med. Biol.* **2018**, *50*, 615–621. [CrossRef] [PubMed]
29. Instituto Brasileiro de Geografia e Estatística (IBGE). *Características étnico-raciais da população: Classificações e identidades*, 2nd ed.; IBGE: Rio de Janeiro, Brazil, 2013; ISBN 978-85-240-4244-7.
30. Chumlea, W.A.; Roche, A.F.; Mukherjee, D. *Nutritional Assessment of the Elderly Through Anthropometry*; Ross Laboratories: Columbus, OH, USA, 1987.
31. Lipschitz, D.A. Screening for nutritional status in the elderly. *Prim. Care* **1994**, *21*, 55–67.
32. National Cholesterol Education Program (NCEP) Expert Panel on Detection, Evaluation; Treatment of High Blood Cholesterol in Adults (Adult Treatment Panel III). Third report of the National Cholesterol Education Program (NCEP) expert panel on detection, evaluation, and treatment of high blood cholesterol in adults (Adult Treatment Panel III) fin. *Circulation* **2002**, *106*, 3143–3421. [CrossRef]

33. Kennerfalk, A.; Ruigómez, A.; Wallander, M.-A.; Wilhelmsen, L.; Johansson, S. Geriatric Drug Therapy and Healthcare Utilization in the United Kingdom. *Ann. Pharmacother.* **2002**, *36*, 797–803. [CrossRef]
34. Cincura, C.; Pontes-Neto, O.M.; Neville, I.S.; Mendes, H.F.; Menezes, D.F.; Mariano, D.C.; Pereira, I.F.; Teixeira, L.A.; Jesus, P.A.P.; de Queiroz, D.C.L.; et al. Validation of the National Institutes of Health Stroke Scale, Modified Rankin Scale and Barthel Index in Brazil: The Role of Cultural Adaptation and Structured Interviewing. *Cerebrovasc. Dis.* **2009**, *27*, 119–122. [CrossRef]
35. De Oliveira Neto, L.; Agrícola, P.M.D.; de Andrade, F.L.J.P.; de Oliveira, L.P.; Lima, K.C.; de Oliveira Neto, L.; Agrícola, P.M.D.; de Andrade, F.L.J.P.; de Oliveira, L.P.; Lima, K.C. What is the impact of the European Consensus on the diagnosis and prevalence of sarcopenia among institutionalized elderly persons? *Rev. Bras. Geriatr. Gerontol.* **2017**, *20*, 754–761. [CrossRef]
36. Katz, S.; Ford, A.B.; Moskowitz, R.W.; Jackson, B.A.; Jaffe, M.W. Studies of illness in the aged. The index of adl: A standardized measure of biological and psychosocial function. *JAMA* **1963**, *185*, 914–919. [CrossRef]
37. Pfeiffer, E. A short portable mental status questionnaire for the assessment of organic brain deficit in elderly patients. *J. Am. Geriatr. Soc.* **1975**, *23*, 433–441. [CrossRef] [PubMed]
38. Martínez de la Iglesia, J.; Dueñas Herrero, R.; Onís Vilches, M.C.; Aguado Taberné, C.; Albert Colomer, C.; Luque Luque, R. Spanish language adaptation and validation of the Pfeiffer's questionnaire (SPMSQ) to detect cognitive deterioration in people over 65 years of age. *Med. Clin.* **2001**, *117*, 129–134.
39. LAVAT Laboratório de Váriaveis Ambientais e Tropicais. Available online: http://www.crn.inpe.br/lavat/ (accessed on 8 May 2019).
40. Nusser, S.M.; Carriquiry, A.L.; Dodd, K.W.; Fuller, W.A. A semiparametric transformation approach to estimating usual daily intake distributions. *J. Am. Stat. Assoc.* **1996**, *91*, 1440–1449. [CrossRef]
41. Willett, W.; Stampfer, M.J. Total energy intake: Implications for epidemiologic analyses. *Am. J. Epidemiol.* **1986**, *124*, 17–27. [CrossRef] [PubMed]
42. Institute of Medicine (US) Subcommittee on Interpretation and Uses of Dietary Reference Intakes; Institute of Medicine (US) Standing Committee on the Scientific Evaluation of Dietary Reference Intakes. *Standing Committee on the Scientific Evaluation of Dietary Reference Intakes. Dietary Reference Intakes: Applications in Dietary Assessment*; National Academies Press: Washington, DC, USA, 2000; ISBN 978-0-309-07183-3.
43. Van Dam, R.M.; Snijder, M.B.; Dekker, J.M.; Stehouwer, C.D.; Bouter, L.M.; Heine, R.J.; Lips, P. Potentially modifiable determinants of vitamin D status in an older population in the Netherlands: The Hoorn Study. *Am. J. Clin. Nutr.* **2007**, *85*, 755–761. [CrossRef] [PubMed]
44. World Health Organization (WHO). *Global Solar UV Index: A Practical Guide. A Joint Recommendation of World Health Organization, World Meteorological Organization, United Nations Environment Programme and the International Commission on Non-Ionizing Radiation Protection*; Library Cataloguing-in-Publication; WHO: Geneva, Switzerland, 2002; ISBN 92 4 159007 6.
45. Lin, T.-C.; Liao, Y.-C. The impact of sunlight exposure on the health of older adults. *J. Nurs.* **2016**, *63*, 116–122.
46. Okan, F.; Okan, S.; Zincir, H. Effect of Sunlight Exposure on Vitamin D Status of Individuals Living in a Nursing Home and Their Own Homes. *J. Clin. Densitom.* **2018**. In Press. [CrossRef]
47. Komar, L.; Nieves, J.; Cosman, F.; Rubin, A.; Shen, V.; Lindsay, R. Calcium Homeostasis of an Elderly Population upon Admission to a Nursing Home. *J. Am. Geriatr. Soc.* **1993**, *41*, 1057–1064. [CrossRef]
48. Oliveira, B.S.; Zunzunegui, M.V.; Quinlan, J.; Batistuzzo de Medeiros, S.R.; Thomasini, R.L.; Guerra, R.O. Lifecourse Adversity and Telomere Length in Older Women from Northeast Brazil. *Rejuvenation Res.* **2018**, *21*, 294–303. [CrossRef]
49. Jääskeläinen, T.; Itkonen, S.T.; Lundqvist, A.; Erkkola, M.; Koskela, T.; Lakkala, K.; Dowling, K.G.; Hull, G.L.; Kröger, H.; Karppinen, J.; et al. The positive impact of general vitamin D food fortification policy on vitamin D status in a representative adult Finnish population: Evidence from an 11-y follow-up based on standardized 25-hydroxyvitamin D data. *Am. J. Clin. Nutr.* **2017**, *105*, 1512–1520. [CrossRef]

Communication

Considerations for the Development of Innovative Foods to Improve Nutrition in Older Adults

Mariane Lutz [1,2,*], **Guillermo Petzold** [1,3] **and Cecilia Albala** [1,4]

1 Thematic Task Force on Healthy Aging, CUECH Research Network, Viña del Mar 2520000, Chile; gpetzold@ubiobio.cl (G.P.); calbala@uchile.cl (C.A.)
2 Interdisciplinary Center for Health Studies, CIESAL, Faculty of Medicine, Universidad de Valparaíso, Angamos 655, Reñaca, Viña del Mar 2520000, Chile
3 Department of Food Engineering, Universidad del Bio-Bio, Andrés Bello 720, Casilla 447, Chillán 3780000, Chile
4 Institute of Nutrition and Food Technology, INTA, Universidad de Chile, El Líbano 5524, Macul, Santiago 7810000, Chile
* Correspondence: mariane.lutz@uv.cl; Tel.: +56-322603095

Received: 29 April 2019; Accepted: 27 May 2019; Published: 5 June 2019

Abstract: The population of older adults is growing globally. This increase has led to an accumulation of chronic illnesses, so-called age-related diseases. Diet and nutrition are considered the main drivers of the global burden of diseases, and this situation applies especially to this population segment. It relates directly to the development of coronary heart disease, hypertension, some types of cancer, and type 2 diabetes, among other diseases, while age-associated changes in body composition (bone and muscle mass, fat, sarcopenia) constitute risk factors for functional limitations affecting health status and the quality of life. Older adults present eating and swallowing problems, dry mouth, taste loss, and anorexia among other problems causing "anorexia of aging" that affects their nutritional status. The strategies to overcome these situations are described in this study. The impact of oral food processing on nutrition is discussed, as well as approaches to improve food acceptance through the design of innovative foods. These foods should supply a growing demand as this group represents an increasing segment of the consumer market globally, whose needs must be fulfilled.

Keywords: older adults; aging; food; nutrition; acceptability

1. Introduction

The increasing life expectancy of the world population, along with decreased mortality, has led to a rapid aging of the population in many countries [1]. The success in improving survival does not necessarily mean that the additional years are healthy or endured with a good quality of life. In fact, as mortality decreases and life expectancy increases, the question about the quality of the years gained, in the different regions of the world, arises. The rapid increase of the population over 60 years old, or older adults (OA), has led to an accumulation of chronic illnesses, the so-called age-related diseases (ARDs). This, together with the decrease in the function of organs and systems, increases vulnerability to a variety of stressors, augmenting functional limitations, disability, and dependency. The situation, however, is not inevitable or irreversible. Although the prevention of chronic diseases and the promotion of health should optimally be carried out throughout the life cycle, many disability-adjusted life years (DALYs) lost can be avoided through proper action initiated in OA [2]. Consequently, the current challenge is to decrease the gap between health expectancy and healthy life expectancy, as expressed by independence, autonomy, and functionality.

Nutrition and physical activity are among the main determinants of health. Diet and nutrition are considered the main drivers of the global burden of diseases [3]. Their global impact is huge, and their association to ARDs such as coronary heart disease, hypertension, some types of cancer, and type 2 diabetes, is well established. On the other hand, age-associated changes in body composition, including a decrease in bone and muscle mass, and a redistribution and increase of body fat, can lead, among others, to impaired immunity, metabolic disorders, frailty, sarcopenia, and osteoporosis, all of which constitute risk factors for functional limitations, falls, fractures, disability, dependency, institutionalization and mortality, affecting health status and the quality of life.

Although the importance of nutrition in OA is well established, undernutrition has not been given sufficient importance. A review of studies carried out mainly in European countries using the Mini Nutritional Assessment (MNA) screening tool found a prevalence of undernutrition of 5.8% in community dwelling, 38.7% in hospitalized patients, 50.5% in rehabilitation care, and 13.8% in institutionalized OA [4]. The Survey on Health, Well-Being, and Aging (SABE) survey, carried out in the capital cities of Latin America and the Caribbean in 1999–2000, revealed that, according to MNA, the prevalence of undernutrition among community-dwelling OA fluctuated between 1% and 8.3% and the risk of malnutrition ranged from 9.1% to 41.8% [5].

Other important physiological changes in OA include eating and swallowing impairment, such as chewing difficulty, dry mouth, taste loss, and loss of appetite, among others. These problems decrease the oral processing capability, which has attracted the attention of the food industry and researchers to provide technological solutions of foods with nutritive properties and attractive sensory attributes [6].

The aim of this publication is to describe some major physiological changes associated with aging and how these changes impact the nutritional status of OA, facts that should be taken into account to prevent the negative impact of aging on their quality of life. Accordingly, the need for innovative foods that consider rheological or texture properties especially directed towards OA is described, as a strategy to improve their acceptability and, consequently, the nutritional status of this increasing population group.

2. Nutritional Status in Aging

Along with the increasing prevalence of certain diseases, the changes in body composition associated to aging influence the nutritional status of OA [7]. Roubenoff [8] distinguishes three interrelated types of changes associated with undernutrition that may occur either as a consequence of the aging process, concomitant diseases, or both—wasting (an unintentional loss of weight, primarily caused by a deficient dietary intake, affecting fat and fat free mass); cachexia (the loss of fat free mass or body cell mass and no initial weight loss, characterized by hypercatabolism and sarcopenia); and loss of muscle mass (which seems to be a condition inherently related to aging, usually with pre-existing cachexia or sarcopenia, with low muscle mass, muscle strength, and physical performance, caused by mechanisms that involve, among others, protein synthesis, proteolysis, neuromuscular integrity and muscle fat content). In many OA the etiology is multi-factorial [9].

The physiological changes that lead to these situations include the loss of appetite [10], mainly due to decreased chemosensory functions and decreased secretions of the hormones that regulate appetite. Cox et al. [11] assessed nine interventional treatment strategies for the anorexia of aging, which aimed to improve appetite, of which food flavor enhancement, oral nutritional supplements, amino acid precursors, fortified foods, and megestrol acetate medication proved to be effective.

Chewing difficulties, swallowing problems, thirst, hunger, and diminished smell and taste are detrimental for the psychological satisfaction and pleasure associated with eating, resulting in a decrease in energy intake. Anorexia may also lead to wasting and sarcopenia (defined as "a syndrome characterized by progressive and generalized loss of skeletal muscle mass and strength with a risk of adverse outcomes such as physical disability, poor quality of life, and death" [9,12–14]), poor endurance, and decreased mobility [15]. Consequently, there are multiple causes of weight loss in the elderly, including the decline of chemosensory function (smell and taste), reduced efficiency of chewing, slowed gastric emptying, and alterations to the neuroendocrine axis (changes in the levels of leptin, cholecystokinin, neuropeptide Y, and other hormones and peptides), which contribute to anorexia [16,17].

3. Anorexia and Malnutrition in OA

Poor nutritional status is one of the main risk factors for frailty, a condition characterized by the inability to respond to stress and preserve homeostasis [18], and associated with both macro- and micronutrients deficiencies [19]. Frail persons are at a high risk of disability, dependency, cognitive impairment, and mortality [20]. In fact, frailty is considered as a state of pre-disability, and has been described as a situation between normal aging and disability (or even death) [21]. Tsutsumimoto et al. [22] investigated whether the anorexia of aging had a significant impact on incident disability and a possible direct association with future disability, or an indirect association with this condition via frailty. The authors showed that OA with anorexia had a higher proportion of frailty and a higher prevalence of disability compared to those without it. In addition, anorexia indirectly affected incident disability via frailty status. The pathophysiology of frailty is complex and multi-factorial, and nutrition is an important factor in its onset and a specific target for treatment [23]. Frailty involves a decrease in dietary intake, coupled with a decline in physical exercise, which leads to a loss of muscle mass, thus making OA more vulnerable to develop complications such as sarcopenia, comorbidities, or disability [18].

Chronic undernutrition (insufficient protein and energy intake) leads to weight loss and sarcopenia (which may, in turn, cause low muscle strength and feelings of exhaustion), while frailty itself may have a negative effect on eating and, consequently, on the nutritional status [24]. When lean body mass is lost, while fat mass is preserved or even increased, the state is called sarcopenic obesity [25]. In this situation, the relationship between age-related reduction of muscle mass and strength is often independent of body mass. It had long been thought that the loss of weight, along with the loss of muscle mass, were major factors affecting muscle weakness in OA [26]. However, changes in muscle composition are also important, e.g., fat infiltration into muscle lowers work performance [27].

Undernutrition involving protein-energy wasting in OA has been extensively described [28,29]. Inadequate food intake, reduced capacity to use available proteins, and a higher need for proteins due to a cumulative physical decline [30] contribute to the alteration of the nutritional state. Efforts have been made to improve protein intake through various strategies, including the use of nutritional supplements [31] and dietary enrichment or food and meal fortification, in which protein intake is increased by augmenting protein density [32]. In a systematic review of clinical studies determining the effects of dietary enrichment with conventional foods on energy and protein intake in OA, Trabal and Farran-Codina [33] concluded that any intervention that increases energy and nutrient density while holding constant or reducing portion sizes, constitutes a desired approach—having observed that low-volume, energy-dense foods increase energy intake without affecting appetite. However, the authors could not get conclusive results, mainly due to the lack of large-scale clinical trials with long-term interventions that allow the establishment of the effects of the treatments reported to address malnutrition in OA. Besides, OA usually exhibit both short- and long-term satiety signals (mostly peripheral) which contrast energy balance and contribute to malnutrition.

4. Protein Needs in OA

OA need more protein due to a series of physiological changes, including a declining anabolic response to protein intake. In fact, OA develop resistance to the positive effects of dietary protein on protein synthesis, limiting muscle accretion and maintenance, in a situation described as "anabolic resistance" [34]. Besides, there is a need to offset the catabolic conditions associated with the multiple chronic and acute diseases that commonly occur in OA, among other situations. Many of these relate to modifications of hormone production and sensitivity, which involves growth hormones (GH), insulin-like growth factor (IGF-I), corticosteroids, androgens, estrogens, and insulin, which affect the anabolic/catabolic state of muscle protein metabolism [35–37]. Metabolic changes associated with aging also include increased splanchnic sequestration and decreased postprandial availability of amino acids (AA), a lower postprandial perfusion of muscle, decreased muscle uptake of dietary AA, reduced anabolic signaling for protein synthesis, a reduced ability to use available protein (insulin resistance, protein anabolic resistance, high splanchnic extraction, immobility), and a reduced digestive capacity [30,38]. On the other hand, the main factors that influence protein use in OA include inadequate intake of protein (anorexia or appetite loss, gastrointestinal disturbances) or a greater need for protein (inflammatory disease, increased oxidative modification of proteins), all of which indicate that protein needs are augmented. A high proportion of inadequate protein intake has been observed in OA [39] and some studies had estimated that 15% to 38% of older men and 27% to 41% of older women consume less protein than recommended [40]. The AA composition of dietary proteins impacts anabolic potency at a muscular level. Leucine is the main regulator of protein turnover in muscle, through the activation of mTOR signaling. Although aged muscle has a reduced anabolic response to small doses of essential AA, 2.5–3 g of leucine are able to reverse this anabolic resistance [41].

The PROT-AGE Study Group established that, in order to maintain and regain muscle, OA should consume an average daily intake in the range of 1.0 to 1.2 g/kg body weight/day. In case of acute or chronic disease, the need of dietary protein increases to 1.2 to 1.5 g/kg body weight/day; and people with severe illness, injury, or marked malnutrition may need 2.0 g/kg body weight/day [30]. In OA, dietary protein or AA supplementation promote protein synthesis and can enhance recovery of physical function [42], improving muscle strength and function more readily than muscle mass [41,43,44]. Observational studies have supported an association between protein intake and muscle strength and mass [45,46], and the ingestion of ~20 g whey protein has been shown to increase muscle protein synthesis rates in healthy OA [47,48]. In fact, it has been recommended that due to the blunted sensitivity of OA muscles to low doses of AA, dietary protein should be distributed to at least 25 to 30 g of high quality protein per meal, containing approximately 2.5 to 2.8 g of leucine, to stimulate muscle protein synthesis [30,49].

An additional approach regarding the sources of proteins to supply the needs of OA takes into consideration their sustainability. A sustainable diet should increase plant protein sources and reduce animal protein intake. The impacts of these more environmentally-friendly diets on the nutritional state of OA are just beginning to be addressed, considering that plant foods are also sources of dietary fiber and a variety of phytochemicals, and may eventually reduce the bioavailability of some nutrients [50]. The recommendations for increased high-quality protein intake in OA should also take into consideration an adequate supply of calcium for preserving bone and muscle mass and, additionally, it is also relevant to reach an adequate energy supply to achieve the optimal protein utilization, with a high P% (proportion of dietary energy derived from proteins) [51].

5. Strategies to Contrast Anorexia and Malnutrition

The most commonly described dietary strategies to deliver proteins and other nutrients to OA include fractioning food intake in small digestible meals, improving taste and flavor, and/or limiting the intake of cholecystokinin (CCK)-stimulating foods such as fats and proteins [10], although in this case the energy and protein densities may be reduced. A strategy that can be used for OA who require increased energy and nutrient intakes is to offer frequent, small servings of food with high energy and nutrient density [52], such as frozen ready-to-eat meals [53]. Besides changes in the amount of food and type of food intake, OA eat fewer snacks between meals [54], experience less cravings for food [55], and feel less hungry and more satiated than younger individuals [56]. Another factor affecting low energy intake and low body weight in OA is small dietary variety, since energy intake is greater when a variety of foods is provided [57,58]. Finally, as highlighted by de Boer et al. [59], the effects of non-physiological anorexia of aging should be considered, including socio-economic factors such as depression, alcoholism, poverty, widowhood, environment changes, social isolation, and loneliness.

6. Foods for OA: The Importance of Texture and Other Sensory Attributes

Good nutrition may help prevent, modulate, or ameliorate age related diseases [60]. However, the physiological changes and dysfunctions in OA cause a series of eating and swallowing problems, considering that the oral food processing (the first step of food consumption) includes not only the intake of food, digestion and absorption of nutrients, and conditioning their bioavailability, but also comprises important sensory attributes. Various approaches have been proposed to improve food acceptance through intelligent design or modifications of the food matrix. Among these, in case of the chewing difficulty of OA due to the lack of functional teeth, an efficient reduction of food size is suggested; while in most of the OA with eating difficulties it is necessary to modify the texture of food as an efficient solution to improve food intake [6].

The development of innovative foods is important to counteract the deficiencies of macro- and micronutrients intake in OA, fortifying the food products with selected ingredients, vitamins, and minerals [61]. In this context, the need of OA for food products with adequate sensory values and optimal nutritional quality is of crucial importance [62]. As mentioned above, an important nutritional concern is to provide OA with sufficiently high-quality proteins [50], while the decline of food chemosensory perception in OA forces the food industry to develop more palatable foods, improving attractive properties such as taste, smell, temperature, color, and texture that positively influence food intake [63]. Accordingly, several strategies have been proposed to make foods for OA more palatable and stimulate their appetite. For example, van der Meij et al. [64] highlight the importance of providing a variety of adapted meals and snacks of different colors. In a similar strategy, Griep et al. [65] showed that intensely flavored products such as a meat substitute (Quorn) and yoghurt increase food intake in OA. On the other hand, an increase in food intake has been observed via flavored additives such as monosodium glutamate [66,67], or the natural flavoring of roast beef, bacon, cheese, citrus or pomegranate byproducts and spices such as rosemary, garlic, paprika, and onion [68,69]. The effect of natural food flavors on food intake in hospitalized OA patients in Hong Kong showed that total energy and protein intakes were increased by 13–26% and 15–28%, respectively, with flavor enhancement [70].

The food industry needs to develop and offer innovative food products with modified texture or rheology, palatable, and nutritious [63] to help overcome aging related anorexia [71]. Texture modified foods are processed products with a soft texture or a reduced particle size, as well as thickened liquids (drinks) oriented towards the market segment of OA with eating dysfunctions [72]. Food textures for the OA population should be soft and moist, while sticky and adhesive textures should be avoided as well as fibrous structures that are not easily disintegrated [73]. Soft texture foods are preferred, because they are easily disintegrated and mixed in the mouth, avoiding mastication [74]. Additionally, OA usually have difficulties forming the food bolus, which in some cases leads to a very long time of chewing before swallowing, which negatively affects the sensory experience associated with that food. Accordingly, Laguna et al. [75] tested the oral processing of foods in OA using gels of different textures (varying in hardness), and reported that not only the texture or consistency (hardness) is important, but also the heterogeneity of the food matrix. The physical characteristics of foods influence their oral processing, affecting mainly the number of chews and time spent in the mouth—major considerations in the design of foods for OA. Moreover, the flow and properties of saliva normally change with age, which can result in dry mouth conditions and taste aberrations [76]. Limited salivation is an important physiological dysfunction, with a 38% drop of the salivary flow in OA and the consequent problems forming the food bolus [77]. Salivation is very important to food processing, involving lubrication and food bolus formation in the mouth, and is consequently also related to the textural experience and overall sensory experience [76]. In this context, Assad-Bustillos et al. [78] reported that soft aerated cereal foods stimulate the salivary flow rate, the food bolus properties and the perception of oral comfort (the oral sensations perceived when eating a food) in OA, priming over the dental status. In addition, Lorieau et al. [79] demonstrated that soft model cheeses tested by OA led to a softer bolus that was more easily formed, the soft cheeses being more comfortable than dryer cheeses as their textures were perceived as soft, fatty and melting.

Another alternative to offering innovative acceptable foods for OA is to modify the culinary processes used to improve oral comfort when they eat. Among these, blade tenderization is an effective technique for improving meat texture. It involves meat perforation with sharp edged blades that are closely spaced to cut muscle fibers and ensure tenderness. Vandenberghe-Descamps et al. [80] demonstrated that easy-to-do culinary processes improve oral comfort, facilitate the formation of a food bolus and ameliorate food texture while eating meat. Regarding roast beef, the cumulative effect of blade tenderization, marinade and low-temperature cooking were the optimal conditions to obtain meat that is easy to chew, humidifies well with saliva, and can be smoothly swallowed, as well as any tender and juicy product.

The food industry should consider the fact that foods for the OA represent an interesting segment of the world consumer market. In the US, this segment holds more than a third of the country's wealth. In Europe, consumption by adults over 50 years old has increased three times as fast as those under this age. Clearly, the stereotype of an OA who is conservatively spending on food and beverages in the face of very limited income is increasingly out of date. In addition, OA constitute the largest percentage of television audiences (>50%) and the largest consumers of printed material, representing an audience prone to receive information about new food products that leverage innovations in the food science and technology area [81].

7. Conclusions

OA experience a series of age-related physiological changes that lead to detrimental nutritional impacts. In order to improve their food intake to accomplish their nutritional needs, overcome changes in their appetite, improve their oral processing of foods, and to increase sensory attributes, the food industry needs to develop intelligent foods through novel design. Among the current alternatives directed towards the growing OA population group, new texture-modified foods represent an efficient, nutritious, and palatable solution to overcome their eating and swallowing difficulties. Innovative foods should supply a growing demand as OA represent an interesting and increasing segment of the consumer market globally, whose needs must be fulfilled (Figure 1).

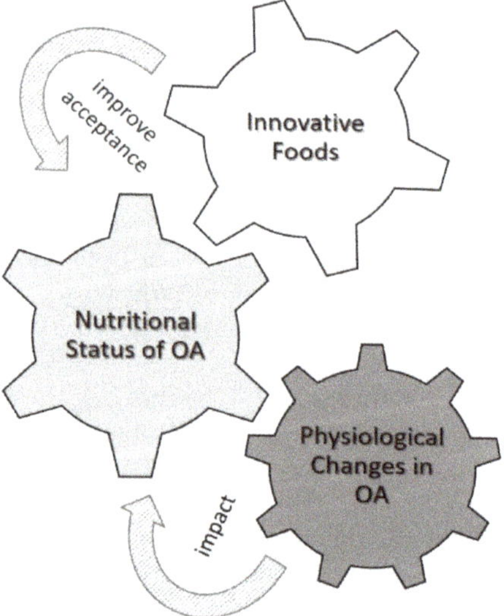

Figure 1. The physiological changes in older adults (OA) impact their nutritional status. Innovative foods with high acceptability must be developed to improve their nutritional status.

Author Contributions: M.L. conceived the subject and aim of the review and wrote the paper; G.P. provided information on sensory and texture aspects of foods; C.A. provided information on physiological changes of aging that affect nutritional status.

Funding: The publication of this review is funded by the Thematic Task Force on Healthy Aging, CUECH Research Network (Association of Public State Universities), Chile.

Conflicts of Interest: The authors declare no conflict of interest.

References

1. World Health Organization. Global Health Observatory Data Repository. Mortality and Global Health Estimates. Available online: http://apps.who.int/gho/data/node.home (accessed on 3 April 2019).
2. Prince, M.J.; Wu, F.; Guo, Y.; Gutierrez Robledo, L.M.; O'Donnell, M.; Sullivan, R.; Yusuf, S. The burden of disease in older people and implications for health policy and practice. *Lancet* **2015**, *385*, 549–562. [CrossRef]
3. Global Nutrition Report. From Promise to Impact: Ending Malnutrition by 2030 Independent Expert Group, Washington, DC, USA. Available online: http://www.ifpri.org/publication/global-nutrition-report-2016-promise-impact-ending-malnutrition-2030 (accessed on 30 March 2019).
4. Kaiser, M.J.; Bauer, J.M.; Rämsch, C.; Uter, W.; Guigoz, Y.; Cederholm, T.; Thomas, D.R.; Anthony, P.S.; Charlton, K.E.; Maggio, M.; et al. Mini Nutritional Assessment International Group. Frequency of malnutrition in older adults: A multinational perspective using the mini nutritional assessment. *J. Am. Geriatr. Soc.* **2010**, *58*, 1734–1738. [CrossRef] [PubMed]
5. Lera, L.; Sánchez, H.; Ángel, B.; Albala, C. Mini Nutritional Assessment short-form: Validation in five Latin American cities. SABE Study. *J. Nutr. Health Aging* **2016**, *20*, 797–805. [CrossRef] [PubMed]
6. Wang, X.; Chen, J. Food oral processing: Recent developments and challenges. *Curr. Opin. Colloid Interface Sci.* **2017**, *28*, 22–30. [CrossRef]
7. Hickson, M.; Frost, G. An investigation into the relationships between quality of life, nutritional status and physical function. *Clin. Nutr.* **2004**, *23*, 213–221. [CrossRef]
8. Roubenoff, R. The pathophysiology of wasting in the elderly. *J. Nutr.* **1999**, *129*, 256S–259S. [CrossRef] [PubMed]
9. Cruz-Jentoft, A.J.; Bahat, G.; Bauer, J.; Boirie, Y.; Bruyère, O.; Cederholm, T.; Cooper, C.; Landi, F.; Rolland, Y.; Sayer, A.A.; et al. Sarcopenia: Revised European consensus on definition and diagnosis. *Age Ageing* **2018**, *48*, 16–31. [CrossRef] [PubMed]
10. Di Francesco, V.; Fantin, F.; Omizzolo, F.; Residori, L.; Bissoli, L.; Bissoli, L.; Bosello, O.; Zamboni, M. The anorexia of aging. *Dig. Dis.* **2007**, *25*, 129–137. [CrossRef]
11. Cox, N.J.; Ibrahim, K.I.; Sayer, A.A.; Robinson, S.M.; Roberts, H.C. Assessment and treatment of the anorexia of aging: A systematic review. *Nutrients* **2019**, *11*, 144. [CrossRef]
12. Fielding, R.A.; Vellas, B.; Evans, W.J.; Bhasin, S.; Morley, J.E.; Newman, A.B.; Abellan van Kan, G.; Andrieu, S.; Bauer, J.; Breuille, D.; et al. Sarcopenia: An undiagnosed condition in older adults. Current consensus definition: Prevalence, etiology, and consequences. International Working Group on Sarcopenia. *J. Am. Med. Dir. Assoc.* **2011**, *12*, 249–256. [CrossRef]
13. Cruz-Jentoft, A.J.; Baeyens, J.P.; Bauer, J.M.; Boirie, Y.; Cederholm, T.; Landi, F.; Martin, F.C.; Michel, J.P.; Rolland, Y.; Schneider, S.M.; et al. Sarcopenia: European consensus on definition and diagnosis. *Age Ageing* **2010**, *39*, 412–423. [CrossRef] [PubMed]
14. Cruz-Jentoft, A.J.; Landi, F.; Schneider, S.M.; Zuñiga, C.; Arai, H.; Boirie, Y.; Chen, L.K.; Fielding, R.A.; Martin, F.C.; Michel, J.P.; et al. Prevalence of and interventions for sarcopenia in ageing adults: A systematic review. Report of the International Sarcopenia Initiative (EWGSOP and IWGS). *Age Ageing* **2014**, *43*, 748–759. [CrossRef] [PubMed]
15. Morley, J.E. Anorexia, sarcopenia, and aging. *Nutrition* **2001**, *17*, 660–663. [CrossRef]
16. Alibhai, S.M.; Greenwood, C.; Payette, H. An approach to the management of unintentional weight loss in elderly people. *Can. Med. Assoc. J.* **2005**, *172*, 773–780. [CrossRef] [PubMed]
17. Wernette, C.M.; White, D.; Zizza, C.A. Signaling proteins that influence energy intake may affect unintentional weight loss in elderly persons. *J. Am. Diet. Assoc.* **2011**, *111*, 864–873. [CrossRef] [PubMed]

18. Fried, L.P.; Tangen, C.M.; Walston, J.; Newman, A.B.; Hirsch, C.; Gottdiener, J.; Seeman, T.; Tracy, R.; Kop, W.J.; Burke, G.; et al. Frailty in older adults: Evidence for a phenotype. *J. Gerontol. A Biol. Sci. Med. Sci.* **2001**, *56*, M146–M156. [CrossRef] [PubMed]

19. Bonnefoy, M.; Berrut, G.; Lesourd, B.; Ferry, M.; Gilbert, T.; Guérin, O.; Hanon, O.; Jeandel, C.; Paillaud, E.; Raynaud-Simon, A.; et al. Frailty and nutrition: Searching for evidence. *J. Nutr. Health Aging* **2015**, *19*, 250–257. [CrossRef] [PubMed]

20. Walston, J.; Hadley, E.C.; Ferrucci, L.; Guralnik, J.M.; Newman, A.B.; Studenski, S.A.; Ershler, W.B.; Harris, T.; Fried, L.P. Research Agenda for Frailty in Older Adults: Toward a better understanding of physiology and etiology: Summary from the American Geriatrics Society/National Institute on Aging Research Conference on Frailty in Older Adults. *J. Am. Geriatrics Soc.* **2006**, *54*, 991–1001. [CrossRef] [PubMed]

21. Van Kan, G.A.; Rolland, Y.M.; Morley, J.E.; Vellas, B. Frailty: Toward a clinical definition. *J. Am. Med. Dir. Assoc.* **2007**, *9*, 71–72. [CrossRef]

22. Tsutsumimoto, K.; Doi, T.; Makizako, H.; Hotta, R.; Nakakubo, S.; Makino, K.; Suzuki, T.; Shimada, H. Aging-related anorexia and its association with disability and frailty. *J. Cachexia Sarcop. Muscle* **2018**, *9*, 834–843. [CrossRef]

23. Clegg, A.; Young, J.; Iliffe, S.; Rikkert, M.O.; Rockwood, K. Frailty in elderly people. *Lancet* **2013**, *381*, 752–762. [CrossRef]

24. Yannakoulia, M.; Ntanasi, E.; Anastasiou, C.A.; Scarmeas, M. Frailty and nutrition: From epidemiological and clinical evidence to potential mechanisms. *Metab. Clin. Exper.* **2017**, *68*, 64–76. [CrossRef]

25. Prado, C.M.; Lieffers, J.R.; McCargar, L.J.; Reiman, T.; Sawyer, M.B.; Sawyer, M.B.; Martin, L.; Baracos, V.E. Prevalence and clinical implications of sarcopenic obesity in patients with solid tumours of the respiratory and gastrointestinal tracts: A population-based study. *Lancet Oncol.* **2008**, *9*, 629–635. [CrossRef]

26. Stenholm, S.; Harris, T.B.; Rantanen, T.; Visser, M.; Kritchevsky, S.B.; Ferrucci, L. Sarcopenic obesity: Definition, cause and consequences. *Curr. Opin. Clin. Nutr. Metab. Care* **2008**, *11*, 693–700. [CrossRef] [PubMed]

27. Visser, M.; Kritchevsky, S.B.; Goodpaster, B.H.; Newman, A.B.; Nevitt, M.; Stamm, E.; Harris, T.B. Leg muscle mass and composition in relation to lower extremity performance in men and women aged 70 to 79: The health, aging and body composition study. *J. Am. Geriatr. Soc.* **2002**, *50*, 897–904. [CrossRef] [PubMed]

28. DiMaria-Ghalili, R.; Amella, E. Nutrition in older adults: Intervention and assessment can help curb the growing threat of malnutrition. *Am. J. Nurs.* **2005**, *105*, 40–50. [CrossRef] [PubMed]

29. Morley, J.E. Undernutrition in older adults. *Fam. Pract.* **2012**, *29*, i89–i93. [CrossRef]

30. Bauer, J.; Biolo, G.; Cederholm, T.; Cesari, M.; Cruz-Jentoft, J.; Morley, J.E.; Phillips, S.; Sieber, C.; Stehle, P.; Teta, D.; et al. Evidence-based recommendations for optimal dietary protein intake in older people: A position paper from the PROT-AGE study group. *J. Am. Med. Dir. Assoc.* **2013**, *14*, 542–559. [CrossRef]

31. Silver, H.J. Oral strategies to supplement older adults' dietary intakes: Comparing the evidence. *Nutr. Rev.* **2009**, *67*, 21–31. [CrossRef]

32. Milne, A.C.; Potter, J.; Vivanti, A.; Avenell, A. Protein and energy supplementation in elderly people at risk from malnutrition. *Cochrane Database Syst. Rev.* **2009**, *2*, CD003288. [CrossRef]

33. Trabal, J.; Farran-Codina, A. Effects of dietary enrichment with conventional foods on energy and protein intake in older adults: A systematic review. *Nutr. Rev.* **2015**, *73*, 624–633. [CrossRef] [PubMed]

34. Burd, N.A.; Gorissen, S.H.; van Loon, L.J. Anabolic resistance of muscle protein synthesis with aging. *Exerc. Sport Sci. Rev.* **2013**, *41*, 169–173. [CrossRef] [PubMed]

35. Chernoff, R. Protein and older adults. *J. Am. Coll. Nutr.* **2004**, *23*, 627S–630S. [CrossRef] [PubMed]

36. Adamo, M.L.; Farrar, R.P. Resistance training, and IGF involvement in the maintenance of muscle mass during the aging process. *Ageing Res. Rev.* **2006**, *5*, 310–331. [CrossRef] [PubMed]

37. Walrand, S.; Guillet, C.; Salles, J.; Cano, N.; Boirie, Y. Physiopathological mechanism of sarcopenia. *Clin. Geriatr. Med.* **2011**, *27*, 365–385. [CrossRef] [PubMed]

38. Deutz, N.E.P.; Bauer, J.M.; Barazzoni, R.; Biolo, G.; Boirie, Y.; Bosy-Westphal, A.; Cederholm, T.; Cruz-Jentoft, A.; Krznariç, Z.; Nair, K.S.; et al. Protein intake and exercise for optimal muscle function with aging: Recommendations from the ESPEN Expert Group. *Clin. Nutr.* **2014**, *33*, 929–936. [CrossRef] [PubMed]

39. Fulgoni, V.L. Current protein intake in America: Analysis of the National Health and Nutrition Examination Survey, 2003–2004. *Am. J. Clin. Nutr.* **2008**, *87*, 1554S–1557S. [CrossRef]

40. Kerstetter, J.E.; O'Brien, K.O.; Insogna, K.L. Low protein intake: The impact on calcium and bone homeostasis in humans. *J. Nutr.* **2003**, *133*, 855S–861S. [CrossRef]

41. Tieland, M.; van de Rest, O.; Dirks, M.L.; van der Zwaluw, N.; Mensink, M.; van Loon, L.J.; de Groot, L.C. Protein supplementation improves physical performance in frail elderly people: A randomized, double-blind, placebo-controlled trial. *J. Am. Med. Dir. Assoc.* **2012**, *13*, 720–726. [CrossRef]

42. Paddon-Jones, D.; Sheffield-Moore, M.; Zhang, X.J.; Volpi, E.; Wolf, S.E.; Aarsland, A.; Ferrando, A.A.; Wolfe, R.R. Amino acid ingestion improves muscle protein synthesis in the young and elderly. *Am. J. Physiol. Endocrinol. Metab.* **2004**, *286*, E321–E328. [CrossRef]

43. Pohlhausen, S.; Uhlig, K.; Kiesswetter, E.; Diekmann, R.; Heseker, H.; Volkert, D.; Stehle, P.; Lesser, S. Energy and protein intake, anthropometrics, and disease burden in elderly home-care receivers—A cross-sectional study in Germany (ErnSIPP Study). *J. Nutr. Health Aging* **2016**, *20*, 361–368. [CrossRef] [PubMed]

44. Kramer, I.F.; Verdijk, L.V.; Hamer, H.M.; Verlaan, S.; Luiking, Y.C.; Kouw, I.W.K.; Senden, J.M.; van Kranenburg, J.; Gijsen, A.P.; Bierau, J.; et al. Both basal and post-prandial muscle protein synthesis rates, following the ingestion of a leucine-enriched whey protein supplement, are not impaired in sarcopenic older males. *Clin. Nutr.* **2017**, *36*, 1440–1449. [CrossRef] [PubMed]

45. Isanejad, M.; Mursu, J.; Sirola, J.; Kröger, H.; Rikkonen, T.; Tuppurainen, M.; Erkkilä, A.T. Dietary protein intake is associated with better physical function and muscle strength among elderly women. *Br. J. Nutr.* **2016**, *115*, 1281–1291. [CrossRef] [PubMed]

46. Landi, F.; Calvani, R.; Tosato, M.; Martone, A.M.; Picca, A.; Ortolani, E.; Savera, G.; Salini, S.; Ramaschi, M.; Bernabei, R.; et al. Animal-derived protein is associated with muscle mass and strength in community-dwellers: Results from the Milan EXPO Survey. *J. Nutr. Health Aging* **2017**, *21*, 1050–1056. [CrossRef] [PubMed]

47. Paddon-Jones, D.; Sheffield-Moore, M.; Katsanos, C.S.; Zhang, X.J.; Wolfe, R.R. Differential stimulation of muscle protein synthesis in elderly humans following isocaloric ingestion of amino acids or whey protein. *Exp. Gerontol.* **2006**, *41*, 215–219. [CrossRef]

48. Kramer, I.F.; Verdijk, L.B.; Hamer, H.M.; Verlaan, S.; Luiking, Y.; Kouw, I.W.; Senden, J.M.; van Kranenburg, J.; Gijsen, A.P.; Poeze, M.; et al. Impact of the macronutrient composition of a nutritional supplement on muscle protein synthesis rates in older men: A randomized, double blind, controlled trial. *J. Clin. Endocrinol. Metab.* **2015**, *100*, 4124–4132.

49. Bauer, J.; Verlaan, S.; Bautmans, I.; Brandt, K.; Donini, L.M.; Maggio, M.; McMurdo, M.E.; Mets, T.; Seal, C.; Wijers, S.L.; et al. Effects of a vitamin D and leucine-enriched whey protein nutritional supplement on measures of sarcopenia in older adults, the PROVIDE Study: A randomized, double-blind, placebo-controlled trial. *J. Am. Med. Dir. Assoc.* **2015**, *16*, 740–747. [CrossRef]

50. Lonnie, M.; Hooker, E.; Brunstrom, J.M.; Corfe, B.M.; Green, M.A.; Watson, A.W.; Williams, E.A.; Stevenson, E.J.; Penson, S.; Johnstone, A.M.; et al. Protein for Life: Review of optimal protein intake, sustainable dietary sources and the effect on appetite in ageing adults. *Nutrients* **2018**, *10*, 360. [CrossRef]

51. De Souza Genaro, P.; Araujo Martini, L. Effect of protein intake on bone and muscle mass in the elderly. *Nutr. Rev.* **2010**, *68*, 616–623. [CrossRef]

52. Nieuwenhuizen, W.F.; Weenen, H.; Rigby, P.; Hetherington, M.M. Older adults and patients in need of nutritional support: Review of current treatment options and factors influencing nutritional intake. *Clin. Nutr.* **2010**, *29*, 160–169. [CrossRef]

53. Höglund, E.; Ekman, S.; Stuhr-Olsson, G.; Lundgren, C.; Albinsson, M.S.; Signäs, M.; Karlsson, C.; Rothenberg, E.; Wendin, K. A meal concept designed for older adults—Small, enriched meals including dessert. *Food Nutr. Res.* **2018**, *62*, 1572–1579. [CrossRef] [PubMed]

54. De Castro, J.M. Age-related changes in spontaneous food intake and hunger in humans. *Appetite* **1993**, *21*, 255–272. [CrossRef] [PubMed]

55. Pelchat, M.L.; Schaefer, S. Dietary monotony and food cravings in young and elderly adults. *Physiol. Behav.* **2000**, *68*, 353–359. [CrossRef]

56. Clarkson, W.K.; Pantano, M.M.; Morley, J.E.; Horowitz, M.; Littlefield, J.M.; Burton, F.R. Evidence for the anorexia of aging: Gastrointestinal transit and hunger in healthy elderly versus young adults. *Am. J. Physiol.* **1997**, *272*, R243–R248.

57. Roberts, S.B.; Hajduk, C.L.; Howarth, N.C.; Russell, R.; McCrory, M.A. Dietary variety predicts low body mass index and inadequate macronutrient and micronutrient intakes in community-dwelling older adults. *J. Gerontol. A Biol. Sci. Med. Sci.* **2005**, *60A*, 613–621. [CrossRef] [PubMed]

58. Hays, N.P.; Roberts, S.B. The anorexia of aging in humans. *Physiol. Behav.* **2006**, *88*, 257–266. [CrossRef] [PubMed]

59. De Boer, A.; Ter Horst, G.J.; Lorist, M.M. Physiological and psychosocial age-related changes associated with reduced food intake in older persons. *Ageing Res. Rev.* **2013**, *12*, 316–328. [CrossRef] [PubMed]

60. Clegg, M.E.; Williams, E.A. Optimizing nutrition in older people. *Maturitas* **2018**, *112*, 34–38. [CrossRef] [PubMed]

61. Baugreet, S.; Hamill, R.M.; Kerry, J.P.; McCarthy, S.N. Mitigating nutrition and health deficiencies in older adults: A role for food innovation? *J. Food Sci.* **2017**, *82*, 848–855. [CrossRef]

62. Schwartz, C.; Vandenberghe-Descamps, M.; Sulmont-Rossé, C.; Tournier, C.; Feron, G. Behavioral and physiological determinants of food choice and consumption at sensitive periods of the life span, a focus on infants and elderly. *Innov. Food Sci. Emerg. Technol.* **2018**, *46*, 91–106. [CrossRef]

63. Aguilera, J.M.; Park, D.J. Texture-modified foods for the elderly: Status, technology and opportunities. *Trends Food Sci. Technol.* **2016**, *57*, 156–164. [CrossRef]

64. Van der Meij, B.S.; Wijnhoven, H.A.; Finlayson, G.S.; Oosten, B.S.; Visser, M. Specific food preferences of older adults with a poor appetite. A forced-choice test conducted in various care settings. *Appetite* **2015**, *90*, 168–175. [CrossRef] [PubMed]

65. Griep, M.I.; Mets, T.F.; Massart, D.L. Effects of flavour amplification of Quorn (R) and yoghurt on food preference and consumption in relation to age, BMI and odour perception. *Br. J. Nutr.* **2000**, *83*, 105–113. [CrossRef]

66. Schiffman, S.S. Intensification of sensory properties of foods for the elderly. *J. Nutr.* **2000**, *130* (Suppl. 4S), 927s–930s. [CrossRef]

67. Dermiki, M.; Prescott, J.; Sargent, L.J.; Willway, J.; Gosney, M.A.; Methven, L. Novel flavours paired with glutamate condition increased intake in older adults in the absence of changes in liking. *Appetite* **2015**, *90*, 108–113. [CrossRef] [PubMed]

68. Smith, J.S.; Ameri, F.; Gadgil, P. Effect of marinades on the formation of heterocyclic amines in grilled beef steaks. *J. Food Sci.* **2008**, *73*, T100–T105. [CrossRef] [PubMed]

69. Best, R.L.; Appleton, K.M. Comparable increases in energy, protein and fat intakes following the addition of seasonings and sauces to an older person's meal. *Appetite* **2011**, *56*, 179–182. [CrossRef]

70. Di Francesco, V.; Zamboni, M.; Dioli, A.; Zoico, E.; Mazzali, G.; Omizzolo, F.; Bissoli, L.; Solerte, S.B.; Benini, L.; Bosello, O. Delayed postprandial gastric emptying and impaired gallbladder contraction together with elevated cholecystokinin and peptide YY serum levels sustain satiety and inhibit hunger in healthy elderly persons. *J. Gerontol.* **2005**, *60*, 1581–1585. [CrossRef] [PubMed]

71. Wysokiński, A.; Sobów, T.; Kłoszewska, I.; Kostka, T. Mechanisms of the anorexia of aging—A review. *Age (Dordr)* **2015**, *37*, 9821. [CrossRef]

72. Cichero, J.A.Y. Texture-modified meals for hospital patients. In *Modifying Food Texture*; Chen, J., Rosenthal, A., Eds.; Woodhead Publishing: Cambridge, UK, 2015; Volume 2, p. 135e162.

73. Cichero, J.A.Y. Adjustment of food textural properties for elderly patients. *J. Text. Stud.* **2016**, *47*, 277–283. [CrossRef]

74. Ishihara, S.; Nakao, S.; Nakauma, M.; Funami, T.; Hori, K.; Ono, T.; Kohyama, K.; Nishinariet, K. Compression test of food gels on artificial tongue and its comparison with human test. *J. Text. Stud.* **2013**, *44*, 104–114. [CrossRef]

75. Laguna, L.; Hetherington, M.M.; Chen, J.; Artigas, G.; Sarkar, A. Measuring eating capability, liking and difficulty perception of older adults: A textural consideration. *Food Qual. Prefer.* **2016**, *53*, 47–56. [CrossRef]

76. Xu, F.; Laguna, L.; Sarkar, A. Aging-related changes in quantity and quality of saliva: Where do we stand in our understanding? *J. Text. Stud.* **2019**, *50*, 27–35. [CrossRef] [PubMed]

77. Vandenberghe-Descamps, M.; Labouré, H.; Prot, A.; Septier, C.; Tournier, C.; Feron, G. Salivary flow decreases in healthy elderly people independently of dental status and drug intake. *J. Text. Stud.* **2016**, *47*, 353–360. [CrossRef]

78. Assad-Bustillos, M.; Tournier, C.; Feron, G.; Guessasma, S.; Reguerre, A.L.; Della Valle, G. Fragmentation of two soft cereal products during oral processing in the elderly: Impact of product properties and oral health status. *Food Hydrocoll.* **2019**, *91*, 153–165. [CrossRef]

79. Lorieau, L.; Septier, C.; Laguerre, A.; Le Roux, L.; Hazart, E.; Ligneul, A.; Famelart, M.H.; Dupont, D.; Floury, J.; Feron, G.; et al. Bolus quality and food comfortability of model cheeses for the elderly as influenced by their texture. *Food Res. Internat.* **2018**, *111*, 31–38. [CrossRef] [PubMed]

80. Vandenberghe-Descamps, M.; Sulmont-Rossé, C.; Septier, C.; Follot, C.; Feron, G.; Labouré, H. Impact of blade tenderization, marinade and cooking temperature on oral comfort when eating meat in an elderly population. *Meat Sci.* **2018**, *145*, 86–93. [CrossRef] [PubMed]

81. Murphy, C.; Vertrees, R. Sensory functioning in older adults: Relevance for food preference. *Curr. Opin. Food Sci.* **2017**, *15*, 56–60. [CrossRef]

 *nutrients*

Article

Dietary Patterns and Their Association with Anxiety Symptoms among Older Adults: The ATTICA Study

Maria F. Masana [1,2,3], Stefanos Tyrovolas [1,2,4,*], Natasa Kollia [4], Christina Chrysohoou [5], John Skoumas [5], Josep Maria Haro [1,2], Dimitrios Tousoulis [5], Charalambos Papageorgiou [6], Christos Pitsavos [5] and Demosthenes B. Panagiotakos [4]

[1] Parc Sanitari Sant Joan de Déu, Universitat de Barcelona, Fundació Sant Joan de Déu, 08007 Barcelona, Spain; mfmasana@gmail.com (M.F.M.); jmharo@pssjd.org (J.M.H.)
[2] Instituto de Salud Carlos III, Centro de Investigación Biomédica en Red de Salud Mental, CIBERSAM, Monforte de Lemos 3-5, Pabellón 11, 28029 Madrid, Spain
[3] Facultat de Medicina, Universitat de Barcelona, Casanova, 143, 08036 Barcelona, Spain
[4] Department of Science of Dietetics and Nutrition, School of Health Science and Education, Harokopio University, 17671 Athens, Greece; natkollia@yahoo.gr (N.K.); dbpanag@hua.gr (D.B.P.)
[5] First Cardiology Clinic, School of Medicine, University of Athens, 15772 Athens, Greece; chrysohoou@usa.net (C.C.); skoumasj@gmail.com (J.S.); tousouli@med.uoa.gr (D.T.); cpitsavo@med.uoa.gr (C.P.)
[6] First Psychiatry Clinic, School of Medicine, University of Athens, 15772 Athens, Greece; cpapage@eginitio.uoa.gr
[*] Correspondence: s.tyrovolas@pssjd.org; Tel.: +34-936-40-63-50

Received: 11 April 2019; Accepted: 30 May 2019; Published: 31 May 2019

Abstract: By 2050, the global population aged 60 years and over is expected to reach nearly 2.1 billion and affective disorders might be also expected to increase. Although nutrition has been related with affective disorders, there is a lack of studies assessing the relation between dietary habits and anxiety among European and Mediterranean older populations. In the present study, we aimed to evaluate the association between dietary habits, energy intake, and anxiety symptoms using data from 1128 Greek older adults (>50 years) without pre-existing cardiovascular disease (CVD) or any other chronic disease who participated in the ATTICA study. Various socio demographic lifestyle, bio-clinical (e.g., blood pressure), and psychological (e.g., depression) characteristics were used, and dietary habits as well as energy intake were calculated using standard procedures. Older people with anxiety were more likely to be sedentary, to be smokers, and to show symptoms of depression. The saturated fat and added sugars (SFAS) dietary pattern was associated with higher anxiety levels (non-standardized b (95% CI): 5.82 (0.03 to 11.61)). No association between energy intake tertiles and anxiety levels pictured in the later regression model. Moreover, female gender, family status, and depression were positively related to anxiety. Therefore, promoting healthy dietary habits could reduce anxiety symptoms of the older adults.

Keywords: snack; sugars; carbohydrate; energy intake; anxiety; older adults; Attica study; Greece

---

1. Introduction

The global population aged 60 years or over is expected to reach nearly 2.1 billion by 2050 due to reductions in fertility and a rise in life expectancy [1] resulting in an increased risk for non-communicable diseases such as cancer, dementia, and affective disorders (anxiety and depression). Particularly, affective disorders might be expected to be increased because older people living alone is continuously rising as a consequence of changes in societies, such as falling family size, family structure and less multigenerational living arrangements [2].

The estimated global prevalence for anxiety disorders was 7.3% (95% CI 4.8–10.9%) and considering socioeconomic factors and culture, the prevalence ranged from 5.3% (3.5–8.1%) to 10.4% (7.0–15.5%) in African and Euro/Anglo cultures, respectively [3]. However, anxiety disorders are frequently under-recognized and misdiagnosed. Consequently, individuals with an undiagnosed or misdiagnosed anxiety disorder may not receive the appropriate treatment, suffering a deterioration of quality of life and social ability [4]. Several risk factors have been related to anxiety, but the emergence of a health and social crisis due to the financial crisis and strict fiscal austerity in Europe (e.g., Greece, Portugal, and Spain) could have contributed to worsen mental health [5]. For instance, in Greece, worse health status was reported in older adults due to unhealthy changes in their lifestyle behaviours [6] and anxiety disorders were also reported to be the fourth cause of years lost due to disability (YLDs) [7].

Depression has shown to be the most common comorbid disorder with anxiety disorders [8], but a bidirectional prospective relationship with one another was also reported [9]. Although the relationship between anxiety and cognition remains unclear, some studies have shown that anxiety was associated with cognitive decline, specifically on spatial and verbal working memory [10], influencing positively in the progression of Mild Cognitive Impairment (MCI) to dementia [11,12]. Some other chronic diseases, such as gastrointestinal disorders (e.g., irritable bowel syndrome) [13], diabetes [14] and thyroid disorders [15] have been also related to anxiety disorders or/and elevated anxiety symptoms. Moreover, patients with cardiovascular disease (CVD) are more likely to show prevalent anxiety disorders compared to the general population, but anxiety disorders have been also reported to increase the CVD risk [16]. For instance, anxiety levels were directly associated with the 10 y CVD incidence (OR (95%CI): 1.03 (1.0–1.1)) in older Greek adults without previous CVD history [17]. Some common pathophysiological mechanisms seem to share anxiety and the interrelated comorbidities, such as increase of corticotropin-releasing factor (CRF) levels, dysregulation of the hypothalamic-pituitary-adrenal (HPA) axis [18], activation of the sympathetic nervous system, and increase of levels of proinflammatory cytokines [19,20].

In recent years, an increasing number of studies are providing evidence for diet as a modifiable risk factor for mental health problems, such as depression and anxiety, however there is a lack of data regarding the older adults, a vulnerable population subgroup that appears to have high prevalence of mental disorders. For instance, a dietary pattern characterized by consumption of vegetables, fruit, beef, lamb, fish, and whole-grain foods has been related to a lower risk of a diagnosis of anxiety, while a western dietary pattern characterized by consumption of foods such as meats and sugar has been associated with more psychiatric symptoms [21]. Later studies also found that an inadequate and poor diet quality (i.e., snack patterns, animal foods) [22] and high intake of alcohol [23] can contribute to an inadequate intake of nutrients leading to mental health problems.

To date, there is a lack of studies assessing the association between dietary habits, energy intake, and anxiety symptoms among older adults and specifically, among European and Mediterranean older populations. Therefore, the aim of the current study was to evaluate the relation between dietary patterns, energy intake, and anxiety in a sample of older people (50+ years old) living in the Attica region.

2. Methods

2.1. Study Population

In brief, ATTICA is a prospective, population-based, cohort study performed in Attica (Athens metropolitan region, Greece), which recruited 3042 non-institutionalized adults (Caucasians; women/men: 1528/1514; age: ≥18 years) without previous CVD. The enrolment of the participants was carried out during 2001–2002, and after that, two follow-up waves followed 5 (in 2006) and 10 years (in 2012) later. Random, multistage sampling based on the age and gender distribution of the reference population, as defined by the Hellenic National Statistical Service Census Survey of 2001, was applied. Sampling procedures anticipated enrolling only one participant per household, while institutionalized individuals were excluded from study participation. All participants underwent detailed baseline assessments which included medical history, physical examination, and blood sampling for biochemical measurements. Baseline CVD was excluded in all participants by the study physicians [24]. For the purposes of the present work, only baseline data for the study's sample of over 50 years old (i.e., $n = 1128$) men and women were used (there was no psychological assessment at follow-up examinations).

2.2. Sociodemographic and Life-Style Variables

The study questionnaire included demographic information such as age, gender, family status (married, divorced, widowed), financial status, and education level. The level of education (as a proxy of social status) was determined by years of schooling and classified into 3 groups: (1) <9 years; (2) up to high school or technical college (10–14 years); (3) university. Mean annual income during the past 3 years was also recorded.

Current smokers were defined as those who smoked at least one cigarette/day, never smokers as those who had never tried a cigarette in their life, and former smokers as those who had stopped smoking for at least one year. Occasional smokers (7 cigarettes/week) were recorded and combined with current smokers because of their small sample size. In order to evaluate more accurately the smoking habits, we calculated the pack-years (cigarette packs/day × years of smoking), adjusted for a nicotine content of 0.8 mg/cigarette.

The physical activity level of each participant was also assessed at baseline using the International Physical Activity Questionnaire (IPAQ; participants reporting no physical activities or exercise on the IPAQ were classified as physically inactive) [25].

2.3. Clinical and Biochemical Assessments

Standardized measurements of anthropometric parameters were performed by trained study researchers, including body weight and height, BMI (kg/m^2), and WC (cm). Resting arterial blood pressure (BP, average of 3 recordings in sitting position) was also measured, and participants exhibiting average BP $\geq$ 140/90 mmHg or taking antihypertensive medication(s) were categorized as hypertensive.

At baseline, fasting blood samples were obtained from all participants after overnight fasting. Triglyceride (TG), total cholesterol (TC) and high-density lipoprotein-cholesterol (HDL-C) levels were measured by a chromatographic enzymatic method using a Technicon automatic analyzer RA-1000 (Dade Behring, Marburg, Germany; corresponding intra- and inter-assay coefficients of variation (CV) were <4%, <9%, and <4%, respectively). Low-density lipoprotein-cholesterol (LDL-C) was calculated by the Friedewald formula [26]. Hypercholesterolemia was defined as TC > 200 mg/dL or treatment with lipid-lowering drug(s). Moreover, fasting blood glucose (FBG) levels were measured by a Beckman glucose analyzer (Beckman Instruments, Fullerton, CA, USA) and subjects with FBG >125 mg/dL or on antidiabetic treatment were classified as having diabetes.

2.4. Dietary Assessment

All participants underwent a detailed baseline dietary evaluation through the EPIC-Greek questionnaire [27], which is a validated semi-quantitative food-frequency questionnaire that was kindly provided by the Unit of Nutrition of Athens Medical School. The energy intake in kcal/day was also calculated based on the participants' responses in this questionnaire. Energy intake tertiles were also created, those in the 1st energy tertile were the participants consuming <1840 kcals/day.

All participants were asked to report the average intake (per week or day) of several food items that they consumed (during the last 12 months). Then, the frequency of consumption was quantified approximately in terms of the number of times a month this food was consumed. Thus, daily consumption multiplied by 30 and weekly consumption multiplied by 4 and a value of 0 was assigned to food items rarely or never consumed. Alcohol consumption was measured in wineglasses

(100 ml) and quantified by ethanol intake (grams per drink). One wineglass was considered equal to 12% ethanol concentration.

A dietary pyramid was developed to describe the Mediterranean dietary pattern [28]. This dietary pattern consist of: (a) daily consumption of non-refined cereals and products (whole grain bread, pasta, brown rice, etc.), vegetables (2–3 servings/day), fruits (6 servings/day), olive oil (as the main added lipid) and dairy products (1–2 servings/day); (b) weekly consumption of fish (4–5 servings/week), poultry (3–4 servings/week), olives, pulses, and nuts (3 servings/week), potatoes, eggs and sweets (3–4 servings/week) and monthly consumption of red meat and meat products (4–5 servings/month). The Mediterranean diet (MedDiet) is also characterized by moderate consumption of wine (1–2 wineglasses/day) and high monounsaturated to saturated fat ratio (>2).

A special diet score that assessed adherence to MedDiet was calculated. In particular, we assigned a score of 0 for rare or no consumption of food items that are close to this dietary pattern, 1 for 1 to 4 times/month, 2 for 5 to 8 times/month, 3 for 9 to 12 times/month, 4 for 13 to 18 times/month and 5 for almost daily consumption. On the other hand, for the consumption of foods that are away from this traditional diet, like meat and meat products, we assigned the opposite scores (i.e. 0 for almost daily consumption to 5 for rare or no consumption). For alcohol, we assigned a score of 5 for consumption of less than 3 wineglasses per day, a score of 0 for consumption of more than 7 wineglasses/day and scores of 1 to 4 for consumption of 3, 4–5, 6 and 7 wine glasses per day.

The calculation of the dietary inflammatory load of the participants' diet was according to the methodology and the rationale of the Dietary Inflammation Index (DII) that has been previously proposed by Shivappa and colleagues [29]. Thus, a Dietary Anti-Inflammation Index (D-AII) was developed based on participants' dietary habits. Detailed information on the D-AII has been reported in a previous ATTICA publication, by Georgousopoulou et al [30].

Aside from the classic indexes of MedDiet and D-AII, additional dietary patterns were defined for the daily dietary intake of the participants. Three main components were extracted using the principal components (PC) methods (Table 1). These were: (A) The lacto-fish-vegetarian (LFV) dietary pattern including the food components of vegetables, cereals, fruits, legumes, fish, and dairy products; (B) The meat-eaters dietary pattern including food components of all kinds of meat, read meat, and poultry; (C) The saturated fat and added sugars (SFAS) dietary pattern including the food components of sweets, soft drinks, nuts, and potatoes.

2.5. Psychological Accessments

Anxiety levels were assessed using the validated Greek translation of the 20 item, self-report State-Trait Anxiety Inventory (STAI) [31], the total score of which ranges from 20 to 80. Higher scores in this scale are indicative of more severe anxiety, and according to Spielberger's criteria, a score of 40 or higher reflects clinically relevant symptoms of anxiety.

Depressive symptomatology was assessed using a translated and validated version of the Zung Self-Rating Depression Scale (ZDRS) [32]. The ZDRS consists of 20 items that cover affective, psychological, and somatic symptoms, and ranges from 20 to 80. A subject with a ZDRS score below 50 is considered normal but with a score of 50 to 59, 60 to 69, or 70 or above is considered to suffer from mild, moderate, or severe depression, respectively.

Cognitive distortion symptomatology was assessed using a validated version of the cognitive distortions scale (CDS) [33]. This tool captures cognitive symptoms or distortions among individuals who have experienced inter-personal victimization: self-criticism, self-blame, helplessness, hopelessness, and preoccupation with danger.

2.6. Bioethics

The ATTICA study was approved by our institutional ethics committee and conformed to the ethical guidelines of the 1975 declaration of Helsinki. All participants were informed about the study protocol, and they provided written signed consent.

2.7. Statistical Analysis

To define the dietary patterns, factor analysis using the PC was applied [34]. Principal components were retained if their eigenvalues were greater than 1.0, a threshold that is commonly used as a cut-off to identify meaningful patterns. The resulting components were interpreted based on the variables with loadings above 0.3. Moreover, the Kaiser-Meyer-Olkin criterion was used as a measure of variables (components) inter-correlation and used as an indicator of internal consistency.

Normally distributed continuous variables are presented as mean values ± SD, and categorical variables as frequencies. Normality was tested using the Shapiro-Wilk criterion; the non-normally distributed variables are presented as median and 1st, 3rd tertile. Associations between categorical variables were tested by the chi-square test, whereas between continuous variables by the Pearson r or Spearman's rho coefficients for normally distributed and skewed variables, respectively. Continuous variables were tested for normality via P-P plots. For normally distributed variables, comparisons were performed by the student's t-test, after controlling for equality of variances by the Levene's test. For continuous variables without normal distribution, comparisons were performed by the non-parametric Mann-Whitney U-test. Previous literature was used as a guide for the selection of variables used for adjustment between dietary patterns and anxiety [35–37]. Multiple linear regression analyses were performed in order to evaluate the association between anxiety as the dependent outcome and a participant´s adherence to dietary patterns, adjusted for multiple confounders. All *p*-values are based on two-sided tests. Statistical analyses were performed with the Statistical Package for Social Sciences version 22 (SPSS Inc., Chicago, IL, USA).

3. Results

In Table 1 the component loadings and the eigenvalues from the factor analysis are presented. The size of eigenvalues strongly suggests the formation of three distinct dietary patterns, explaining 51.78% of the total variance of the information. The first dietary pattern could be characterized as the "lacto-fish-vegetarian", the second as the "meat-eaters" and the third as the "saturated fat and added sugars".

Table 1. Factor loadings of the food components included in the extracted dietary patterns.

Food Components	LFV Dietary Pattern	Meat-Eaters Dietary Pattern	SFAS Dietary Pattern
Vegetables	0.756	–	–
Fruit	0.653	–	–
Cereals	0.622	–	–
Legumes	0.609	–	–
Fish	0.579	–	–
Dairy Products	0.435	–	–
Total Meat	–	0.941	–
Red Meat	–	0.828	–
Poultry	–	0.708	–
Sweets	–	–	0.675
Soft Drinks	–	–	0.649
Nuts	–	–	0.648
Potatoes	–	–	0.523
Cumulative Variance explained (%)	25.675	41.609	51.788

Abbreviations: LFV, lacto-fish-vegetarian; and, SFAS, saturated fat and added sugars. In each factor only the components with values > 0.30 are included in the table.

The sociodemographic, lifestyle, and clinical characteristics of participants are presented by anxiety status in Table 2. Compared to the participants without anxiety, those with anxiety were more likely to be sedentary (*p* = 0.001), smokers (*p* = 0.027), and suffered depressive symptoms (*p* < 0.001). The rest of measured variables (i.e., age, gender, education, income, family status, diet, energy intake, BMI, hypertension, diabetes, hypercholesterolemia, and cognitive distortion) were not significantly associated with the level of anxiety.

Table 2. Baseline characteristics of the study sample 50+ years old, by anxiety status.

Characteristic	Total	No Anxiety	Anxiety	p
Sample size (n)	758	149	609	
Age (Mean ± SD)	59.67 ± 8.28	54.11 ± 4.31	54.68 ± 3.39	0.372
Sex				0.286
Female	41.6	37.3	45.9	
Male	58.4	62.7	54.1	
Education status (years of school) (Mean ± SD)	10.53 ± 3.96	12.43 ± 3.65	11.36 ± 4.15	0.100
Financial status (%)				0.151
Bad	12.8	8.0	17.6	
Poor	17.4	18.7	16.2	
Good	47.7	45.3	50.0	
Very good	22.1	28.0	16.2	
Family status (%)				0.580
Never married	10.7	10.7	10.08	
Married	78.5	80.0	77.0	
Divorced	7.4	8.0	6.8	
Widowed	3.4	1.3	5.4	
Current smoking (%)	46.3	37.3	55.4	0.027
Physically active (%)				0.001
Sedentary	52.3	38.7	66.2	
Physically active	47.7	61.3	33.8	
MedDiet (0–55)	23.02 ± 6.17	24.27 ± 5.25	24.03 ± 6.36	0.808
SFAS dietary pattern (Mean ± SD)	−0.32 ± 0.86	−0.189 ± 0.756	−0.33 ± 0.81	0.337
Meat-eaters dietary pattern (Mean ± SD)	−0.24 ± 0.76	−0.35 ± 0.55	−0.22 ± 0.77	0.294
LFV dietary pattern (Mean ± SD)	0.25 ± 0.98	0.29 ± 0.89	0.146 ± 1.09	0.434
Energy intake (kcal/day)	2409.38 ± 1019.19	2120.10 ± 693.15	2048.51 ± 855.46	0.617
D-AII (10–77)	32.61 ± 6.34	32.81 ± 6.04	32.44 ± 6.62	0.72
BMI (kg/m^2)	27.60 ± 4.25	26.87 ± 4.11	26.48 ± 3.29	0.532
Waist circumference (cm)	94.64 ± 18.87	95.58 ± 21.12	93.70 ± 16.70	0.57
Hypertension (%)	45.9	43.1	48.6	0.498
Diabetes (%)	10.1	8.0	12.2	0.399
Hypercholesterolemia (%)	51.0	46.7	55.4	0.286
Zung score (Mean ± SD)	35.17 ± 7.57	31.65 ± 6.30	38.54 ± 6.88	<0.001
Cognitive distortion score (Mean ± SD)	17.84 ± 24.35	11.36 ± 15.11	24.48 ± 30.35	0.073

Abbreviations: MedDiet, mediterranean diet; BMI, body mass index; LFV, lacto-fish-vegetarian; SFAS, saturated fat and added Sugars; diet anti-inflammatory index (D-AII).

As a first step, results of multi-adjusted analysis assessing the energy intake by tertiles on anxiety levels are presented in Table 3, without adjusting for dietary patterns. Initially the association between energy intake (kcals/day) as a continuous variable and anxiety levels was tested. Based on this analysis, it was observed that the higher levels of energy intake were positively related with higher anxiety levels (non-standardized b (95% CI): 0.01 (0.003 to 0.2)), after various adjustments (i.e., sex, smoking habits, physical activity, etc.) (data shown only in text). When the analysis was applied by energy intake tertiles, it was shown that the 1st energy intake tertile had an independent inverse association with anxiety levels as compared with the highest one (3rd tertile) (non-standardized b (95% CI): −11.65 (−22.83 to −0.48), p = 0.04). The 2nd energy intake tertile as compared with the 3rd tertile, was no related with anxiety levels (non-standardized b (95% CI): −7.61 (−18.55 to 3.34), p = 0.16). In addition, female gender (non-standardized b (95% CI): 11.96 (0.53 to 23.38), p = 0.04), family status (non-standardized b (95% CI): 18.17 (4.56 to 31.77), p = 0.012), and depression (non-standardized b (95% CI): 1.07 (0.27 to 1.87), p = 0.01) were positively related to higher anxiety levels. Cognitive distortion (non-standardized b (95% CI): 0.20 (0.09 to 0.32), p = 0.001) was also observed to be associated with the presence of anxiety symptoms, while physical activity was inversely related with anxiety levels (non-standardized b (95% CI): −8.58 (−15.90 to −1.25), p = 0.02). Finally, the results remained similar when basic metabolic rate was inserted as a confounding variable in the model.

Table 3. Correlates of anxiety among adults aged ≥ 50 years estimated by multivariable linear regression, in the ATTICA study, *n* = 758.

Variable	Non-Standardized b	95% CI
Female vs. male sex	11.96 *	0.53, 23.38
Education status (per 1 year)	0.87	−0.10, 1.85
Family status (married vs. other status)	18.17 *	4.56, 31.77
Financial status (high vs. low-medium)	−9.66 *	−19.32, −0.002
Physical activity (active vs. sedentary)	−8.58 *	−15.90, −1.25
Current smoking (yes vs. no)	−1.50	−9.40, 6.41
Energy intake (kcals)		
1st vs. 3rd tertile	−11.65 *	−22.83, −0.48
2nd vs. 3rd tertile	−7.61	−18.55, 3.34
BMI (per 1 kg/m^2)	0.91	−0.24, 2.06
Hypertension (yes vs. no)	1.68	−5.31, 8.68
Diabetes (yes vs. no)	1.88	−22.32, 26.09
Hypercholesterolemia (yes vs. no)	1.95	−5.16, 9.06
Depression (Zung scale)	1.07 *	0.27, 1.87
Cognitive distortion scale	0.20 **	0.09, 0.32

Data are presented as non-standardized coefficients (b) and their 95% confidence intervals (CIs). * $p < 0.05$, ** $p < 0.01$. Abbreviations: BMI, body mass index. Model is adjusted for all the covariates in the table.

As a next step, Table 4 illustrates the results from multiple linear regression analysis that evaluated the association between dietary patterns, energy intake and anxiety (models I and II). After adjusting for sociodemographic, lifestyle and clinical characteristics, as in Table 3, plus for confounding due to obesity, central obesity (waist circumference), and energy intake, the dietary pattern characterized by the consumption of saturated fats and added sugars (SFAS dietary pattern) was consistently associated with higher anxiety levels (non-standardized b (95% CI): 5.82 (0.03 to 11.61), $p = 0.04$) (Model II). Moreover, anxiety was positively associated with female gender (non-standardized b (95% CI): 21.06 (3.19 to 38.94), $p = 0.02$), and family status (non-standardized b (95% CI): 19.83 (2.47 to 37.19), $p = 0.02$). In addition, depressive symptomatology was related with higher level of anxiety (non-standardized b (95% CI): 1.88 (0.48 to 3.28), $p = 0.01$). Finally, LFV dietary pattern as well as meat-eaters and SFAS dietary patterns were replaced with the D-AII, a dietary index that is picturing pro- and anti-inflammatory dietary habits. In this additional analysis, there was not association between the D-AII and the anxiety levels ($p = 0.94$) (data shown only in text).

Table 4. Correlates of anxiety among adults aged ≥ 50 years estimated by additive multivariable linear regression (Model I and Model II), in the ATTICA study, *n* = 758.

Variable	Model I		Model II	
	Non-Standardized b	95% CI	Non-Standardized b	95% CI
Female vs. male sex	21.77	7.25, 36.28	21.06	3.19, 38.94
Education status (per 1 year)	1.92	0.29, 3.56	1.78	−0.07, 3.65
Family status (married vs. other status)	20.72	4.81, 36.63	19.83	2.47, 37.19
Financial status (high vs. low-medium)	−9.33	−19.61, 0.95	−9.84	−21.22, 1.53
Physical activity (active vs. sedentary)	−4.32	−13.35, 4.69	−4.00	−14.42, 6.41
Current smoking (yes vs. no)	−3.54	−12.30, 5.21	−2.42	−12.50, 7.66
LFV dietary pattern	−1.96	−9.05, 5.12	−1.20	−9.29, 6.88
Meat-eaters dietary pattern	3.41	−3.80, 10.62	3.33	−4.92, 11.59
SFAS dietary pattern	5.62	1.21, 10.02	5.82	0.03, 11,61
Obesity (yes vs. no)	19.22	0.51, 37.94	18.00	−4.52, 40.54
Waist circumference (cm)	−0.17	−0.45, 0.12	−0.16	−0.51, 0.18
Hypertension (yes vs. no)	5.70	−2.47, 13.87	5.17	−3.73, 14.08
Diabetes (yes vs. no)	−1.36	−26.13, 23.39	1.71	−26.53, 29.96

Table 4. Cont.

Variable	Model I		Model II	
	Non-Standardized b	95% CI	Non-Standardized b	95% CI
Hypercholesterolemia (yes vs. no)	1.96	−5.60, 9.54	2.67	−6.04, 11.40
Depression (Zung scale)	1.97	0.76, 3.17	1.88	0.48, 3.28
Cognitive distortion scale	0.14	−0.04, 0.37	0.16	−0.07, 0.39
Energy intake (kcals)	–	–		
1st vs. 3rd tertile	–	–	−0.48	−18.98, 18.02
2nd vs. 3rd tertile	–	–	−2.97	−16.80, 10.86

Data are presented as non-standardized coefficients (b) and their 95% confidence intervals (CIs). Abbreviations: LFV, lacto-fish-vegetarian; SFAS, saturated fat and added sugars. Model I and Model II are adjusted for all the respective covariates in the table.

4. Discussion

While anxiety is multi-factorial and mostly related with psychological, social and biological aspects [38], little is known about the contribution of holistic dietary patterns as preventive means of anxiety in older populations. Through factor analysis, three distinct dietary patterns were pictured (the LFV, the meat-eaters and the SFAS dietary patterns) among the older adults (50+ years old) of the ATTICA study. The present study reported that at an initial step, energy intake was related with anxiety levels among older adults. However, when the abovementioned dietary patterns were inserted in the analysis, it was reported that the SFAS dietary pattern was associated with higher levels of anxiety among older adults, while energy intake was no longer significant. Notably, this relation was independent of well-established risk factors such as depression, physical activity, and cognitive distortion. To the best of our knowledge, this is among the first studies to evaluate the association between dietary patterns, energy intake, and anxiety levels using a non-CVD, European, older population. These results point to the importance of specific dietary patterns in relation with anxiety and the public health actions needed to be taken with western older populations in which the burden of cognitive and mental disorders is increasing at alarming rates.

Older participants with anxiety from the Attica region in Greece were more likely to be smokers, sedentary and presented depressive symptoms, and this results are in agreement with the results of studies found in the literature. In particular, systematic reviews have reported that smoking [39] and depression [9] might be risk factors for the development of anxiety, but a bidirectional relationship has also been observed. Moreover, previous studies have suggested that a sedentary behavior (i.e., sitting time and screen time) is related to an increased risk for anxiety [40], and higher levels of anxiety were associated with increased diabetes risk among women but not among men [41].

Multi-adjusting analysis revealed an association between socio-economic and lifestyle factors and mental health problems with anxiety levels in older adults. In previous studies, associations between socio-economic and lifestyle factors have also reported. For instance, women with anxiety have reported lower socioeconomic status and income compared to men [42], and other risk factors such as stressful life events, chronic diseases, physical inactivity, depression, insomnia, and lower cognitive function have also been related to anxiety [43]. Regular physical activity has been related to reduced anxiety due to its impact on some pathological mechanisms, such as modulation of HPA and an increase of brain-derived neurotrophic factor (BDNF) levels and β-endorphins [44]. Smoking have been reported to activate the HPA axis increasing adrenocorticotropic hormone (ACTH) and cortisol [45], which have been related with anxiety [46]. However, in our fully adjusted regression analysis, physical activity, financial status and smoking habits did not remain significant. Anxiety has been also related to worse performance in some cognitive domains, such as spatial and verbal working memory [10], and bidirectional associations between anxiety symptomology and processing speed and attention have also reported in older adults [47]. Literature suggests that anxiety is highly comorbid with depression due to the overlap of symptomatology of both disorders increasing its severity and chronicity, and consequently, a worsening in the quality of life [48], but an inverse relationship between anxiety and reduced cognitive function have been also observed and explained by comorbid depression in non-demented elderly general population [49]. Therefore, older adults with healthy habits such as regular physical activity, healthy sleeping routine, and no smoking could contribute to maintain a good mental health reducing affective disorders, and consequently, reducing the deterioration of cognitive function.

Dietary habits are usually estimated using PC analysis and are reflecting the real behavior of the population [50]. To date various dietary patterns have been identified among older populations. Healthy dietary patterns (i.e., higher consumption of vegetables, fruits, poultry and fish), have been related with better quality of life, self-reported health, lower burden of morbidity as well as higher survival among the older adults [51,52]. However, until today, research on the effect of dietary patterns on the anxiety levels among older adults is sparse. In our analysis we extracted three major dietary patters (the LFV, the meat-eaters and the SFAS dietary patterns). Of them, only the SFAS dietary pattern characterized by the consumption of added sugars and saturated fats was related with higher anxiety levels, while no association was reported between the LFV and the meat-eaters dietary patterns. Furthermore, no relation was reported between anti-inflammatory dietary patterns (D-AII) and anxiety. These associations are partly in line with the literature, where mixed results are reported. For example, a previous ATTICA work focused only in women reported that increased consumption of sweets and meat products were associated with higher anxiety levels [35]. Other recent studies report that diets high in sugars, processed foods (i.e., meat products), and/or fats are related with higher anxiety levels through alterations of glucose, protein and energy homeostasis, and increases in inflammatory cytokines (i.e., IL-6, IL-1β, TNFα) and corticosterone [53,54]. Inverse associations have also been reported between anti-inflammatory dietary patterns [36], high intake of fruit and vegetables [37], and anxiety. Our analysis between energy intake tertiles and anxiety referred no association in the final regression model, although, some animal studies report an anxiolytic caloric restriction (CR) effect [55] contributing beneficially to brain microstructure integrity [56]. Further studies, especially with longitudinal data, are needed on dietary habits and anxiety levels among older adults to confirm our findings.

Anxiety disorders in high-income countries for the population over 70 years old in 2017 account for 1.41% of total YLDs (https://vizhub.healthdata.org/gbd-compare/). These associations combined with the fact that Western societies are ageing along with the increased epidemic of depression and cognitive problems raise major concerns about the need for early non-pharmacological measures (and promotion of lifestyle changes) in order to prevent anxiety among the older populations.

Strengths and limitations

The present study has several strengths since it is one of the few studies to evaluate the relation between dietary habits, energy intake, and anxiety in Mediterranean older adults. However, some limitations deserve mentioning. It should be acknowledged that the baseline/entry study examination was conducted once and hence, may be susceptible to a certain degree of measurement error. This observational study may have potential recall bias due to its cross-sectional design, and consequently, the nature of the study limits the possibility for etiological conclusions between dietary habits and anxiety. The assessment of dietary habits using as a tool the FFQ is known for certain bias. Furthermore, there might be several misinterpretations since the components extracted from a PC are sometimes subjective.

5. Conclusions

Our results report that a dietary pattern characterized by saturated fats and added sugars (SFAS) seems to be related to higher anxiety in older adults living in Greece while energy intake was not related with anxiety in the fully adjusted model. Further studies and longitudinal data analysis are needed to better address this interesting question on the role of dietary habits in the symptomatology of anxiety among older populations.

Author Contributions: Conceptualization, M.F.M. and S.T. Data curation, N.K. Formal analysis, M.F.M., S.T., N.K., and D.B.P. Methodology, M.F.M., S.T., and N.K. Writing—original draft, M.F.M. and S.T. Writing—review and editing, M.F.M., S.T., N.K., D.B.P., C.C., J.S., C.P. (Charalambos Papageorgiou), J.M.H., D.T., and C.P. (Christos Pitsavos).

Funding: This work was supported by "Ageing Trajectories of Health: Longitudinal Opportunities and Synergies—ATHLOS Project" (Grant Agreement n° 635316) funded by the European Commission. The ATTICA study is supported by research grants from the Hellenic Cardiology Society (HCS2002) and the Hellenic Atherosclerosis Society (HAS2003). Stefanos Tyrovolas was supported by the Foundation for Education and European Culture, the Miguel Servet Programme (reference CP18/00006), and the Fondos Europeos de Desarrollo Regional.

Conflicts of Interest: The authors declare no conflict of interest.

References

1. World Population Ageing [Highlights]. Available online: http://www.un.org/en/development/desa/population/publications/pdf/ageing/WPA2017_Highlights.pdf (accessed on 20 January 2019).
2. World Report on Ageing and HeAltH. Available online: www.who.int (accessed on 20 January 2019).
3. Baxter, A.J.; Scott, K.M.; Vos, T.; Whiteford, H.A. Global prevalence of anxiety disorders: A systematic review and meta-regression. *Psychol. Med.* **2013**, *43*, 897–910. [CrossRef] [PubMed]
4. Kasper, S. Anxiety disorders: Under-diagnosed and insufficiently treated. *Int. J. Psychiatry Clin. Pract.* **2006**, *10*, 3–9. [CrossRef] [PubMed]
5. Karanikolos, M.; Mladovsky, P.; Cylus, J.; Thomson, S.; Basu, S.; Stuckler, D.; Mackenbach, J.P.; McKee, M. Financial crisis, austerity, and health in Europe. *Lancet.* **2013**, *381*, 1323–1331. [CrossRef]
6. Foscolou, A.; Tyrovolas, S.; Soulis, G.; Mariolis, A.; Piscopo, S.; Valacchi, G.; Anastasiou, F.; Lionis, C.; Zeimbekis, A.; Tur, J.-A.; et al. The Impact of the Financial Crisis on Lifestyle Health Determinants Among Older Adults Living in the Mediterranean Region: The Multinational MEDIS Study (2005–2015). *J. Prev. Med. Public Health* **2017**, *50*, 1–9. [CrossRef] [PubMed]
7. Tyrovolas, S.; Kassebaum, N.J.; Stergachis, A.; Abraha, H.N.; Alla, F.; Androudi, S.; Car, M.; Chrepa, V.; Fullman, N.; Fürst, T.; et al. The burden of disease in Greece, health loss, risk factors, and health financing, 2000–2016: An analysis of the Global Burden of Disease Study 2016. *Lancet Public Health* **2018**, *3*, e395–e406. [CrossRef]
8. Prina, A.M.; Ferri, C.P.; Guerra, M.; Brayne, C.; Prince, M. Co-occurrence of anxiety and depression amongst older adults in low- and middle-income countries: Findings from the 10/66 study. *Psychol. Med.* **2011**, *41*, 2047–2056. [CrossRef] [PubMed]
9. Jacobson, N.C.; Newman, M.G. Anxiety and depression as bidirectional risk factors for one another: A meta-analysis of longitudinal studies. *Psychol. Bull.* **2017**, *143*, 1155–1200. [CrossRef] [PubMed]
10. Vytal, K.E.; Cornwell, B.R.; Letkiewicz, A.M.; Arkin, N.E.; Grillon, C. The complex interaction between anxiety and cognition: Insight from spatial and verbal working memory. *Front. Hum. Neurosci.* **2013**, *7*, 93. [CrossRef] [PubMed]
11. Gulpers, B.; Ramakers, I.; Hamel, R.; Köhler, S.; Oude Voshaar, R.; Verhey, F. Anxiety as a Predictor for Cognitive Decline and Dementia: A Systematic Review and Meta-Analysis. *Am. J. Geriatr. Psychiatry* **2016**, *24*, 823–842. [CrossRef]
12. Li, X.-X.; Li, Z. The impact of anxiety on the progression of mild cognitive impairment to dementia in Chinese and English data bases: A systematic review and meta-analysis. *Int. J. Geriatr. Psychiatry* **2018**, *33*, 131–140. [CrossRef]
13. Mayer, E.A.; Craske, M.; Naliboff, B.D. Depression, anxiety, and the gastrointestinal system. *J. Clin. Psychiatry* **2001**, *62*, 28–36. [PubMed]
14. Smith, K.J.; Béland, M.; Clyde, M.; Gariépy, G.; Pagé, V.; Badawi, G.; Rabasa-Lhoret, R.; Schmitz, N. Association of diabetes with anxiety: A systematic review and meta-analysis. *J. Psychosom. Res.* **2013**, *74*, 89–99. [CrossRef] [PubMed]
15. Fischer, S.; Ehlert, U. Hypothalamic-pituitary-thyroid (HPT) axis functioning in anxiety disorders. A ystematic review. *Depress. Anxiety* **2018**, *35*, 98–110. [CrossRef] [PubMed]
16. Tully, P.J.; Harrison, N.J.; Cheung, P.; Cosh, S. Anxiety and Cardiovascular Disease Risk: A Review. *Curr. Cardiol. Rep.* **2016**, *18*, 120. [CrossRef] [PubMed]

17. Kyrou, I.; Kollia, N.; Panagiotakos, D.; Georgousopoulou, E.; Chrysohoou, C.; Tsigos, C.; Randeva, H.S.; Yannakoulia, M.; Stefanadis, C.; Papageorgiou, C.; et al. Association of depression and anxiety status with 10-year cardiovascular disease incidence among apparently healthy Greek adults: The ATTICA Study. *Eur. J. Prev. Cardiol.* **2017**, *24*, 145–152. [CrossRef]

18. Boyer, P. Do anxiety and depression have a common pathophysiological mechanism? *Acta Psychiatr. Scand.* **2000**, *102*, 24–29. [CrossRef]

19. Myers, B.; Van Meerveld, B.G. Role of anxiety in the pathophysiology of irritable bowel syndrome: Importance of the amygdala. *Front. Neurosci.* **2009**, *3*, 2. [CrossRef]

20. Olafiranye, O.; Jean-Louis, G.; Zizi, F.; Nunes, J.; Vincent, M. Anxiety and cardiovascular risk: Review of Epidemiological and Clinical Evidence. *Mind Brain* **2011**, *2*, 32–37.

21. Jacka, F.N.; Pasco, J.A.; Mykletun, A.; Williams, L.J.; Hodge, A.M.; O'Reilly, S.L.; Nicholson, G.C.; Kotowicz, M.A.; Berk, M. Association of Western and Traditional Diets with Depression and Anxiety in Women. *Am. J. Psychiatry* **2010**, *167*, 305–311. [CrossRef]

22. Weng, T.-T.; Hao, J.-H.; Qian, Q.-W.; Cao, H.; Fu, J.-L.; Sun, Y.; Huang, L.; Tao, F.-B. Is there any relationship between dietary patterns and depression and anxiety in Chinese adolescents? *Public Health Nutr.* **2012**, *15*, 673–682. [CrossRef]

23. Velten, J.; Lavallee, K.L.; Scholten, S.; Meyer, A.H.; Zhang, X.-C.; Schneider, S.; Margraf, J. Lifestyle choices and mental health: A representative population survey. *BMC Psychol.* **2014**, *2*, 58. [CrossRef] [PubMed]

24. Pitsavos, C.; Panagiotakos, D.B.; Chrysohoou, C.; Stefanadis, C. Epidemiology of cardiovascular risk factors in Greece: Aims, design and baseline characteristics of the ATTICA study. *BMC Public Health* **2003**, *3*, 32. [CrossRef] [PubMed]

25. Papathanasiou, G.; Georgoudis, G.; Papandreou, M.; Spyropoulos, P.; Georgakopoulos, D.; Kalfakakou, V.; Evangelou, A. Reliability measures of the short International Physical Activity Questionnaire (IPAQ) in Greek young adults. *Hellenic J. Cardiol.* **2009**, *50*, 283–294. [PubMed]

26. Friedewald, W.T.; Levy, R.I.; Fredrickson, D.S. Estimation of the concentration of low-density lipoprotein cholesterol in plasma, without use of the preparative ultracentrifuge. *Clin. Chem.* **1972**, *18*, 499–502. [PubMed]

27. Katsouyanni, K.; Rimm, E.B.; Gnardellis, C.; Trichopoulos, D.; Polychronopoulos, E.; Trichopoulou, A. Reproducibility and relative validity of an extensive semi-quantitative food frequency questionnaire using dietary records and biochemical markers among Greek schoolteachers. *Int. J. Epidemiol.* **1997**, *26*, S118–S127. [CrossRef] [PubMed]

28. Willett, W.C.; Sacks, F.; Trichopoulou, A.; Drescher, G.; Ferro-Luzzi, A.; Helsing, E.; Trichopoulos, D. Mediterranean diet pyramid: A cultural model for healthy eating. *Am. J. Clin. Nutr.* **1995**, *61*, 1402S–1406S. [CrossRef]

29. Shivappa, N.; Steck, S.E.; Hurley, T.G.; Hussey, J.R.; Hébert, J.R. Designing and developing a literature-derived, population-based dietary inflammatory index. *Public Health Nutr.* **2014**, *17*, 1689–1696. [CrossRef]

30. Georgousopoulou, E.N.; Kouli, G.-M.; Panagiotakos, D.B.; Kalogeropoulou, A.; Zana, A.; Chrysohoou, C.; Tsigos, C.; Tousoulis, D.; Stefanadis, C.; Pitsavos, C. Anti-inflammatory diet and 10-year (2002–2012) cardiovascular disease incidence: The ATTICA study. *Int. J. Cardiol.* **2016**, *22*, 473–478. [CrossRef]

31. Fountoulakis, K.; Papadopoulou, M.; Kleanthous, S.; Papadopoulou, A.; Bizeli, V.; Nimatoudis, I.; Iacovides, A.; Kaprinis, G.S. Reliability and psychometric properties of the Greek translation of the State-Trait Anxiety Inventory form Y: Preliminary data. *Ann. Gen. Psychiatry.* **2006**, *5*, 2. [CrossRef]

32. Zung, W.W.K. A self-rating depression scale. *Arch. Gen. Psychiatry* **1965**, *12*, 63–70. [CrossRef]

33. Briere, J. *Cognitive Distortions Scale (CDS) Professional Manual*; Psychological Assessment Resources: Odessa, FL, USA, 2000.

34. Floyd, F.J.; Widaman, K.F. Factor analysis in the development and refinement of clinical assessment instruments. *Psychol. Assess.* **1995**, *7*, 286–299. [CrossRef]

35. Yannakoulia, M.; Panagiotakos, D.B.; Pitsavos, C.; Tsetsekou, E.; Fappa, E.; Papageorgiou, C.; Stefanadis, C. Eating habits in relations to anxiety symptoms among apparently healthy adults. A pattern analysis from the ATTICA Study. *Appetite* **2008**, *51*, 519–525. [CrossRef] [PubMed]

36. Phillips, C.M.; Shivappa, N.; Hébert, J.R.; Perry, I.J. Dietary inflammatory index and mental health: A cross-sectional analysis of the relationship with depressive symptoms, anxiety and well-being in adults. *Clin. Nutr.* **2018**, *37*, 1485–1491. [CrossRef] [PubMed]

37. Saghafian, F.; Malmir, H.; Saneei, P.; Keshteli, A.H.; Hosseinzadeh-Attar, M.J.; Afshar, H.; Siassi, F.; Esmaillzadeh, A.; Adibi, P. Consumption of fruit and vegetables in relation with psychological disorders in Iranian adults. *Eur. J. Nutr.* **2018**, *57*, 2295–2306. [CrossRef] [PubMed]

38. Vink, D.; Aartsen, M.J.; Schoevers, R.A. Risk factors for anxiety and depression in the elderly: A review. *J. Affect. Disord.* **2008**, *106*, 29–44. [CrossRef]

39. Moylan, S.; Jacka, F.N.; Pasco, J.A.; Berk, M. Cigarette smoking, nicotine dependence and anxiety disorders: A systematic review of population-based, epidemiological studies. *BMC Med.* **2012**, *10*, 123. [CrossRef] [PubMed]

40. Teychenne, M.; Costigan, S.A.; Parker, K. The association between sedentary behaviour and risk of anxiety: A systematic review. *BMC Public Health* **2015**, *15*, 513. [CrossRef]

41. Demmer, R.T.; Gelb, S.; Suglia, S.F.; Keyes, K.M.; Aiello, A.E.; Colombo, P.C.; Galea, S.; Uddin, M.; Koenen, K.C.; Kubzansky, L.D. Sex Differences in the Association Between Depression, Anxiety, and Type 2 Diabetes Mellitus. *Psychosom. Med.* **2015**, *77*, 467–477. [CrossRef]

42. Mwinyi, J.; Pisanu, C.; Castelao, E.; Stringhini, S.; Preisig, M.; Schiöth, H.B. Anxiety Disorders are Associated with Low Socioeconomic Status in Women but Not in Men. *Women's Health Issues* **2017**, *27*, 302–307. [CrossRef]

43. Kang, H.-J.; Bae, K.-Y.; Kim, S.-W.; Shin, I.-S.; Yoon, J.-S.; Kim, J.-M. Anxiety symptoms in Korean elderly individuals: A two-year longitudinal community study. *Int. Psychogeriatr.* **2016**, *28*, 423–433. [CrossRef]

44. Anderson, E.; Shivakumar, G. Effects of exercise and physical activity on anxiety. *Front. Psychiatry* **2013**, *4*, 27. [CrossRef] [PubMed]

45. Rohleder, N.; Kirschbaum, C. The hypothalamic–pituitary–adrenal (HPA) axis in habitual smokers. *Int. J. Psychophysiol.* **2006**, *59*, 236–243. [CrossRef] [PubMed]

46. Vreeburg, S.A.; Zitman, F.G.; van Pelt, J.; DeRijk, R.H.; Verhagen, J.C.M.; van Dyck, R.; Hoogendijk, W.J.G.; Smit, J.H.; Penninx, B.W.J.H. Salivary Cortisol Levels in Persons with and without Different Anxiety Disorders. *Psychosom. Med.* **2010**, *72*, 340–347. [CrossRef] [PubMed]

47. Petkus, A.J.; Reynolds, C.A.; Wetherell, J.L.; Kremen, W.S.; Gatz, M. Temporal dynamics of cognitive performance and anxiety across older adulthood. *Psychol. Aging* **2017**, *32*, 278–292. [CrossRef] [PubMed]

48. Hirschfeld, R.M.A. The Comorbidity of Major Depression and Anxiety Disorders: Recognition and Management in Primary Care. *Prim. Care Companion J. Clin. Psychiatry* **2001**, *3*, 244–254. [CrossRef] [PubMed]

49. Biringer, E.; Mykletun, A.; Dahl, A.A.; Smith, A.D.; Engedal, K.; Nygaard, H.A.; Lund, A. The association between depression, anxiety, and cognitive function in the elderly general population–the Hordaland Health Study. *Int. J. Geriatr. Psychiatry* **2005**, *20*, 989–997. [CrossRef] [PubMed]

50. Hu, F.B. Dietary pattern analysis: A new direction in nutritional epidemiology. *Curr. Opin. Lipidol.* **2002**, *13*, 3–9. [CrossRef]

51. Anderson, A.L.; Harris, T.B.; Tylavsky, F.A.; Perry, S.E.; Houston, D.K.; Hue, T.F.; Strotmeyer, E.S.; Sahyoun, N.R. Dietary patterns and survival of older adults. *J. Am. Diet. Assoc.* **2011**, *111*, 84–91. [CrossRef]

52. Govindaraju, T.; Sahle, B.W.; McCaffrey, T.A.; McNeil, J.J.; Owen, A.J. Dietary Patterns and Quality of Life in Older Adults: A Systematic Review. *Nutrients* **2018**, *10*, 971. [CrossRef]

53. Bakhtiyari, M.; Ehsanpouoh, E.; Enayati, N.; Joodi, G.; Sadr, S.; Delpisheh, A.; Alihaydari, J.; Homayounfar, R. Anxiety as a consequence of modern dietary pattern in adults in Tehran–Iran. *Eat. Behav.* **2013**, *14*, 107–112. [CrossRef]

54. Dutheil, S.; Ota, K.T.; Wohleb, E.S.; Rasmussen, K.; Duman, R.S. High-Fat Diet Induced Anxiety and Anhedonia: Impact on Brain Homeostasis and Inflammation. *Neuropsychopharmacology* **2016**, *41*, 1874–1887. [CrossRef] [PubMed]

55. Kenny, R.; Dinan, T.; Cai, G.; Spencer, S.J. Effects of mild calorie restriction on anxiety and hypothalamic-pituitary-adrenal axis responses to stress in the male rat. *Physiol. Rep.* **2014**, *2*, e00265. [CrossRef] [PubMed]

56. Willette, A.A.; Coe, C.L.; Birdsill, A.C.; Bendlin, B.B.; Colman, R.J.; Alexander, A.L.; Allison, D.B.; Weindruch, R.H.; Johnson, S.C. Interleukin-8 and interleukin-10, brain volume and microstructure, and the influence of calorie restriction in old rhesus macaques. *Age* **2013**, *35*, 2215–2227. [CrossRef] [PubMed]

Article

Food Insecurity Is Associated with Mild Cognitive Impairment among Middle-Aged and Older Adults in South Africa: Findings from a Nationally Representative Survey

Ai Koyanagi [1,2,3,*], Nicola Veronese [4,5], Brendon Stubbs [6,7,8], Davy Vancampfort [9,10], Andrew Stickley [11,12], Hans Oh [13], Jae Il Shin [14,15,16], Sarah Jackson [17], Lee Smith [18] and Elvira Lara [2,19]

[1] Research and Development Unit, Parc Sanitari Sant Joan de Déu, Universitat de Barcelona, Fundació Sant Joan de Déu, 08830 Barcelona, Spain

[2] Instituto de Salud Carlos III, Centro de Investigación Biomédica en Red de Salud Mental, CIBERSAM, 28029 Madrid, Spain; elvira.lara@uam.es

[3] ICREA, Pg. Lluis Companys 23, 08010 Barcelona, Spain

[4] Aging Branch, Neuroscience Institute, National Research Council, 35128 Padova, Italy; ilmannato@gmail.com

[5] National Institute of Gastroenterology "S. De Bellis" Research Hospital, Castellana Grotte, 70013 Bari, Italy

[6] Physiotherapy Department, South London and Maudsley NHS Foundation Trust, Denmark Hill, London SE5 8AZ, UK; brendon.stubbs@kcl.ac.uk

[7] Department of Psychological Medicine, Institute of Psychiatry, Psychology and Neuroscience, King's College London, London WC2R 2LS, UK

[8] Faculty of Health, Social Care and Education, Anglia Ruskin University, Chelmsford CM1 1SQ, UK

[9] Department of Rehabilitation Sciences, KU Leuven, 3001 Leuven, Belgium; davy.vancampfort@kuleuven.be

[10] University Psychiatric Center KU Leuven, KU Leuven, 3000 Leuven, Belgium

[11] Department of Preventive Intervention for Psychiatric Disorders, National Institute of Mental Health, National Center of Neurology and Psychiatry, Kodaira, Tokyo 187-8553, Japan; amstick66@gmail.com

[12] The Stockholm Center for Health and Social Change (SCOHOST), Södertörn University, 141 89 Huddinge, Sweden

[13] School of Social Work, University of Southern California, CA 90015, USA; hansoh@usc.edu

[14] Department of Pediatrics, Yonsei University College of Medicine, Yonsei-ro 50, Seodaemun-gu, C.P.O. Box 8044, Seoul 03722, Korea; SHINJI@yuhs.ac

[15] Division of Pediatric Nephrology, Severance Children's Hospital, Seoul 03722, Korea

[16] Institute of Kidney Disease Research, Yonsei University College of Medicine, Seoul 03722, Korea

[17] Department of Behavioural Science and Health, University College London, London WC1E 7HB, UK; s.e.jackson@ucl.ac.uk

[18] The Cambridge Centre for Sport and Exercise Sciences, Anglia Ruskin University, Cambridge CB1 1PT, UK; Lee.Smith@anglia.ac.uk

[19] Department of Psychiatry, Hospital Universitario de La Princesa, Instituto de Investigación Sanitaria Princesa (IIS-Princesa), 28006 Madrid, Spain

[*] Correspondence: a.koyanagi@pssjd.org; Tel.: +34-936-002-685

Received: 27 February 2019; Accepted: 26 March 2019; Published: 30 March 2019

Abstract: There are no studies on the association between food insecurity and mild cognitive impairment (MCI). Thus, cross-sectional, community-based data on individuals aged ≥50 years from the World Health Organization's Study on Global AGEing and Adult Health (SAGE) conducted in South Africa (2007–2008) were analyzed to assess this association. The definition of MCI was based on the National Institute on Ageing-Alzheimer's Association criteria. Past 12-month food insecurity was assessed with two questions on frequency of eating less and hunger due to lack of food. Multivariable logistic regression analysis was conducted. The sample consisted of 3,672 individuals aged ≥50 years [mean (SD) age 61.4 (18.3); 56% females]. The prevalence of MCI was

8.5%, while 11.0% and 20.8% experienced moderate and severe food insecurity, respectively. After adjustment for potential confounders, moderate and severe food insecurity were associated with 2.82 (95%CI = 1.65–4.84) and 2.51 (95%CI = 1.63–3.87) times higher odds for MCI compared with no food insecurity, respectively. The OR for those aged $\geq$65 years with severe food insecurity was particularly high (OR = 3.87; 95%CI = 2.20–6.81). In conclusion, food insecurity was strongly associated with MCI among South African older adults. Future longitudinal research is required to assess whether addressing food insecurity may reduce risk of MCI and subsequent dementia.

Keywords: mild cognitive impairment; food insecurity; South Africa; epidemiology

1. Introduction

Dementia is a debilitating syndrome that results in the deterioration of memory, thinking, behavior, and the ability to conduct daily activities. It is one of the main causes of disability and dependency among the older adult population globally [1]. The 2016 Global Burden of Disease (GBD) study showed that 28.8 million disability-adjusted life years (DALYs) are attributed to dementia, and that dementia is the fifth leading cause of death globally [2]. Worldwide, it has been estimated that approximately 50 million people have dementia, of which about 60% live in low- and middle-income countries (LMICs), while there are nearly 10 million new cases each year [3]. As a consequence of global aging, the total number of people with dementia is expected to triple from its current figure by 2050 and reach 152 million, with this increase being largely attributable to rising numbers in LMICs [3]. Despite the overwhelming burden of dementia, especially in the years to come, there are no truly disease-modifying treatments for dementia [4]. Thus, there is a crucial need to identify modifiable risk factors for the preclinical transitional stages of dementia such as mild cognitive impairment (MCI). MCI has a high progression rate to dementia (12%, 20%, and 50% at 1, 3, and 5 years, respectively [5]), and is increasingly being considered as an important stage for intervention to prevent or delay the onset of dementia.

Currently, there is increasing evidence that food insecurity is associated with cognitive decline [6–8]. Food insecurity is defined as "limited or uncertain availability of nutritionally adequate and safe foods or limited or uncertain ability to acquire food in socially acceptable ways" [9]. It has been hypothesized that food insecurity may increase risk of cognitive decline via stress, depression, or poor nutritional intake [8]. However, there are no studies specifically on the association between food insecurity and MCI, or studies on food insecurity and any form of cognitive impairment from LMICs despite the fact that food insecurity is more common in this setting [10].

South Africa is an apposite setting in which to examine the association between food insecurity and MCI as the level of food insecurity has been reported to be high in this country [11], which may be attributable to the extremely high levels of absolute poverty compared with other middle-income countries [12]. Furthermore, rapid urbanization, the HIV epidemic, and increasing food prices are also likely to be implicated in the high rate of food insecurity in South Africa [11–13]. Finally, an extremely high prevalence of obesity and hypertension have been reported in South Africa [14], and these factors can potentially contribute to an upward trend in dementia in the future as these conditions are known to be risk factors for dementia [15]. Thus, the aim of the current study was to assess whether food insecurity is associated with MCI using data from a nationally representative sample of older adults in South Africa collected as part of the Study on Global AGEing and Adult Health (SAGE).

2. Materials and Methods

Data from the SAGE survey conducted in South Africa between 2007–2008 were analyzed. This dataset is publically available to all interested researchers via the WHO website (http://www.who.int/healthinfo/sage/en/) upon request. Detailed sampling information can be found in the above-mentioned WHO website. Briefly, a stratified multistage cluster sampling design was used to obtain a nationally representative sample. Strata were defined by the nine provinces (Eastern Cape, Free State, Gauteng, Kwa-Zulu Natal, Limpopo, Mpumalanga, North West, Northern Cape and Western Cape), locality (urban or rural), and predominant racial group (African/Black, White, Colored and Indian/Asian). Enumeration areas (EAs) constituted the primary sampling units (PSUs) and were selected with probability proportional to size: the measure of size being the number of individuals aged 50 or over in the EA. A different questionnaire was administered to a proxy respondent if the selected individual could not participate in the survey due to limited cognitive function. Information from proxies was not used in the current study. The survey response rate was 75%. Household weights were post-stratified by province and locality according to the South African Community Survey 2007. Individual weights were post-stratified by province, sex and age groups according to the 2009 Medium Mid-Year population estimates from Statistics South Africa. Ethical approval was obtained from the WHO Ethical Review Committee and Human Sciences Research Council, Pretoria, South Africa. All participants provided written informed consent.

2.1. Mild Cognitive Impairment (MCI)

We used the recommendations of the National Institute on Aging-Alzheimer's Association to define MCI [16]. The exact same algorithms used in past SAGE publications were applied [17,18]. Briefly, MCI was defined as fulfilling all of the following conditions:

(a) Concern regarding cognitive changes: This condition referred to replying 'bad' or 'very bad' to the question "How would you best describe your memory at present?" and/or answering 'worse' to the question "Compared to 12 months ago, would you say your memory is now better, the same or worse than it was then?"

(b) Objective evidence of impairment in at least one cognitive domain: The following performance tests were used to assess cognitive function: word list immediate and delayed verbal recall based on the Consortium to Establish a Registry for Alzheimer's Disease [19], which evaluated learning and episodic memory; digit span forward and backwards based on the Weschler Adult Intelligence Scale [20], that assessed working and attention memory; and the animal naming task [19], which evaluated verbal fluency. Individuals with a level of performance that was below -1 SD after adjustment for education and age on at least one of these tests were considered to have this condition.

(c) Preserved independence in functional abilities: Questions on self-reported past-30-day difficulties with basic activities of daily living (ADL) were used to assess this condition [21]. Specifically, the questions were: "How much difficulty did you have in getting dressed?" and "How much difficulty did you have with eating (including cutting up your food)?" The response options included none, mild, moderate, severe, and extreme (cannot do). Independence in functional activities was considered to be preserved if the participant answered either none, mild, or moderate to both of these questions. All other participants were omitted from the analysis (83 individuals aged ≥50 years).

(d) Absence of dementia: Individuals who could not participate in the survey due to severe cognitive impairment were excluded from the current study.

2.2. Food Insecurity

Food insecurity was defined with the use of the two following questions: "In the last 12 months, how often did you ever eat less than you felt you should because there wasn't enough food?" and "In the last 12 months, were you ever hungry, but didn't eat because you couldn't afford enough food?" Both of these questions had as response options: every month (coded = 1); almost every month (coded = 2); some months, but not every month (coded = 3); only in 1 or 2 months (coded = 4); never (coded = 5). These items were based on similar items found in food security questionnaires such as the US Household Food Security Survey Module and National Health and Nutrition Examination Survey (NHANES) Food Security module. As in a previous SAGE study, those who answered 1 through 3 to both questions or answered 1 to either item were categorized as severely food insecure. Those who did not fulfill the criteria for severe food insecurity but answered 2 through 4 for either question were coded as moderately food insecure. Those who answered 5 to both items were categorized as food secure [22].

2.3. Control Variables

Past literature was used as a guide to select the control variables [7]. Specifically, these included sex, age (years), years of education, wealth quintiles based on income, race (Black, White, other), low physical activity, smoking (never, past, current), alcohol use in the past 30 days, past 12 month DSM-IV depression, body mass index (BMI) based on measured weight and height [<18.5 (underweight), 18.5–24.9 (normal), 25.0–29.9 (overweight), $\geq$30 kg/m^2 (obese)], and chronic physical conditions (diabetes, hypertension, stroke). Stroke and diabetes were based only on lifetime self-reported diagnosis. Hypertension referred to having self-reported diagnosis, systolic blood pressure $\geq$140 mmHg, or diastolic blood pressure $\geq$90 mmHg. The Global Physical Activity Questionnaire was used to assess the level of physical activity [23]. Low physical activity was defined as <150 minutes of moderate-to-vigorous physical activity in a typical week [24]. The endorsement of DSM-IV depression was based on the World Mental Health Survey version of the Composite International Diagnostic Interview [25].

2.4. Statistical Analysis

The statistical analysis was done with Stata 14.1 (Stata Corp LP, College station, TX, USA). The analysis was limited to those aged $\geq$50 years. Middle-aged individuals were also included in the current study because cognitive dysfunction can appear up to 10 years prior to dementia diagnosis [26], and intervening in mid-life is now considered crucial [27]. The difference in sample characteristics by level of food insecurity was tested by Chi-squared tests and Student's *t*-tests for categorical and continuous variables, respectively.

We conducted multivariable logistic regression analysis to assess the association between food insecurity (exposure) and MCI (outcome) in the overall sample (i.e., age $\geq$50 years) and by age group (50–64 and $\geq$65 years) as the risk factors for MCI may differ between mid-life and late-life [28]. The regression analysis was adjusted for age, sex, education, wealth, race, physical activity, smoking, alcohol use, BMI, diabetes, stroke, hypertension, and depression. Given that some authors have suggested that depression may be an important mediator in the association between food insecurity and cognitive decline [8], we also constructed a model without adjustment for depression using the overall sample to assess the degree to which the association between food insecurity and MCI is explained by depression. All variables were included in the regression analysis as categorical variables with the exception of age and years of education (continuous variables). The sample weighting and the complex study design were taken into account in the analyses. Results from the regression analyses are presented as odds ratios (ORs) with 95% confidence intervals (CIs). The level of statistical significance was set at $p < 0.05$.

3. Results

The final analytical sample comprised 3,672 individuals aged $\geq$50 years [mean (SD) age 61.4 (18.3) years; 56% females]. The proportion of those aged 50–64 and $\geq$65 years was 66.8% and 33.2%, respectively. Overall, the prevalence of MCI was 8.5% (95%CI = 6.9–10.3), while the prevalence of moderate (unweighted n = 332) and severe (unweighted n = 677) food insecurity was 11.0% (95%CI = 9.0–13.4) and 20.8% (95%CI = 17.9–23.9), respectively. The sample characteristics are shown in Table 1. Compared to respondents without food insecurity, those who were food insecure were significantly more likely to have less education and wealth, and lower levels of physical activity, and be of Black race.

Table 1. Sample characteristics (overall and by food insecurity).

Characteristic		Overall	Food Insecurity			*p*-Value [a]
			None	Moderate	Severe	
Sex	Female	56.0	54.4	56.5	62.8	0.047
	Male	44.0	43.5	37.2	43.6	
Age (years)	Mean (SD)	61.4 (18.3)	61.7 (18.9)	62.5 (17.4)	60.1 (17.1)	0.074
Education (years)	Mean (SD)	6.1 (10.1)	6.8 (11.0)	4.5 (8.1)	4.7 (7.3)	<0.001
Wealth	Poorest	20.7	14.7	27.7	32.7	<0.001
	Poorer	19.9	17.9	20.8	26.4	
	Middle	18.5	17.7	25.1	19.8	
	Richer	19.8	22.1	20.3	13.2	
	Richest	21.1	27.6	6.0	7.9	
Race	Black	74.3	68.2	80.1	88.9	<0.001
	White	9.2	12.9	2.0	1.5	
	Other	16.5	19.0	17.9	9.6	
Low physical activity	No	50.0	54.1	39.5	41.9	0.002
	Yes	50.0	45.9	60.5	58.1	
Smoking	Never	66.6	66.1	69.0	66.7	0.162
	Past	23.8	23.3	20.2	27.3	
	Current	9.7	10.7	10.8	6.0	
Alcohol consumption	No	86.0	86.8	91.1	80.7	0.007
	Yes	14.0	13.2	8.9	19.3	
BMI (kg/m^2)	Normal	24.0	23.3	21.5	25.8	0.382
	Overweight	26.5	27.8	22.1	25.5	
	Obese	46.6	46.4	51.9	45.1	
	Underweight	2.9	2.5	4.5	3.6	
Diabetes	No	90.9	91.3	88.1	90.5	0.465
	Yes	9.1	8.7	11.9	9.5	
Stroke	No	96.6	96.6	96.7	96.3	0.951
	Yes	3.4	3.4	3.3	3.7	
Hypertension	No	21.4	21.8	16.7	20.5	0.278
	Yes	78.6	78.2	83.3	79.5	
Depression	No	97.1	97.7	97.6	95.0	0.064
	Yes	2.9	2.3	2.4	5.0	

Abbreviation: SD Standard deviation; BMI Body mass index; Data are percentage unless otherwise stated. [a] *p*-value was calculated by Chi-squared tests and Student's *t*-tests for categorical and continuous variables, respectively.

The prevalence of MCI was higher among those with food insecurity than in those without food insecurity (Figure 1). For example, overall, the prevalence of MCI among those without food insecurity was 5.9% but this increased to 14.8% and 13.5% among those with moderate and severe food insecurity, respectively.

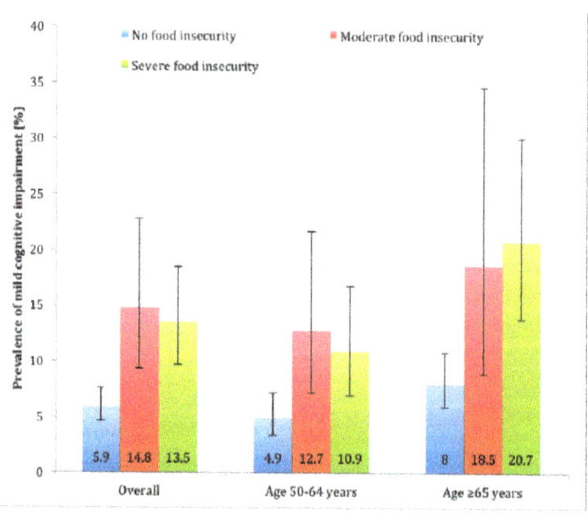

Figure 1. Prevalence of mild cognitive impairment by food insecurity status. Estimates are based on weighted sample. Bars denote 95% confidence interval.

The association between food insecurity and MCI estimated by multivariable logistic regression is shown in Table 2. After adjustment for a variety of sociodemographic and behavioral factors as well as physical health conditions and depression, compared to no food insecurity, moderate and severe food insecurity were associated with 2.82 (95%CI = 1.65–4.84) and 2.51 (95%CI = 1.63–3.87) times higher odds for MCI, respectively, in the overall sample. These estimates were not substantially different from the model that did not include depression with the corresponding figures being 2.83 (95%CI = 1.65–4.85) and 2.58 (95%CI = 1.68–3.97), respectively (data shown only in text). The OR among those aged ≥65 years with severe food insecurity was particularly elevated (OR = 3.87; 95%CI = 2.20–6.81).

Table 2. Association of food insecurity and other covariates with mild cognitive impairment (outcome) estimated by multivariable logistic regression.

Characteristic	Category	Overall		50–64 years		≥65 years	
		OR	95%CI	OR	95%CI	OR	95%CI
Food insecurity	None	1.00		1.00		1.00	
	Moderate	2.82 ***	[1.65,4.84]	2.84 **	[1.41,5.69]	2.76 *	[1.19,6.41]
	Severe	2.51 ***	[1.63,3.87]	1.98 *	[1.06,3.72]	3.87 ***	[2.20,6.81]
Sex	Female	1.00		1.00		1.00	
	Male	0.56 **	[0.37,0.85]	0.59	[0.32,1.06]	0.54	[0.26,1.10]
Age (years)		1.05 ***	[1.03,1.07]	1.05	[0.99,1.11]	1.04 *	[1.01,1.08]
Education (years)		0.98	[0.93,1.03]	0.95	[0.89,1.02]	1.01	[0.94,1.09]
Wealth	Poorest	1.00		1.00		1.00	
	Poorer	1.08	[0.64,1.83]	0.68	[0.34,1.36]	2.32	[0.99,5.40]
	Middle	1.51	[0.82,2.80]	1.06	[0.52,2.16]	2.95	[0.98,8.87]
	Richer	1.82	[0.96,3.46]	1.54	[0.73,3.27]	2.45	[0.87,6.93]
	Richest	1.35	[0.60,3.04]	1.12	[0.37,3.42]	2.11	[0.63,7.09]
Race	Black	1.00		1.00		1.00	
	White	0.23 *	[0.07,0.74]	0.18	[0.03,1.02]	0.24 *	[0.06,0.96]
	Other	0.87	[0.50,1.50]	0.92	[0.46,1.80]	0.58	[0.25,1.34]
Low physical activity	No	1.00		1.00		1.00	
	Yes	0.79	[0.52,1.20]	0.72	[0.40,1.31]	0.75	[0.45,1.23]
Smoking	Never	1.00		1.00		1.00	
	Past	1.11	[0.67,1.82]	1.45	[0.76,2.77]	0.71	[0.36,1.37]
	Current	1.15	[0.63,2.11]	1.16	[0.54,2.48]	1.1	[0.41,2.93]
Alcohol consumption	No	1.00		1.00		1.00	
	Yes	2.44 **	[1.39,4.28]	2.90 **	[1.49,5.64]	1.57	[0.71,3.48]
BMI (kg/m^2)	Normal	1.00		1.00		1.00	
	Overweight	0.95	[0.53,1.69]	1.69	[0.82,3.49]	0.40 *	[0.18,0.89]
	Obese	1.17	[0.74,1.85]	1.63	[0.85,3.11]	0.61	[0.31,1.21]
	Underweight	0.89	[0.40,2.01]	0.92	[0.30,2.84]	0.76	[0.20,2.80]
Diabetes	No	1.00		1.00		1.00	
	Yes	0.9	[0.46,1.76]	0.97	[0.39,2.43]	0.87	[0.33,2.29]
Stroke	No	1.00		1.00		1.00	
	Yes	4.64 ***	[1.91,11.29]	5.02 **	[1.81,13.91]	2.92 *	[1.02,8.39]
Hypertension	No	1.00		1.00		1.00	
	Yes	0.9	[0.55,1.46]	1.17	[0.58,2.34]	0.56	[0.30,1.03]
Depression	No	1.00		1.00		1.00	
	Yes	1.77	[0.68,4.58]	2.41	[0.92,6.31]	0.58	[0.06,6.00]

Abbreviation: OR Odds ratio; CI Confidence interval; BMI Body mass index. Models are mutually adjusted for all variables in the Table. * $p < 0.05$, ** $p < 0.01$, *** $p < 0.001$.

4. Discussion

In our study on community-dwelling adults aged ≥50 years in South Africa, we found a high prevalence of food insecurity (31.8%). After adjustment for potential confounders, moderate and severe food insecurity were associated with significant 2.82 and 2.51 times higher odds for MCI, while the OR was particularly elevated for severe food insecurity among those aged ≥65 years (OR = 3.87). Depression was not a major explanatory factor in this association.

To the best of our knowledge, this is the first study to specifically examine the relationship between food insecurity and MCI. Our findings are in line with previous studies that have examined the association between food insecurity and cognitive function in the USA although these studies were on general cognitive function and not MCI. One community-based cross-sectional study conducted among a representative sample of 1358 Puerto Ricans aged 45–75 years living in Massachusetts found that the adjusted difference in the Mini-Mental State Examination (MMSE) score was 0.90 lower among those with very low food security as compared to those who were food secure [8]. A longitudinal follow-up study among 597 participants aged 40–75 years based on the same study population found that food insecurity at baseline was associated with lower global cognitive function after two years of follow-up [7]. Finally, one cross-sectional US study using data from the NHANES found that among 1851 adults between 60 and 85 years, food insecurity was associated with poor cognitive function [6].

The exact mechanisms linking food insecurity and MCI are unknown, but there are several hypotheses. It is possible that stress resulting from food insecurity may increase the risk of MCI. For example, prolonged elevation of cortisol (an HPA axis response to chronic stress) can lead to alterations in brain structure and function (e.g., in the hippocampus) and subsequent cognitive decline [29]. Also, increases in pro-inflammatory cytokines induced by stress [30] may lead to an increased risk of dementia [31]. Furthermore, food insecurity often compromises diet quality, as people tend to switch to more affordable but less nutritious food when food is scarce (e.g., high fat and carbohydrates, low vitamins and micronutrients) [32]. Poor diet has been associated with increased risk of cognitive decline. For example, vitamin B, C, and E have all been shown to have a protective effect against dementia [33]. Furthermore, randomized clinical trials have shown that supplementation of n-3 polyunsaturated fatty acids [34] and folic acid [35] may improve cognitive function in older people with MCI. Finally, high carbohydrate intake has been associated with higher risk of MCI [36].

Our finding that a particularly strong association was observed among older individuals concurs with that of the study by Gao and colleagues [8], which found that the difference in MMSE scores between very low food insecurity and food secure were −1.75 and −0.48 for those aged ≥60 and <60 years, respectively [8]. This age difference may be related to increased vulnerability of the brain among the older population. For example, older adults have smaller hippocampal volumes and this may increase susceptibility to cognitive deficits when exposed to stress [37]. The fact that people aged ≥65 years had 3.87 times higher odds for MCI compared with those without food insecurity is an important finding as the risk of dementia is particularly elevated in this age group [38].

Strengths of the study include the large sample size and the use of nationally representative data. However, some limitations should be taken into consideration when interpreting the findings. First, participants with mild forms of dementia could have been included in our study sample as the study did not include a clinical assessment of dementia. Second, there is no consensus regarding the acceptable level of functional impairment in MCI [39]. The definition of preservation of independence in functional abilities used in our study, which has been used in previous publications [17,18,28,40], was rather conservative. This was done to avoid the omission of MCI cases with disability not related to their cognitive ability. Despite these potential limitations, it is worth noting that the prevalence of MCI in our study was consistent with previously reported figures [41]. Third, data on biomarkers such as cerebrospinal fluid Aβ and tau were not available. These data could have provided a better understanding on how food insecurity affects brain pathology [4]. Fourth, the estimates in the analyses stratified by age should be interpreted with caution as the sample size was small and some estimates had wide confidence intervals. Next, we lacked information on HIV infection, which has been reported to be associated with a higher risk of food insecurity [42] and impaired cognitive function [43]. However, given that the link between food insecurity and HIV infection is likely to be mainly explained by poverty [42], we believe that the adjustment for wealth in our study is likely to have minimized the potential for residual confounding due to HIV infection. Furthermore, our measure of food insecurity was based on two questions and did not constitute a comprehensive food insecurity measure. Finally, because this was a cross-sectional study, causality cannot be inferred. For example, it is possible that people with cognitive impairments have difficulty in utilizing social safety net services, and this might have led to food insecurity.

5. Conclusions

In conclusion, we found that the prevalence of food insecurity is high in South African older adults, and that food insecurity is associated with MCI, with this association being particularly pronounced among individuals aged $\geq$65 years. Although biological plausibility suggests that food insecurity may lead to MCI, future studies with a longitudinal and experimental design are warranted to assess causality and the utility of addressing food insecurity as a preventive strategy for MCI and subsequent dementia.

Author Contributions: Conceptualization, A.K. and E.L.; methodology, A.K., N.V.; formal analysis, A.K., N.V., E.L; writing—original draft preparation, A.K.; writing—review and editing, all authors.

Funding: This research received no external funding.

Acknowledgments: Ai Koyanagi's work is supported by the PI15/00862 project, integrated into the National R + D + I and funded by the ISCIII - General Branch Evaluation and Promotion of Health Research - and the European Regional Development Fund (ERDF-FEDER).

Conflicts of Interest: The authors declare no conflict of interest.

References

1. Wimo, A.; Guerchet, M.; Ali, G.C.; Wu, Y.T.; Prina, A.M.; Winblad, B.; Jonsson, L.; Liu, Z.; Prince, M. The worldwide costs of dementia 2015 and comparisons with 2010. *Alzheimers Dement.* **2017**, *13*, 1–7. [CrossRef] [PubMed]
2. GBD 2016 Dementia Collaborators. Global, regional, and national burden of Alzheimer's disease and other dementias, 1990–2016: A systematic analysis for the Global Burden of Disease Study 2016. *Lancet Neurol.* **2019**, *18*, 88–106. [CrossRef]
3. World Health Organization. Available online: https://www.who.int/news-room/fact-sheets/detail/dementia (accessed on 20 February 2019).
4. Livingston, G.; Sommerlad, A.; Orgeta, V.; Costafreda, S.G.; Huntley, J.; Ames, D.; Ballard, C.; Banerjee, S.; Burns, A.; Cohen-Mansfield, J.; et al. Dementia prevention, intervention, and care. *Lancet* **2017**, *390*, 2673–2734. [CrossRef]
5. Solfrizzi, V.; Panza, F.; Colacicco, A.M.; D'Introno, A.; Capurso, C.; Torres, F.; Grigoletto, F.; Maggi, S.; Del Parigi, A.; Reiman, E.M.; et al. Vascular risk factors, incidence of MCI, and rates of progression to dementia. *Neurology* **2004**, *63*, 1882–1891. [CrossRef]
6. Frith, E.; Loprinzi, P.D. Food insecurity and cognitive function in older adults: Brief report. *Clin. Nutr.* **2017**. [CrossRef]
7. Wong, J.C.; Scott, T.; Wilde, P.; Li, Y.G.; Tucker, K.L.; Gao, X. Food Insecurity Is Associated with Subsequent Cognitive Decline in the Boston Puerto Rican Health Study. *J. Nutr.* **2016**, *146*, 1740–1745. [CrossRef] [PubMed]
8. Gao, X.; Scott, T.; Falcon, L.M.; Wilde, P.E.; Tucker, K.L. Food insecurity and cognitive function in Puerto Rican adults. *Am. J. Clin. Nutr.* **2009**, *89*, 1197–1203. [CrossRef] [PubMed]
9. Bickel, G.; Nord, M.; Price, C.; Hamilton, W.; Cook, J. *Guide to Measuring Household Food Security*; Department of Agriculture, Food and Nutrition Service: Alexandria, VA, USA, 2000.
10. FAO; IFAD; UNICEF; WFP; WHO. *The State of Food Security and Nutrition in the World 2017*; Building Resilience for Peace and Food Security; FAO: Rome, Italy, 2017.
11. Misselhorn, A.; Hendriks, S.L. A systematic review of sub-national food insecurity research in South Africa: Missed opportunities for policy insights. *PLoS ONE* **2017**, *12*, e0182399. [CrossRef]

12. Altman, M.; Hart, T.G.; Jacobs, P.T. Household food security status in South Africa. *Agrekon* **2009**, *48*, 345–361. [CrossRef]

13. Misselhorn, A.A. What drives food insecurity in southern Africa? A meta-analysis of household economy studies. *Glob. Environ. Chang.* **2005**, *15*, 33–43. [CrossRef]

14. Stubbs, B.; Vancampfort, D.; Veronese, N.; Schofield, P.; Lin, P.Y.; Tseng, P.T.; Solmi, M.; Thompson, T.; Carvalho, A.F.; Koyanagi, A. Multimorbidity and perceived stress: A population-based cross-sectional study among older adults across six low- and middle-income countries. *Maturitas* **2018**, *107*, 84–91. [CrossRef]

15. World Alzheimer Report 2015. Available online: https://www.alz.co.uk/research/world-report-2015 (accessed on 31 January 2019).

16. Albert, M.S.; DeKosky, S.T.; Dickson, D.; Dubois, B.; Feldman, H.H.; Fox, N.C.; Gamst, A.; Holtzman, D.M.; Jagust, W.J.; Petersen, R.C.; et al. The diagnosis of mild cognitive impairment due to Alzheimer's disease: Recommendations from the National Institute on Aging-Alzheimer's Association workgroups on diagnostic guidelines for Alzheimer's disease. *Alzheimers Dement.* **2011**, *7*, 270–279. [CrossRef]

17. Koyanagi, A.; Lara, E.; Stubbs, B.; Carvalho, A.F.; Oh, H.; Stickley, A.; Veronese, N.; Vancampfort, D. Chronic Physical Conditions, Multimorbidity, and Mild Cognitive Impairment in Low- and Middle-Income Countries. *J. Am. Geriatr. Soc.* **2018**. [CrossRef] [PubMed]

18. Koyanagi, A.; Oh, H.; Vancampfort, D.; Carvalho, A.F.; Veronese, N.; Stubbs, B.; Lara, E. Perceived Stress and Mild Cognitive Impairment among 32,715 Community-Dwelling Older Adults across Six Low- and Middle-Income Countries. *Gerontology* **2019**, *65*, 155–163. [CrossRef]

19. Morris, J.C.; Heyman, A.; Mohs, R.C.; Hughes, J.P.; van Belle, G.; Fillenbaum, G.; Mellits, E.D.; Clark, C. The Consortium to Establish a Registry for Alzheimer's Disease (CERAD). Part I. Clinical and neuropsychological assessment of Alzheimer's disease. *Neurology* **1989**, *39*, 1159–1165.

20. The Psychological Corporation. *The Psychological Corporation: The WAIS III-WMS III Updated Technical Manual*; The Psychological Corporation: San Antonio, TX, USA, 2002.

21. Katz, S.; Ford, A.B.; Moskowitz, R.W.; Jackson, B.A.; Jaffe, M.W. Studies of illness in the aged. The index of ADL: A standardized measure of biological and psychosocial function. *JAMA* **1963**, *185*, 914–919. [CrossRef]

22. Schrock, J.M.; McClure, H.H.; Snodgrass, J.J.; Liebert, M.A.; Charlton, K.E.; Arokiasamy, P.; Naidoo, N.; Kowal, P. Food insecurity partially mediates associations between social disadvantage and body composition among older adults in india: Results from the study on global AGEing and adult health (SAGE). *Am. J. Hum. Biol.* **2017**, *29*. [CrossRef] [PubMed]

23. Bull, F.C.; Maslin, T.S.; Armstrong, T. Global physical activity questionnaire (GPAQ): Nine country reliability and validity study. *J. Phys. Act. Health* **2009**, *6*, 790–804. [CrossRef] [PubMed]

24. Vancampfort, D.; Stubbs, B.; Lara, E.; Vandenbulcke, M.; Swinnen, N.; Koyanagi, A. Mild cognitive impairment and physical activity in the general population: Findings from six low- and middle-income countries. *Exp. Gerontol.* **2017**, *100*, 100–105. [CrossRef] [PubMed]

25. Vancampfort, D.; Stubbs, B.; Mugisha, J.; Firth, J.; Schuch, F.B.; Koyanagi, A. Correlates of sedentary behavior in 2,375 people with depression from 6 low- and middle-income countries. *J. Affect. Disord.* **2018**, *234*, 97–104. [CrossRef]

26. Amieva, H.; Jacqmin-Gadda, H.; Orgogozo, J.M.; Le Carret, N.; Helmer, C.; Letenneur, L.; Barberger-Gateau, P.; Fabrigoule, C.; Dartigues, J.F. The 9 year cognitive decline before dementia of the Alzheimer type: A prospective population-based study. *Brain* **2005**, *128 Pt 5*, 1093–1101. [CrossRef]

27. World Alzheimer Report 2014. Available online: https://www.alz.co.uk/research/world-report-2014 (accessed on 29 January 2019).

28. Lara, E.; Koyanagi, A.; Olaya, B.; Lobo, A.; Miret, M.; Tyrovolas, S.; Ayuso-Mateos, J.L.; Haro, J.M. Mild cognitive impairment in a Spanish representative sample: Prevalence and associated factors. *Int. J. Geriatr. Psychiatry* **2016**, *31*, 858–867. [CrossRef]

29. Pruessner, J.C.; Dedovic, K.; Pruessner, M.; Lord, C.; Buss, C.; Collins, L.; Dagher, A.; Lupien, S.J. Stress regulation in the central nervous system: Evidence from structural and functional neuroimaging studies in human populations—2008 Curt Richter Award Winner. *Psychoneuroendocrinology* **2010**, *35*, 179–191. [CrossRef]

30. Leonard, B.E. The HPA and immune axes in stress: The involvement of the serotonergic system. *Eur. Psychiatry* **2005**, *20* (Suppl. 3), S302–S306. [CrossRef]

31. Lai, K.S.P.; Liu, C.S.; Rau, A.; Lanctot, K.L.; Kohler, C.A.; Pakosh, M.; Carvalho, A.F.; Herrmann, N. Peripheral inflammatory markers in Alzheimer's disease: A systematic review and meta-analysis of 175 studies. *J. Neurol. Neurosurg. Psychiatry* **2017**. [CrossRef] [PubMed]

32. Pilgrim, A.; Barker, M.; Jackson, A.; Ntani, G.; Crozier, S.; Inskip, H.; Godfrey, K.; Cooper, C.; Robinson, S. Does living in a food insecure household impact on the diets and body composition of young children? Findings from the Southampton Women's Survey. *J. Epidemiol. Community Health* **2012**, *66*, e6. [CrossRef]

33. Fenech, M. Vitamins Associated with Brain Aging, Mild Cognitive Impairment, and Alzheimer Disease: Biomarkers, Epidemiological and Experimental Evidence, Plausible Mechanisms, and Knowledge Gaps. *Adv. Nutr.* **2017**, *8*, 958–970. [CrossRef] [PubMed]

34. Bo, Y.; Zhang, X.; Wang, Y.; You, J.; Cui, H.; Zhu, Y.; Pang, W.; Liu, W.; Jiang, Y.; Lu, Q. The n-3 Polyunsaturated Fatty Acids Supplementation Improved the Cognitive Function in the Chinese Elderly with Mild Cognitive Impairment: A Double-Blind Randomized Controlled Trial. *Nutrients* **2017**, *9*, 54. [CrossRef] [PubMed]

35. Ma, F.; Li, Q.; Zhou, X.; Zhao, J.; Song, A.; Li, W.; Liu, H.; Xu, W.; Huang, G. Effects of folic acid supplementation on cognitive function and Abeta-related biomarkers in mild cognitive impairment: A randomized controlled trial. *Eur. J. Nutr.* **2017**. [CrossRef]

36. Roberts, R.O.; Roberts, L.A.; Geda, Y.E.; Cha, R.H.; Pankratz, V.S.; O'Connor, H.M.; Knopman, D.S.; Petersen, R.C. Relative intake of macronutrients impacts risk of mild cognitive impairment or dementia. *J. Alzheimers Dis.* **2012**, *32*, 329–339. [CrossRef] [PubMed]

37. Lupien, S.J.; Maheu, F.; Tu, M.; Fiocco, A.; Schramek, T.E. The effects of stress and stress hormones on human cognition: Implications for the field of brain and cognition. *Brain Cogn.* **2007**, *65*, 209–237. [CrossRef]

38. Jorm, A.F.; Jolley, D. The incidence of dementia: A meta-analysis. *Neurology* **1998**, *51*, 728–733. [CrossRef] [PubMed]

39. Lindbergh, C.A.; Dishman, R.K.; Miller, L.S. Functional Disability in Mild Cognitive Impairment: A Systematic Review and Meta-Analysis. *Neuropsychol. Rev.* **2016**, *26*, 129–159. [CrossRef]

40. Lara, E.; Koyanagi, A.; Domenech-Abella, J.; Miret, M.; Ayuso-Mateos, J.L.; Haro, J.M. The Impact of Depression on the Development of Mild Cognitive Impairment over 3 Years of Follow-Up: A Population-Based Study. *Dement. Geriatr. Cogn. Disord.* **2017**, *43*, 155–169. [CrossRef]

41. Petersen, R.C. Mild Cognitive Impairment. *Continuum (Minneap Minn)* **2016**, *22*, 404–418. [CrossRef]

42. Masa, R.; Chowa, G.; Nyirenda, V. Prevalence and Predictors of Food Insecurity among People Living with HIV Enrolled in Antiretroviral Therapy and Livelihood Programs in Two Rural Zambian Hospitals. *Ecol. Food Nutr.* **2017**, *56*, 256–276. [CrossRef] [PubMed]

43. Kanmogne, G.D.; Fonsah, J.Y.; Tang, B.; Doh, R.F.; Kengne, A.M.; Umlauf, A.; Tagny, C.T.; Nchindap, E.; Kenmogne, L.; Franklin, D.; et al. Effects of HIV on executive function and verbal fluency in Cameroon. *Sci. Rep.* **2018**, *8*, 17794. [CrossRef] [PubMed]

 *nutrients*

Article

The Effect of an Infant Formula Supplemented with AA and DHA on Fatty Acid Levels of Infants with Different FADS Genotypes: The COGNIS Study

Isabel Salas Lorenzo [1,2], Aida M. Chisaguano Tonato [3], Andrea de la Garza Puentes [1,2,4,*], Ana Nieto [5,6], Florian Herrmann [5,6], Estefanía Dieguez [5,6], Ana I. Castellote [1,2,7], M. Carmen López-Sabater [1,2,7,*], Maria Rodríguez-Palmero [8] and Cristina Campoy [5,6,9]

[1] Department of Nutrition, Food Sciences and Gastronomy, Faculty of Pharmacy and Food Sciences, University of Barcelona, Av. Joan XXIII 27-31, E-08028 Barcelona, Spain; salas.lorenzo.i@gmail.com (I.S.L.); aicastellote@ub.edu (A.I.C.)

[2] Institut de Recerca en Nutrició i Seguretat Alimentària de la UB (INSA-UB), 08921 Barcelona, Spain

[3] Nutrition, Faculty of Health Sciences, University of San Francisco de Quito, Quito 170157, Ecuador; achisaguano@usfq.edu.ec

[4] Parc Sanitari Sant Joan de Déu, Fundació Sant Joan de Déu, Institut de Recerca Sant Joan de Déu, 08830 Sant Boi de Llobregat, Spain

[5] Centre of Excellence for Paediatric Research EURISTIKOS, University of Granada, 18071 Granada, Spain; ananietoruiz@gmail.com (A.N.); herrmann@florian-herrmann.de (F.H.); estefaniadieguez@ugr.es (E.D.); ccampoy@ugr.es (C.C.)

[6] Department of Paediatrics, University of Granada, 18071 Granada, Spain

[7] CIBER Physiopathology of Obesity and Nutrition CIBERobn, Institute of Health Carlos III, 28029 Madrid, Spain

[8] Basic Research Department. Ordesa Laboratories, 08830 Barcelona, Spain; Maria.Rodriguez@ordesa.es

[9] CIBER Epidemiology and Public Health CIBEResp, Institute of Health Carlos III, 28029 Madrid, Spain

[*] Correspondence: adelagarza@ub.edu (A.d.l.G.P.); mclopez@ub.edu (M.C.L.-S.); Tel.: +34-934-024-512 (A.d.l.G.P.); +34-934-024-512 (M.C.L.-S.)

Received: 13 December 2018; Accepted: 5 March 2019; Published: 12 March 2019

Abstract: Polymorphisms in the fatty acid desaturase (FADS) genes influence the arachidonic (AA) and docosahexaenoic (DHA) acid concentrations (crucial in early life). Infants with specific genotypes may require different amounts of these fatty acids (FAs) to maintain an adequate status. The aim of this study was to determine the effect of an infant formula supplemented with AA and DHA on FAs of infants with different FADS genotypes. In total, 176 infants from the COGNIS study were randomly allocated to the Standard Formula (SF; n = 61) or the Experimental Formula (EF; n = 70) group, the latter supplemented with AA and DHA. Breastfed infants were added as a reference group (BF; n = 45). FAs and FADS polymorphisms were analyzed from cheek cells collected at 3 months of age. FADS minor allele carriership in formula fed infants, especially those supplemented, was associated with a declined desaturase activity and lower AA and DHA levels. Breastfed infants were not affected, possibly to the high content of AA and DHA in breast milk. The supplementation increased AA and DHA levels, but mostly in major allele carriers. In conclusion, infant FADS genotype could contribute to narrow the gap of AA and DHA concentrations between breastfed and formula fed infants.

Keywords: fatty acids; omega 6; omega 3; breast milk; infant formula; fatty acid desaturases; early life nutrition; control formula; intervention formula; exclusive breastfeeding

---

1. Introduction

Long chain polyunsaturated fatty acids (LCPUFA) have an important role in the immune system regulation, blood clots, neurotransmitters, cholesterol metabolism, and in the structure of membrane phospholipids in the brain and the retina [1]. Attention has been devoted especially to arachidonic acid (AA) and docosahexaenoic acid (DHA), due to their key role for optimal health, cognition and development during fetal and early postnatal life [2]. Breastmilk is usually the only external source of AA and DHA for infants during the first months of life [3,4]. When breastfeeding is not possible, infants require breast milk substitutes [5], which are usually supplemented with nutrients to match the breast milk content [6]. However, there is still debate and different opinions in regards of DHA and AA supplementation in infant formulas. The European Food Safety Authority (EFSA) [7] and the Commission Delegated Regulation (EU) 2016/127 [8] have proposed DHA supplementation as mandatory for infant formulas, while no minimum amount of AA was determined to be necessary, setting AA supplementation as an optional ingredient. However, it has been observed that when infants receive both AA and DHA supplementation they have better outcomes in cognitive performance than receiving DHA alone [9]. Other authors have even found that DHA alone did not influenced cognitive development at all [10]. Therefore, whether DHA should be supplemented alone or with AA remains controversial. Additionally, there is no agreed specific dose for supplementation, and there is little evidence of the long-term effect.

LCPUFAs can also be endogenously synthesized from the essential fatty acids (FAs): linoleic acid (LA) and alpha-linolenic acid (ALA). LCPUFA synthesis requires desaturation and elongation reactions. D6 and D5 desaturases (D6D and D5D) are two key enzymes that catalyze the synthesis by introducing *cis* double bonds at specific positions. Fatty acid desaturase genes FADS1 and FADS2 encode D5D and D6D, respectively, making them a rate limiting factor in LCPUFA conversion [11]. However, the endogenous synthesis of AA and DHA from their precursors is limited in humans as Demmelmair et al. observed in women that only 1.2% of the AA is directly derived from LA intake [12]. Likewise, both blood and tissue levels of PUFAs are influenced to a large extent by genetic heritability [13]. There are many studies showing the major impact of gene variants of the FADS gene cluster on the FA composition of blood, tissues and human milk [14–16]. For instance, single nucleotide polymorphisms (SNPs) in the FADS gene modulate the capacity for endogenous synthesis of LCPUFAs by compromising the desaturase activity of the involved enzymes [17–21]. It has been observed that up to 28% of variation of AA blood levels is due to FADS genetic variants, while LA is affected in 9% [13]. Furthermore, Baylin et al. observed that an impaired desaturase activity may induce an unbalanced proportion of n3 (e.g., the FADS2 deletion could prevent the conversion of the precursor ALA into LCPUFAs) [22]. Likewise, Schaeffer et al. observed that variants in the FADS1 and FADS2 genes showed strong associations with levels of n6 (e.g., LA, gamma-linolenic acid (GLA), dihomo-gamma-linolenic acid (DGLA), AA, and adrenic acid (AdA)), and n3 FAs (ALA, eicosapentaenoic acid (EPA), docosapentaenoic acid (DPA) [13], and DHA [23]). It has also been observed that variants in the FADS cluster can also influence total LDL, and DHL cholesterol, triglycerides, phospholipids, C-reactive protein, proinflamatory eicosanoids and cardiovascular disease endpoints [24,25].

Even though there is wide evidence of the effect of FADS genetic variants in FA concentrations of different biological tissues, there is little evidence for infants during their first year of life [19,26–28], which is a critical period of early life programming in which LCPUFAs play an important role [29]. Studying this could identify vulnerable groups in the pediatric population and contribute to the refinement of current recommendations and legislations in regards to infant AA and DHA supplementation. Given the low rates of exclusive breastfeeding in Europe, and while efforts are still being made to promote it, it is important to secure an appropriate source of LCPUFAs for infants who are not breastfed to promote their development. Therefore, the aim of this study was to determine the effect of an infant formula supplemented with AA and DHA on fatty acid levels of three-month-old infants with different FADS genotypes.

2. Materials and Methods

2.1. Ethics Statement

This study was carried out in accordance with the ethical standards established by the Declaration of Helsinki (2004), the Good Clinical Practice recommendations of the EEC (document 111/3976/88 July 1990) and the current Spanish legislation governing clinical research in humans (Royal Decree 561/1993 on clinical trials). In addition, the study was approved by the San Cecilio University Hospital Ethics Committee and the Faculty of Medicine at the University of Granada.

2.2. Study Population and Design

We analyzed 176 infants from the total of 220 participants in the COGNIS study (A Neurocognitive and Immunological Study of a New Formula for Healthy Infants), which is an interventional, randomized, and double-blinded study registered at www.ClinicalTrials.gov (NCT02094547). Full-term infants were recruited at the University Hospitals (Clinical San Cecilio and Mother-Infant Hospital) in the city of Granada (Spain), where samples and data were also collected. Each parent or legal guardian signed a written informed consent before the recruitment. To be eligible for enrolment of COGNIS study, infants had to meet the following inclusion criteria: 0–2 month old full-term infants, adequate birth weight for gestational age, normal Apgar score, umbilical pH $\geq$ 7.10 (normal range), availability to continue throughout the entire study period, signed informed consent, and for infants in the SF and EF groups, a maximum of 30 days of exclusive breastfeeding was considered and a minimum of 70% of infant formula consumption was required afterwards. Participants were excluded if they were participating in other studies, if they had nervous system or gastrointestinal disorders, and if the mother had a disease history or had received harmful drug treatment during pregnancy. Questionnaires and medical records were used to obtain maternal characteristics, including maternal age, gestational age, pre-pregnancy BMI, pre-pregnancy weight, smoking status during pregnancy, educational level, and Edinburgh scale. Likewise, infant characteristics such as gender, birth weight, birth length, and birth head circumference were obtained. The study design and information of the COGNIS participants are given in Figure 1.

After inclusion, infants were randomly allocated to the Standard Formula (SF) or the Experimental Formula (EF) group. Later, a third group was added with infants who were Exclusively Breastfed (BF) for at least 2 months to function as the control group.

After infant randomization, the research team decided to withdraw from the trial those infants who met the following criteria: Infants fed with infant formulas unrelated to COGNIS study, breastfed infants with formula intake >25% before 6 months, formula fed infants with human milk intakes higher than 25% beyond the 3rd month of life, any adverse event that could interfere with study follow-up, cow's milk protein allergy/intolerance or lactose intolerance, infant formula intake rejection or neurological disorder.

All infants from COGNIS study were followed up at 2, 3, 6, 12, 18 months of life and 2.5 years of age. During the follow-up visits, and depending on the subject age, different assessment procedures and data collection were carried out; however, those will be explored elsewhere. Nevertheless, for the purpose to obtain a better control of records avoiding possible twists on data due to factors that involved other time lines, such as mixed feeding in the 1st month of life or initiation of complementary feeding at 6 months, this study only takes into account infants from the second visit (3 months of age).

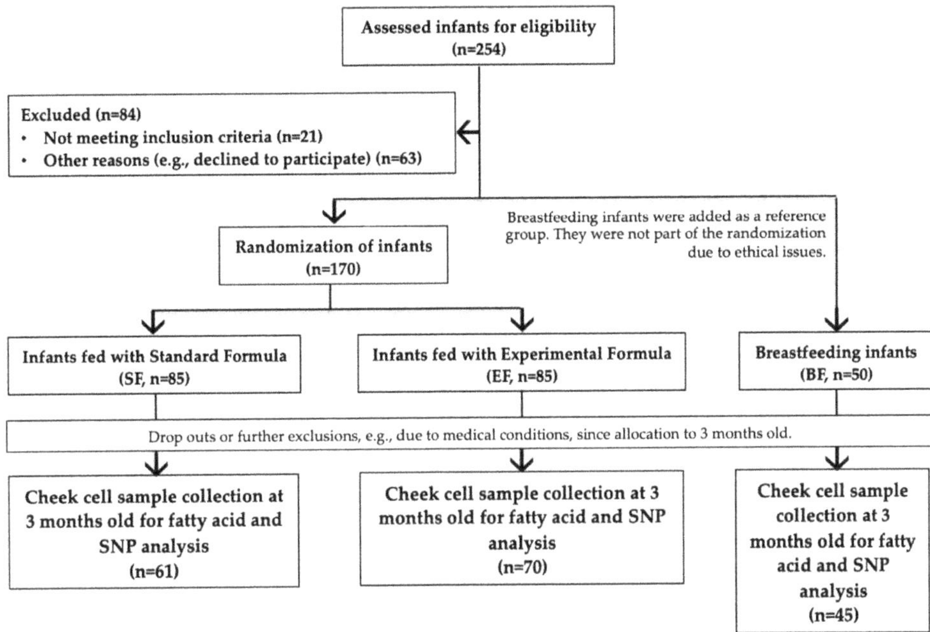

Figure 1. Participants in the COGNIS study and classification following type of feeding.

2.3. Formulas

Infant formulas from SF and EF were based on cow's milk and were provided by ORDESA Laboratories, S.L., Barcelona, Spain. The experimental formula was characterized by the presence of LCPUFAs AA (*Mortierella alpine*) and DHA (fish oil), milk fat globule membrane (MFGM) components {10% of total protein content (wt:wt}, symbiotics, gangliosides, nucleotides and sialic acid (Nutriexpert® factor). Fat blend, OMEGA FATS (palm, palm-kemel, rapeseed, sunflower, oleic sunflower fatty acids) and BETAPOL (Palm, palm-kemel, sunflower and rapeseed oils) were present in both formulas. For more information regarding the lipid profile, consult the Supplementary Materials (Table S2). Additionally, both formulas followed the guidelines of the Committee on Nutrition of the European Society for Pediatric Gastroenterology, Hepatology and Nutrition (ESPGHAN), and the international and national recommendations for the composition of infant formulas. Nutritional composition of infant formulas is shown in Table 1.

Table 1. Standard and Experimental Infant Formula Nutrition Facts per 100 mL.

	Standard Formula	Experimental Formula
	100 mL (13.5%)	100 mL (13.5%)
Energy (kcal/kJ)	69/288	68/285
Proteins * (g)	1.35	1.35
Casein/whey (%)	40/60	40/60
Carbohydrates (g)	7.97	7.56
Lactose (g)	7.17	6.82
Maltodextrin (g)	0.8	0.7
Fat # (g)	3.5	3.5
Linoleic acid (LA, mg)	579	569
α-Linolenic acid (ALA, mg)	49	49
Arachidonic acid (AA, mg)	-	15.8
Docosahexaenoic acid (DHA, mg)	-	11.2

Powder diluted in water (13.5%) * With alpha-lactoalbumin (15% of total protein) and with Immunoglobulins.
With 22% of palmitic acid in beta position.

2.4. Cheek Cell Sample Collection

Cheek cell samples were collected at 3 months of age to analyze FAs and genotype FADS SNPs. Samples were collected 1 h after feeding by scraping the inside of the cheeks with a Rovers® EndoCervex-Brush®. The tip of the brush was transferred and jolted in a cryotube with distilled water, shaking the tip before removing the brush. After centrifugation, the supernatant was carefully discarded. The cell pellets were stored at $-80\,°C$ until analysis.

2.5. Cheek Cell Fatty Acid Analysis

A modified version of the method described by de la Garza et al. [30] was used to analyze FAs from the glycerophospholipid fraction. Methanol with butylated hydroxytoluene (BHT) was used for lipid extraction. FA reactions with sodium methylate in methanol (25 wt% in methanol) and boron trifluoride methanol solution (14% v/v) were used to obtain FA methyl esters (FAMEs). Next, the FAs were separated by rapid gas chromatography following the method developed by Bondia et al. [31]. The system consisted of a Shimadzu GC-2010 gas chromatograph (Kyoto, Japan) equipped with a "split-splitless" injector, an automatic injector with AOC-20i-AOC-20s sampler and a flame ionization detector (FID). The separation of the methyl esters from the FAs was carried out with a fast capillary column of fused silica VF-23ms (10 m × 0,10 mm internal diameter, 0,10 µm film thickness) coated with a stationary phase 100% cyanopropyl-phenyl-methyl-polysiloxane of varian (Palo Alto, CA, USA). The methyl esters of the FAs were identified by comparison with the retention times of standards, FAME-37 and PUFA-2 animal. Quantification was done by normalization, expressing the results in relative amounts (percentage). Enzyme activities were estimated as product:precursor indexes of individual FAs as follows: GLA:LA and DGLA:LA indexes for D6D enzyme activity, and the AA:DGLA index for D5D enzyme activity. Additionally, AA:LA and eicosapentaenoic acid (EPA):ALA indexes were analyzed.

2.6. SNP Selection and Genotyping

SNPs within the FADS gene were selected if they were documented in previous studies for comparison purposes [19,27,32–38] and if their minor allele frequency (MAF) was higher than 10%. DNA material was extracted from infant cheek cells. FADS1 (rs174537, rs174545, rs174546, rs174548, and rs174553) and FADS2 (rs1535, rs174570, and rs2072114) SNPs were genotyped from 2.5 µl of DNA mixed with 2.5 µL of 2X TaqMan® OpenArray® Genotyping Master Mix. Analysis was then performed with 4 µL of the mixture in a microplate using the TaqMan® OpenArray® genotyping technology. Analyses were carried out at the *Autonomous University of Barcelona* (UAB) using the QuantStudio 12 k Flex® instrument (ThermoFisher) and the corresponding OpenArray® SNP Genotyping Analysis software.

2.7. Statistical Analysis

SPSS statistical software package for Windows (version 20.0; SPSS Inc., Chicago, IL, USA) was used to perform the statistical analyses. Data were tested for normality using the Kolmogorov-Smirnov test and non-normal data were log transformed. This exploratory study evaluated the associations between SNPs and PUFAs within the study groups using a linear regression analysis. We decided to analyze each SNP individually to provide more evidence about their effects on LCPUFAs levels in the first stage of life, given the lack of literacy in this period of life. Heterozygotes and minor allele homozygotes were analyzed together as one group to improve sample size. However, this codification implies an additive and dominant model. SNPs were studied as a numeric variable by coding them according to the minor allele count; 0 for major homozygotes and 1 for heterozygotes and minor allele homozygotes. We also tested the analyses with the three allele groups and confirmed that the results showed the same tendency. The Hardy-Weinberg equilibrium and genotype distribution were analyzed with the x^2-test (Supplementary Materials Table S1). We used a multivariate general linear model (GLM) to compare FA levels (mean ± standard deviation) between the study groups and according to FADS genotype. FAs were expressed as the percentage of total FAs. The analyses were corrected for potential confounders such as maternal characteristics (pre-pregnancy body mass index, age, education and smoking status), and gender of the child. The *p*-value cut-off has been reconsidered and changed according to Bonferroni correction (0.05/8 SNPs × 3 groups = 24) and assuming a moderate correlation of 30% between SNPs. The significance cut-off values resulted at <0.005 and this has been applied to each trait.

3. Results

3.1. Sample Characteristics

The population has the characteristics shown in Table 2. Mothers from the BF group were more likely to have a higher level of education and were older than mothers from the EF group. The EF group had more male infants than the BF group. No differences were observed in infant anthropometric data.

Table 2. Characteristics of the population.

Characteristics	SF		EF		BF		*p*
	N	Mean ± SD	*N*	Mean ± SD	*N*	Mean ± SD	
Maternal characteristics							
Age (years)	61	30.31 [ab] ± 6.53	70	29.97 ± 6.22 [a]	45	33.24 ± 5.39 [b]	**0.014**
Gestational age (months)	61	39.52 ± 1.29	70	39.26 ± 1.44	45	39.4 ± 1.3	0.55
Pre-pregnancy weight (kg)	56	65.43 ± 12.92	63	64.26 ± 12.52	43	65.66 ± 10.4	0.81
Pre-pregnancy BMI (kg/m^2) (%)							0.76
Underweight	3	5.45	8	12.7	5	11.63	
Normal weight	29	52.73	29	46.03	23	53.49	
Overweight	14	25.45	15	23.81	11	25.58	
Obesity	9	16.36	11	17.46	4	9.3	
Education (%)							**<0.001**
Primary	14	22.58	15	21.43	1	2.22	
Secondary	18	29.03	25	35.71	4	8.89	
Professional	12	19.35	16	22.86	13	28.89	
Bachelor degree	18	29.03	14	20	27	60	
Smoking during pregnancy (Yes, %)	10	20.83	10	15.87	2	5.13	0.11
Edinburgh Scale (%)							0.42
No depression	49	79.03	53	76.81	39	86.67	
Probable depression	13	20.97	16	23.19	6	13.33	
Infant characteristics							
Sex, male (%)	37	59.68 [ab]	44	62.86 [a]	18	40.00 [b]	**0.042**
Birth weight (kg)	61	3.34 ± 0.41	70	3.32 ± 0.5	45	3.35 ± 0.42	0.91
Birth length (cm)	61	50.67 ± 2.01	68	50.72 ± 2.1	45	50.62 ± 2.39	0.97
WAZ	61	0.04 ± 0.86	68	0.02 ± 0.97	44	0.13 ± 0.86	0.81
LAZ	61	0.58 ± 1.03	68	0.59 ± 1.08	44	0.71 ± 0.98	0.79
BMIZ	61	−0.39 ± 0.93	68	−0.39 ± 1.04	44	−0.37 ± 0.95	0.99

The presented values are means and proportions. Different superscript letters indicate differences among study groups according to ANOVA and Bonferroni post-hoc test. A chi-square test was applied to qualitative variables. Significance level was established at *p* < 0.05. SF, Standard Formula; EF, Experimental Formula; BF, Breastfeeding; WAZ, weight for age z-score; LAZ, length for age z-score; HAZ, height for age z-score.

3.2. Associations of FADS SNPs with Fatty Acids

Table 3 shows nominal and significant associations between PUFAs and FADS minor alleles after adjusting for maternal age, maternal education, maternal smoking habit, and sex of infant.

The most significant associations (*p* < 0.005) were found in the EF group, where minor allele carriership of rs174537 was negatively associated with LA, AA and DHA (βc −0.376, −0.440, and −0.415, respectively), and FADS minor allele carriership of rs2072114 was negatively associated with AA and the AA:LA index (βc −0.522 and −0.450, respectively).

Within the SF group, nominal negative associations were found after correcting for confounders, while the BF group showed none.

The complete analysis can be found in Supplementary Materials Table S3.

Table 3. Associations between FADS genes and fatty acids levels in infants.

Fatty Acids and Gene	SNP	M/m	Standard Formula (n = 46) (n = 46)				Experimental Formula (n = 56) (n = 56)				Breastfeeding (n = 33) (n = 33)			
			β	P	βc	Pc	β	P	βc	Pc	β	P	βc	Pc
C18:2*n*6 (LA)														
FADS1	rs174537	G/T	−0.035	0.818	−0.132	0.408	−0.361	**0.006**	−0.376	**0.005 ***	−0.027	0.880	−0.094	0.687
FADS1	rs174545	C/G	−0.035	0.818	−0.132	0.408	−0.322	**0.015**	−0.351	**0.008**	0.026	0.888	−0.092	0.688
FADS1	rs174546	C/T	−0.035	0.818	−0.132	0.408	−0.322	**0.015**	−0.351	**0.008**	−0.027	0.880	−0.094	0.687
FADS1	rs174553	A/G	−0.035	0.818	−0.132	0.408	−0.322	**0.015**	−0.351	**0.008**	−0.027	0.880	−0.094	0.687
FADS2	rs1535	A/G	−0.035	0.818	−0.132	0.408	−0.346	**0.008**	−0.357	**0.008**	−0.027	0.880	−0.094	0.687
FADS2	rs174570	C/T	0.155	0.304	0.137	0.406	−0.272	**0.043**	−0.207	0.134	−0.106	0.558	−0.162	0.464
C20:3*n*6 (DGLA)														
FADS2	rs174570	C/T	−0.276	0.063	−0.230	0.167	−0.327	**0.014**	−0.288	**0.034**	−0.123	0.495	−0.046	0.815
FADS2	rs2072114	A/G	−0.056	0.709	−0.079	0.632	−0.368	**0.005 ***	−0.342	**0.014**	−0.090	0.618	−0.145	0.446
C20:4*n*6 (AA)														
FADS1	rs174537	G/T	−0.297	**0.045**	−0.224	0.155	−0.396	**0.002 ***	−0.440	**0.001 ***	0.023	0.897	0.035	0.876
FADS1	rs174545	C/G	−0.297	**0.045**	−0.224	0.155	−0.351	**0.007**	−0.375	**0.006**	0.046	0.803	0.036	0.873
FADS1	rs174546	C/T	−0.297	**0.045**	−0.224	0.155	−0.351	**0.007**	−0.375	**0.006**	0.023	0.897	0.035	0.876
FADS1	rs174548	C/G	−0.118	0.436	−0.073	0.653	−0.360	**0.006**	−0.367	**0.007**	0.052	0.773	0.034	0.878
FADS1	rs174553	A/G	−0.297	**0.045**	−0.224	0.155	−0.351	**0.007**	−0.375	**0.006**	0.023	0.897	0.035	0.876
FADS2	rs1535	A/G	−0.297	**0.045**	−0.224	0.155	−0.340	**0.010**	−0.374	**0.007**	0.023	0.897	0.035	0.876
FADS2	rs174570	C/T	−0.412	**0.004 ***	−0.347	**0.030**	−0.262	0.051	−0.237	0.096	−0.257	0.148	−0.187	0.379
FADS2	rs2072114	A/G	−0.077	0.613	−0.049	0.761	−0.502	**<0.001 ***	−0.522	**<0.001 ***	−0.059	0.746	−0.054	0.797
C22:4*n*6 (AdA)														
FADS1	rs174537	G/T	−0.350	**0.017**	−0.408	**0.010**	−0.346	**0.009**	−0.365	**0.006**	0.015	0.933	−0.042	0.855
FADS1	rs174545	C/G	−0.350	**0.017**	−0.408	**0.010**	−0.333	**0.011**	−0.330	**0.014**	0.005	0.979	−0.042	0.856
FADS1	rs174546	C/T	−0.350	**0.017**	−0.408	**0.010**	−0.333	**0.011**	−0.330	**0.014**	0.015	0.933	−0.042	0.855
FADS1	rs174548	C/G	−0.280	0.059	−0.372	**0.022**	−0.332	**0.012**	−0.317	**0.016**	−0.027	0.883	−0.140	0.542
FADS1	rs174553	A/G	−0.350	**0.017**	−0.408	**0.010**	−0.333	**0.011**	−0.330	**0.014**	0.015	0.933	−0.042	0.855
FADS2	rs1535	A/G	−0.350	**0.017**	−0.408	**0.010**	−0.348	**0.008**	−0.354	**0.009**	0.015	0.933	−0.042	0.855
FADS2	rs174570	C/T	−0.222	0.138	−0.244	0.147	−0.374	**0.004 ***	−0.337	**0.013**	−0.046	0.799	0.029	0.897
FADS2	rs2072114	A/G	0.060	0.693	0.025	0.883	−0.362	**0.006**	−0.302	**0.032**	0.011	0.953	−0.019	0.931
C22:5*n*6 (DPAn6)														
FADS1	rs174545	C/G	−0.111	0.463	−0.054	0.739	−0.255	0.056	−0.288	**0.038**	0.102	0.577	0.068	0.742
FADS1	rs174546	C/T	−0.111	0.463	−0.054	0.739	−0.255	0.056	−0.288	**0.038**	0.100	0.578	0.068	0.741
FADS1	rs174553	A/G	−0.111	0.463	−0.054	0.739	−0.255	0.056	−0.288	**0.038**	0.100	0.578	0.068	0.741
C18:3*n*3 (ALA)														
FADS2	rs174570	C/T	0.303	**0.040**	0.279	0.079	−0.040	0.772	−0.042	0.765	−0.061	0.736	−0.089	0.665
C20:5*n*3 (EPA)														
FADS1	rs174537	G/T	−0.287	0.053	−0.315	0.057	−0.249	0.065	−0.331	**0.017**	0.113	0.530	0.269	0.243
FADS1	rs174545	C/G	−0.287	0.053	−0.315	0.057	−0.225	0.093	0.310	**0.025**	0.162	0.374	0.274	0.236
FADS1	rs174546	C/T	−0.287	0.053	−0.315	0.057	−0.225	0.093	0.310	**0.025**	0.113	0.530	0.269	0.243
FADS1	rs174548	C/G	−0.238	0.112	−0.284	0.093	−0.247	0.064	−0.303	**0.026**	0.158	0.380	0.296	0.198
FADS1	rs174553	A/G	−0.287	0.053	−0.315	0.057	−0.225	0.093	−0.310	**0.025**	0.113	0.530	0.269	0.243
GLA:LA (D6D)														
FADS2	rs174570	C/T	−0.394	**0.007**	−0.338	**0.037**	−0.049	0.722	−0.127	0.361	−0.078	0.670	0.007	0.973
DGLA:LA (D6D)														
FADS1	rs174537	G/T	−0.24	**0.010**	−0.17	0.28	0.06	0.62	−0.01	0.94	0.09	0.62	0.19	0.38
FADS1	rs174545	C/G	−0.24	**0.010**	−0.17	0.28	0.04	0.72	−0.03	0.81	0.07	0.71	0.19	0.40
FADS1	rs174546	C/T	−0.24	**0.010**	−0.17	0.28	0.04	0.72	−0.03	0.81	0.09	0.62	0.19	0.38
FADS1	rs174553	A/G	−0.24	**0.010**	−0.17	0.28	0.04	0.72	−0.03	0.81	0.09	0.62	0.19	0.38
FADS2	rs174570	C/T	−0.40	**0.006**	−0.34	**0.032**	−0.21	0.12	−0.20	0.15	−0.08	0.64	0.01	0.93
FADS2	rs2072114	A/G	−0.18	0.22	−0.14	0.36	−0.29	**0.026**	−0.26	0.06	−0.09	0.59	−0.14	0.47
AA:LA (D6D + D5D)														
FADS1	rs174537	G/T	−0.36	**0.013**	−0.28	0.06	−0.22	0.09	−0.26	0.06	0.04	0.81	0.09	0.66
FADS1	rs174545	C/G	−0.36	**0.013**	−0.28	0.06	−0.20	0.13	−0.21	0.13	0.04	0.83	0.09	0.67
FADS1	rs174546	C/T	−0.36	**0.013**	−0.28	0.06	−0.20	0.13	−0.21	0.13	0.04	0.81	0.09	0.66
FADS1	rs174548	C/G	−0.18	0.24	−0.15	0.36	−0.27	**0.045**	−0.26	0.06	−0.01	0.95	−0.05	0.83
FADS1	rs174553	A/G	−0.36	**0.013**	−0.28	0.06	−0.20	0.13	−0.21	0.13	0.04	0.81	0.09	0.66
FADS2	rs1535	A/G	−0.36	**0.013**	−0.28	0.06	−0.17	0.18	−0.20	0.14	0.04	0.81	0.09	0.66
FADS2	rs174570	C/T	−0.47	**0.001 ***	−0.41	**0.007**	−0.13	0.32	−0.14	0.32	−0.24	0.16	−0.12	0.54
FADS2	rs2072114	A/G	−0.16	0.27	−0.12	0.43	−0.43	**0.001 ***	−0.450	**0.001 ***	−0.07	0.67	−0.05	0.79
AA:DGLA (D5D)														
FADS1	rs174548	C/G	−0.226	0.132	−0.160	0.333	−0.269	**0.043**	−0.285	**0.041**	−0.133	0.462	−0.108	0.636
FADS2	rs1535	A/G	−0.195	0.195	−0.141	0.386	−0.261	**0.050**	−0.283	**0.049**	−0.072	0.691	−0.071	0.757
C22:6*n*3 (DHA)														
FADS1	rs174537	G/T	−0.257	0.085	−0.303	0.054	−0.393	**0.003 ***	−0.415	**0.002 ***	0.057	0.753	0.176	0.432
FADS1	rs174545	C/G	−0.257	0.085	−0.303	0.054	−0.341	**0.010**	−0.339	**0.013**	0.089	0.629	0.177	0.429
FADS1	rs174546	C/T	−0.257	0.085	−0.303	0.054	−0.341	**0.010**	−0.339	**0.013**	0.057	0.753	0.176	0.432
FADS1	rs174548	C/G	−0.144	0.339	−0.211	0.192	−0.338	**0.010**	−0.328	**0.015**	0.101	0.577	0.164	0.463
FADS1	rs174553	A/G	−0.257	0.085	−0.303	0.054	−0.341	**0.010**	−0.339	**0.013**	0.057	0.753	0.176	0.432
FADS2	rs1535	A/G	−0.257	0.085	−0.303	0.054	−0.342	**0.009**	−0.354	**0.010**	0.057	0.753	0.176	0.432
FADS2	rs174570	C/T	−0.233	0.120	−0.258	0.114	−0.265	**0.048**	−0.238	0.092	−0.352	**0.045**	−0.274	0.196
FADS2	rs2072114	A/G	0.071	0.637	0.056	0.732	−0.289	**0.029**	−0.244	0.093	−0.081	0.654	−0.017	0.934

Table 3. *Cont.*

Fatty Acids and Gene	SNP	M/m	Standard Formula (*n* = 46) (*n* = 46)				Experimental Formula (*n* = 56) (*n* = 56)				Breastfeeding (*n* = 33) (*n* = 33)			
			β	*P*	βc	Pc	β	*P*	βc	Pc	β	*P*	βc	Pc
EPA:ALA (D6D + D5D)														
FADS1	rs174537	G/T	−0.37	**0.010**	−0.35	**0.022**	−0.13	0.32	−0.19	0.15	0.08	0.65	−0.04	0.83
FADS1	rs174545	C/G	−0.37	**0.010**	−0.35	**0.022**	−0.08	0.52	−0.14	0.30	0.06	0.71	−0.04	0.84
FADS1	rs174546	C/T	−0.37	**0.010**	−0.35	**0.022**	−0.08	0.52	−0.14	0.30	0.08	0.65	−0.04	0.83
FADS1	rs174553	A/G	−0.37	**0.010**	−0.35	**0.022**	−0.08	0.52	−0.14	0.30	0.08	0.65	−0.04	0.83
FADS2	rs1535	A/G	−0.37	**0.010**	−0.35	**0.022**	−0.07	0.61	−0.13	0.34	0.08	0.65	−0.04	0.83
FADS2	rs174570	C/T	−0.37	**0.010**	−0.37	**0.019**	−0.06	0.61	−0.07	0.58	−0.14	0.43	−0.09	0.65

Associations between SNPs and FAs were determined using linear regression analysis. βc and Pc are values corrected for potential confounders such as maternal age, maternal education, smoking and infant gender. SNPs were coded according to minor allele count and analyzed as a numeric variable. "β" = beta per minor allele standardized per the major allele. *p*-values < 0.05 are highlighted in bold and significant associations that persisted after Bonferroni corrections are additionally denoted by asterisks (* *p* < 0.005). M: Major allele; m: minor allele; SNP, single nucleotide polymorphism; LA: Linoleic Acid; GLA: gamma-linolenic acid; DGLA: dihomo-gamma-linolenic acid; AA: Arachidonic Acid; AdA: adrenic acid; DPA*n*6: docosapentaenoic acid *n*6; ALA: alpha-linolenic Acid; EPA: eicosapentaenoic acid; DHA: docosahexaenoic Acid.

3.3. Fatty Acid Comparison by FADS Genotype among Feeding Practice Groups

LCPUFA levels were statistically different among the feeding practice groups when classifying infants by FADS genotype (Table 4). Among major homozygotes, both the BF and EF groups had a higher AA level than the SF group, but when infants carried minor alleles, those with breastfeeding showed a higher AA level than both the EF and SF groups. DHA levels also exhibited differences. Among major homozygotes, infants in the BF and EF groups showed a higher DHA level than infants in the SF group. On the other hand, among minor allele carriers, the EF group presented a higher DHA level than the SF group, but the BF group had the highest DHA level of all. The complete analysis can be found in Supplementary Materials Table S4.

Table 4. Fatty acids according to infant SNPs and study group.

Fatty Acids and Gene	SNP	M/m	MM							p	Mm + mm							p
			Standard Formula		Experimental Formula		Breastfeeding				Standard Formula		Experimental Formula		Breastfeeding			
			N	Mean ± SD	N	Mean ± SD	N	Mean ± SD			N	Mean ± SD	N	Mean ± SD	N	Mean ± SD		
C20:4n6 (AA)																		
FADS1	rs174537	G/T	30	2.0000 ± 0.64 [a]	41	2.4900 ± 0.53 [b]	18	2.9200 ± 0.79 [b]	<0.001 *		31	1.7300 ± 0.53 [a]	25	2.0000 ± 0.62 [a]	20	2.7300 ± 0.73 [b]	<0.001 *
FADS1	rs174545	C/G	30	2.0000 ± 0.64 [a]	40	2.4900 ± 0.54 [b]	17	2.9000 ± 0.81 [b]	<0.001 *		31	1.7300 ± 0.53 [a]	27	2.0500 ± 0.63 [a]	20	2.7300 ± 0.73 [b]	<0.001 *
FADS1	rs174546	C/T	30	2.0000 ± 0.64 [a]	40	2.4900 ± 0.54 [b]	18	2.9200 ± 0.79 [b]	<0.001 *		31	1.7300 ± 0.53 [a]	27	2.0500 ± 0.63 [a]	20	2.7300 ± 0.73 [b]	<0.001 *
FADS1	rs174548	C/G	33	1.9200 ± 0.63 [a]	40	2.5000 ± 0.54 [b]	21	2.8700 ± 0.78 [b]	<0.001 *		28	1.8000 ± 0.56 [a]	27	2.0500 ± 0.62 [a]	17	2.7500 ± 0.73 [b]	<0.001 *
FADS1	rs174553	A/G	30	2.0000 ± 0.64 [a]	40	2.4900 ± 0.54 [b]	18	2.9200 ± 0.79[b]	<0.001 *		31	1.7300 ± 0.53 [a]	27	2.0500 ± 0.63 [a]	20	2.7300 ± 0.73 [b]	<0.001 *
FADS2	rs1535	A/G	30	2.0000 ± 0.64 [a]	40	2.4900 ± 0.54 [b]	18	2.9200 ± 0.79 [b]	<0.001 *		31	1.7300 ± 0.53 [a]	27	2.0600 ± 0.63 [a]	20	2.7300 ± 0.73 [b]	<0.001 *
FADS2	rs174570	C/T	44	1.9600 ± 0.60 [a]	53	2.4100 ± 0.58 [b]	28	2.9500 ± 0.75 [c]	<0.001 *		17	1.6200 ± 0.52 [a]	13	1.8900 ± 0.57 [ab]	10	2.4300 ± 0.64 [b]	0.006
FADS2	rs2072114	A/G	47	1.8700 ± 0.61 [a]	54	2.4400 ± 0.53 [b]	30	2.8500 ± 0.76 [b]	<0.001 *		14	1.8400 ± 0.56 [a]	13	1.8100 ± 0.69 [a]	8	2.6800 ± 0.75[b]	0.007
C22:6n3 (DHA)																		
FADS1	rs174537	G/T	30	0.5100 ± 0.24 [a]	41	0.9900 ± 0.22 [b]	18	1.2300 ± 0.42 [b]	<0.001 *		31	0.4500 ± 0.25 [a]	25	0.7900 ± 0.27 [b]	20	1.1800 ± 0.46 [c]	<0.001 *
FADS1	rs174545	C/G	30	0.5100 ± 0.24 [a]	40	0.9900 ± 0.22 [b]	17	1.2000 ± 0.43 [b]	<0.001 *		31	0.4500 ± 0.25 [a]	27	0.8200 ± 0.28 [b]	20	1.1800 ± 0.46 [c]	<0.001 *
FADS1	rs174546	C/T	30	0.5100 ± 0.24 [a]	40	0.9900 ± 0.22 [b]	18	1.2300 ± 0.42 [b]	<0.001 *		31	0.4500 ± 0.25 [a]	27	0.8200 ± 0.28 [b]	20	1.1800 ± 0.46 [c]	<0.001 *
FADS1	rs174548	C/G	33	0.4900 ± 0.24 [a]	40	0.9900 ± 0.22 [b]	21	1.2000 ± 0.43[b]	<0.001 *		28	0.4600 ± 0.26 [a]	27	0.8200 ± 0.29 [b]	17	1.2000 ± 0.45 [c]	<0.001 *
FADS1	rs174553	A/G	30	0.5100 ± 0.24 [a]	40	0.9900 ± 0.22 [b]	18	1.2300 ± 0.42 [b]	<0.001 *		31	0.4500 ± 0.25 [a]	27	0.8200 ± 0.28 [b]	20	1.1800 ± 0.46 [c]	<0.001 *
FADS2	rs1535	A/G	30	0.5100 ± 0.24 [a]	40	0.9900 ± 0.22 [b]	18	1.2300 ± 0.42 [b]	<0.001 *		31	0.4500 ± 0.25 [a]	27	0.8200 ± 0.28 [b]	20	1.1800 ± 0.46 [c]	<0.001 *
FADS2	rs174570	C/T	44	0.5000 ± 0.25 [a]	53	0.9500 ± 0.22 [b]	28	1.3000 ± 0.42 [c]	<0.001 *		17	0.4200 ± 0.21 [a]	13	0.7700 ± 0.33 [b]	10	0.9200 ± 0.35 [b]	<0.001 *
FADS2	rs2072114	A/G	47	0.4700 ± 0.24 [a]	54	0.9600 ± 0.25 [b]	30	1.2300 ± 0.45 [b]	<0.001 *		14	0.5100 ± 0.28 [a]	13	0.7600 ± 0.26 [b]	8	1.0900 ± 0.38 [c]	<0.001 *

Values are means ± standard deviations. The general linear model and Bonferroni post-hoc test were applied. The analysis was corrected for potential confounders such as pre-gestational IMC, smoking, education and age of mother and infant gender. *p*-values < 0.05 are highlighted in bold and significant differences that persisted after Bonferroni corrections are additionally denoted by asterisks (* *p* < 0.005). Different superscript letter indicate which groups are different from the others. M: Major allele; m: minor allele; SNP, single nucleotide polymorphism; AA: Arachidonic Acid; DHA: docosahexaenoic Acid.

4. Discussion

The present study analyzed the influence of an infant formula, supplemented with AA + DHA, on LCPUFA levels in infants with different FADS genotype. A number of studies have investigated the association between variants in the FADS gene cluster and FA levels in human tissue [13,14,22–24]; however, little information is available on the neonatal population. Other studies have analyzed the effect of LCPUFA supplementation in early life [9,39–41], but no association with FADS SNPs has been investigated whatsoever. Our study contributes to generating new evidence in this matter.

FADS minor alleles have shown to decrease the desaturase activity [13,16,17,42–46], compromising LCPUFA production. Some authors have demonstrated that FADS SNPs lower proportions of GLA, AA, and EPA, and accumulate the LCPUFA precursors LA and ALA [11,46,47]. Others have demonstrated that FADS minor allele carriers exhibit lower AA:DGLA and EPA:ALA indexes [48,49]. As mentioned before, some studies include indexes of AA to LA, as well as EPA and DHA to ALA in order to use them as markers of the activity of fatty acids desaturation mediated by the enzymes D5D and D6D, respectively; in our study, we use GLA:AA and DGLA:LA to indicate D6D activity, whereas D5D activity is reflected by the ratio of AA to DGLA, and we consider it relevant to record if an infant formula supplemented with LCPUFA can influence the enzymatic activity of infants compared with non-supplemented, to allow a possible comparison for future studies. Moreover, it has been established that the AA is the most severely affected FA [13]. In our study, infants from the BF group carrying FADS minor alleles were not associated with FAs, which suggests that desaturase activity is not affected by FADS genotype when infants are exclusively breastfed.

However, infants carrying FADS minor alleles in the EF and SF groups were associated with decreased D5D and D6D activities, which is in line with the previous studies [46,50–52]. The most affected group was the EF, where infants carrying minor alleles of rs173547 (FADS1) had decreased levels of LA, AA and DHA, and minor allele carriers of rs2072114 (FADS2) were negatively associated with AA and the D5D and D6D desaturases activity (AA:LA index) ($p < 0.005$). Accordingly, the nominal associations ($p < 0.05$) were also inclined to decrease the desaturase activities when infants presented FADS minor alleles. In the EF group this was observed with n6 PUFAs (LA, DGLA, AA, AdA, DPAn6, AA:DGLA), EPA and DHA. Moreover, in the SF group the SNP rs174570 (FADS2) stood out by associating with decreased levels of AA, GLA:LA, DGLA:LA, AA:LA and EPA:ALA. These results suggest that the infants without AA and DHA supplementation were the least affected in terms of FA levels by FADS genetic variants. To make a proper interpretation of the results, it is of interest to take into account the different effect size among the study groups. For instance, the strongest association was found within the EF group, where minor allele carriers of rs2072114 showed the highest negative association with AA (βc -0.522, $p < 0.001$).

In our study, AA and DHA levels, in spite of the genotype, showed a gradient of SF < EF < BF. More specifically, among major homozygotes of FADS SNPs (except rs174570), AA and DHA concentrations were closer between the EF and the BF groups, and higher compared to SF infants, which shows the expected effect of the infant formula supplementation. Nevertheless, when carrying minor alleles, the EF group did not reach similar levels to the BF infants. In summary, Table 4 should be interpreted with caution due to effect size differences among study groups (e.g., the difference between the FA means of EF and BF groups is narrower when infants carry major homozygotes, and these differences increase among minor allele carriers).

This study exposes minor allele carriers as a potential vulnerable group since the same supplementation might not be enough for them, especially in the case of AA that showed the same levels than the SF group, whereas the DHA was at least higher than them.

Increasing the supplemented dose of preformed LCPUFAs may compensate for the effect of FADS minor alleles. According to Miklavcic et al., increasing the formula supplementation to AA 34 mg/100 kcal and DHA 17 mg/100 kcal prevents the reduction in AA by minor alleles [53]. However, in our study, the EF group received a supplementation of 23.2 mg/100 kcal of AA and 15.5 mg/100 kcal of DHA, which could be a reason for their suggested high SNP influence. As for the SF infants, their

AA and DHA levels were almost unaffected by minor alleles. Since previous evidence suggested that high supplementation levels of AA and DHA could lessen the influence of FADS SNPs on LCPUFA levels, we somewhat expected a gradient effect of SNPs of SF > EF > BF. However, this expectation was not determined since the SF group was not supplemented at all. The gradient resulted in EF > SF > BF. It is interesting how the LCPUFA levels of the SF group were less influenced by the FADS SNPs than the EF group. We could speculate some kind of protection effect (e.g., a more effective synthesis in the absence of preformed AA and DHA intake [54]) against FADS SNPs when infants have low LCPUFA levels. It is important to remember that, even though the EF group was more influenced by FADS SNPs than the SF group, infants from the EF group still presented higher AA and DHA levels. Corresponding to the theory of the higher the LCPUFA supply the less SNP influence, AA and DHA levels of BF infants were not perceptibly affected by genotype, possibly because they had the highest LCPUFA concentrations related to the high content in breast milk. One should consider that mean fat content of human milk may vary considerably between individuals as well as between study populations from affluent or developing countries [55]. Maternal factors, such as diet and weight gain during pregnancy, and sampling procedures have distinct impact on fat levels [56]. Additionally, DHA levels in breast milk are quite sensitive to maternal diet [4] and maternal FADS genotypes are associated with breast-milk AA concentration and this might therefore influence the supply of breast milk FAs [42].

Several authors have demonstrated that FADS SNPs affect FA levels and are associated with health conditions [13,39,57–62]. The major contribution of our study is providing evidence in infants below 6 months of age, identifying a potential vulnerable group of infants who could benefit from more personalized nutrition, and contributing to defining the ideal AA and DHA supplementation of infant formulas. This line of research merits further attention because early life nutrient exposure can program future health. We acknowledge some limitations, such as the lack of information on maternal dietary intake and supplementation, and the relatively small sample size of our population. We should also take into consideration the estimation of effect size to interpret the influence of SNPs on LCPUFA levels according to the study group. Now that we have evidence for each individual SNP, it is of interest to perform a FADS haplotype analysis to compare the results. The strengths of our study include the double-blinded cohort study design and the participation of three infant groups with different feeding practices. It is important to mention that the infant formulas were created previous to the implementation of the Commission Delegated Regulation (EU) 2016/127 [8] when DHA supplementation was not mandatory.

5. Conclusions

In conclusion, formula fed infants with FADS minor alleles, especially those with AA and DHA supplementation, were associated with decreased desaturase activity and lower AA and DHA levels. Breastfed infants were not affected, possibly due to the high LCPUFA content in breast milk. The AA and DHA supplementation of the infant formula provided the infants carrying major allele homozygotes with closer levels to those obtained with breastfeeding. This exposes minor alleles as a potential factor of vulnerability since the same supplementation might not be enough for them. Considering infant FADS genotype to meet the individual needs could contribute to narrow the gap of AA and DHA concentrations between breastfed and formula fed infants. A new LCPUFA supplementation dose should be explored to determine if an increased supplementation of AA and DHA might prevent the reduction of these FAs observed in the presence of FADS minor alleles. However, when breastfeeding is not possible, supplemented formulas should be considered as the second choice since they provide better AA and DHA concentrations compared to infants without supplementation.

Supplementary Materials: The following are available online at http://www.mdpi.com/2072-6643/11/3/602/s1, Table S1: The genetic variants studied within the FADS genes, Table S2: Fatty acid content in infant formulas, Table S3: Associations between FADS genes and fatty acid levels in infants, Table S4: Fatty acids and enzymatic indexes according to infant SNPs and study group.

Author Contributions: Cristina Campoy designed and coordinated the project; A.N., F.H. and E.D. performed the follow-up of the children, collected data and prepared the samples for later analysis; I.S.L. and A.d.l.G.P. performed the experimental procedures; M.C.L.-S., A.I.C., M.R.-P., I.S.L., A.M.C.T. and A.d.l.G.P. analyzed the data and designed the manuscript; A.M.C.T., I.S.L. and A.d.l.G.P. conducted the statistical analysis, interpretation of results and drafting of the manuscript. All authors read and approved the final manuscript.

Funding: This research was funded by ORDESA Laboratories, S.L., Spanish Ministry of Economy, Industry and Competitiveness, NEOBEFOOD Project (2010–2013) and SMARTFOODS Project (2014–2018)—CIEN Strategy (Ministry of Innovation and Science-CDTI) through 2 different contracts established between Ordesa Laboratories and the University of Granada General Foundation (ref. n° 3349 and n° 4003, respectively) and between Ordesa Laboratories and the Bosch Gimpera Foundation/University of Barcelona (ref. n° 306811 and 308516). The project was partially funded by EU Project DynaHEALTH (HORIZON 2020-GA No.633595).

Acknowledgments: The authors want to acknowledge the parents and children who participated in the study and also Ordesa laboratories, S.L. Barcelona, Spain. We are also grateful to personnel, scientists, staff and all people involved in the COGNIS team who have made this research possible.

Conflicts of Interest: Maria Rodríguez-Palmero is an employee of Ordesa Laboratories, S.L. She contributed to the manuscript design, analyzed the results and reviewed the final draft. The funders had no other role in the design of the study; in the collection, analyses, or interpretation of data; in the writing of the manuscript, or in the decision to publish the results.

References

1. Abedi, E.; Sahari, M.A. Long-chain polyunsaturated fatty acid sources and evaluation of their nutritional and functional properties. *Food Sci. Nutr.* **2014**, *2*, 443–463. [CrossRef]

2. Richard, C.; Lewis, E.D.; Field, C.J. Evidence for the essentiality of arachidonic and docosahexaenoic acid in the postnatal maternal and infant diet for the development of the infant's immune system early in life. *Appl. Physiol. Nutr. Metab.* **2016**, *41*, 461–475. [CrossRef] [PubMed]

3. Koletzko, B. Human milk lipids. *Ann. Nutr. Metab.* **2017**, *69*, 28–40. [CrossRef] [PubMed]

4. Grote, V.; Verduci, E.; Scaglioni, S.; Vecchi, F.; Contarini, G.; Giovannini, M.; Koletzko, B.; Agostoni, C. Breast milk composition and infant nutrient intakes during the first 12 months of life. *Eur. J. Clin. Nutr.* **2015**, *70*, 1–7. [CrossRef]

5. World Health Organization How to Prepare Powdered Infant Formula in Care Settings Formula Is Not Sterile. Available online: http://www.who.int/foodsafety/publications/micro/PIF_Care_en.pdf (accessed on 20 August 2008).

6. Brenna, J.T.; Varamini, B.; Jensen, R.G.; Diersen-schade, D.A.; Boettcher, J.A., Arterburn, L.M. Docosahexaenoic and arachidonic acid concentrations in humanbreast milk worldwide. *Am. J. Clin. Nutr.* **2007**, *85*, 1457–1464. [CrossRef]

7. Scientific opinion on the essential composition of infant and follow-on formulae. *EFSA J.* **2014**, *1212*. [CrossRef]

8. European Union Law Reglamento Delegado (UE) 2016/127 de la Comisión, de 25 de Septiembre de 2015, que Complementa el Reglamento (UE) n° 609/2013 del Parlamento Europeo y del Consejo en lo que Respecta a Los Requisitos Específicos de Composición e Información Aplicables a lo. Available online: https://eur-lex.europa.eu/legal-content/es/TXT/?uri=CELEX:32016R0127 (accesed on 11 March 2019).

9. Liao, K.; Mccandliss, B.D.; Carlson, S.E.; Colombo, J.; Shaddy, D.J.; Kerling, E.H.; Lepping, R.J.; Sittiprapaporn, W.; Cheatham, C.L.; Gustafson, K.M. Event-related potential differences in children supplemented with long-chain polyunsaturated fatty acids during infancy. *Dev. Sci.* **2016**, 1–16. [CrossRef]

10. Birch, E.E.; Garfield, S.; Hoffman, D.R.; Uauy, R.; Birch, D.G. A randomized controlled trial of early dietary supply of long-chain polyunsaturated fatty acids and mental development in term infants. *Dev. Med. Child Neurol.* **2000**, *42*, 174–181. [CrossRef]

11. CI, J.; Kiliaan, A.J. Long-chain polyunsaturated fatty acids (LCPUFA) from genesis to senescence: The influence of LCPUFA on neural development, aging, and neurodegeneration. *Prog. Lipid Res.* **2014**, *53*, 1–17. [CrossRef]

12. Demmelmair, H.; Baumheuer, M.; Koletzko, B.; Dokoupil, K.; Kratl, G. Metabolism of U13C-labeled linoleic acid in lactating women. *J. Lipid Res.* **1998**, *39*, 1389–1396. [PubMed]

13. Schaeffer, L.; Gohlke, H.; Müller, M.; Heid, I.M.; Palmer, L.J.; Kompauer, I.; Demmelmair, H.; Illig, T.; Koletzko, B.; Heinrich, J. Common genetic variants of the FADS1 FADS2 gene cluster and their reconstructed haplotypes are associated with the fatty acid composition in phospholipids. *Hum. Mol. Genet.* **2006**, *15*, 1745–1756. [CrossRef] [PubMed]

14. Glaser, C.; Heinrich, J.; Koletzko, B. Role of FADS1 and FADS2 polymorphisms in polyunsaturated fatty acid metabolism. *Metabolism.* **2010**, *59*, 993–999. [CrossRef] [PubMed]

15. Glaser, C.; Lattka, E.; Rzehak, P.; Steer, C.; Koletzko, B. Genetic variation in polyunsaturated fatty acid metabolism and its potential relevance for human development and health. *Matern. Child Nutr.* **2011**, *7*, 27–40. [CrossRef]

16. Lattka, E.; Illig, T.; Koletzko, B.; Heinrich, J. Genetic variants of the FADS1 FADS2 gene cluster as related to essential fatty acid metabolism. *Curr. Opin. Lipidol.* **2010**, *21*, 64–69. [CrossRef] [PubMed]

17. Rzehak, P.; Thijs, C.; Standl, M.; Mommers, M.; Glaser, C.; Jansen, E.; Klopp, N.; Koppelman, G.H.; Singmann, P.; Postma, D.S.; et al. Variants of the FADS1 FADS2 gene cluster, blood levels of polyunsaturated fatty acids and eczema in children within the first 2 years of life. *PLoS ONE* **2010**, *5*, e13261. [CrossRef]

18. Zhang, J.Y.; Qin, X.; Liang, A.; Kim, E.; Lawrence, P.; Park, W.J.; Kothapalli, K.S.D.; Thomas Brenna, J. Fads3 modulates docosahexaenoic acid in liver and brain. *Prostaglandins Leukot Essent Fat. Acids* **2017**, *123*, 25–32. [CrossRef] [PubMed]

19. Fahmida, U.; Htet, M.K.; Adhiyanto, C.; Kolopaking, R.; Yudisti, M.A.; Maududi, A.; Suryandari, D.A.; Dillon, D.; Afman, L.; Müller, M. Genetic variants of FADS gene cluster, plasma LC-PUFA levels and the association with cognitive function of under-two-year-old Sasaknese Indonesian children. *Asia Pac. J. Clin. Nutr.* **2015**, *24*, 323–328. [CrossRef]

20. Morales, E.; Bustamante, M.; Gonzalez, J.R.; Guxens, M.; Torrent, M.; Mendez, M.; Garcia-Esteban, R.; Julvez, J.; Forns, J.; Vrijheid, M.; et al. Genetic variants of the FADS gene cluster and ELOVL gene family, colostrums LC-PUFA levels, breastfeeding, and child cognition. *PLoS ONE* **2011**, *6*, e17181. [CrossRef] [PubMed]

21. Lattka, E.; Illig, T.; Heinrich, J.; Koletzko, B. Do FADS genotypes enhance our knowledge about fatty acid related phenotypes? *Clin. Nutr.* **2010**, *29*, 277–287. [CrossRef] [PubMed]

22. Baylin, A.; Ruiz-narvaez, E.; Kraft, P.; Campos, H. α-Linolenic acid, Δ6-desaturase gene polymorphism, and the risk of nonfatal myocardial infarction. *Am. J. Clin. Nutr.* **2007**, *85*, 554–560. [CrossRef] [PubMed]

23. Moltó-Puigmartí, C.; Plat, J.; Mensink, R.P.; Müller, A.; Jansen, E.; Zeegers, M.P.; Thijs, C. FADS1 FADS2 gene variants modify the association between fish intake and the docosahexaenoic acid proportions in human milk. *Am. J. Clin. Nutr.* **2010**, *91*, 1368–1376. [CrossRef] [PubMed]

24. Gieger, C.; Geistlinger, L.; Altmaier, E.; De Angelis, M.H.; Kronenberg, F.; Meitinger, T.; Mewes, H.W.; Wichmann, H.E.; Weinberger, K.M.; Adamski, J.; et al. Genetics meets metabolomics: A genome-wide association study of metabolite profiles in human serum. *PLoS Genet.* **2008**, *4*, e1000282. [CrossRef] [PubMed]

25. Aulchenko, Y.S.; Ripatti, S.; Lindqvist, I.; Boomsma, D.; Iris, M. Europe PMC funders group loci influencing lipid levels and coronary heart disease risk in 16 european population cohorts. *Nat Genet.* **2009**, *41*, 47–55. [CrossRef] [PubMed]

26. Jensen, H.A.R.; Harsløf, L.B.S.; Nielsen, M.S.; Christensen, L.B.; Ritz, C.; Michaelsen, K.F.; Vogel, U.; Lauritzen, L. FADS single-nucleotide polymorphisms are associated with behavioral outcomes in children, and the effect varies between sexes and is dependent on PPAR genotype. *Am. J. Clin. Nutr.* **2014**, *100*, 826–832. [CrossRef] [PubMed]

27. Molto-Puigmarti, C.; Jansen, E.; Heinrich, J.; Standl, M.; Mensink, R.P.; Plat, J.; Penders, J.; Mommers, M.; Koppelman, G.H.; Postma, D.S.; et al. Genetic variation in FADS genes and plasma cholesterol levels in 2-year-old infants: KOALA birth cohort study. *PLoS ONE* **2013**, *8*, e61671. [CrossRef]

28. Standl, M.; Sausenthaler, S.; Lattka, E.; Koletzko, S.; Bauer, C.P.; Wichmann, H.E.; von Berg, A.; Berdel, D.; Krämer, U.; Schaaf, B.; et al. FADS gene variants modulate the effect of dietary fatty acid intake on allergic diseases in children. *Clin. Exp. Allergy* **2011**, *41*, 1757–1766. [CrossRef] [PubMed]

29. Lepping, R.J.; Honea, R.A.; Martin, L.E.; Liao, K.; Choi, I.-Y.; Lee, P.; Papa, V.B.; Brooks, W.M.; Shaddy, D.J.; Carlson, S.E.; et al. Long-chain polyunsaturated fatty acid supplementation in the first year of life affects brain function, structure, and metabolism at age nine years. *Dev. Psychobiol.* **2018**, 1–12. [CrossRef]

30. de la Garza Puentes, A.; Montes Goyanes, R.; Chisaguano Tonato, A.M.; Castellote, A.I.; Moreno-Torres, R.; Campoy Folgoso, C.; López-Sabater, M.C. Evaluation of less invasive methods to assess fatty acids from phospholipid fraction: Cheek cell and capillary blood sampling. *Int. J. Food Sci. Nutr.* **2015**, *66*, 936–942. [CrossRef] [PubMed]

31. Bondia, E.M.; Castellote, A.I.; Lopez, M.; Rivero, M. Determination of plasma fatty acid composition gas chromatography in neonates by gas chromatography. *J. Chromatogr. B* **1994**, *4347*, 369–374. [CrossRef]

32. Hong, S.H.; Kwak, J.H. Association of polymorphisms in FADS gene with age-related changes in serum phospholipid polyunsaturated fatty acids and oxidative stress markers in middle-aged nonobese men. *Clin. Interv. Aging* **2013**, *13*, 585–596.

33. Cormier, H.; Rudkowska, I.; Thifault, E.; Lemieux, S.; Couture, P.; Vohl, M.C. Polymorphisms in Fatty Acid Desaturase (FADS) gene cluster: Effects on glycemic controls following an omega-3 Polyunsaturated Fatty Acids (PUFA) supplementation. *Genes (Basel)* **2013**, *4*, 485–498. [CrossRef] [PubMed]

34. Harsløf, L.B.S.; Larsen, L.H.; Ritz, C.; Hellgren, L.I.; Michaelsen, K.F.; Vogel, U.; Lauritzen, L. FADS genotype and diet are important determinants of DHA status: A cross-sectional study in Danish infants. *Am. J. Clin. Nutr.* **2013**, *97*, 1403–1410. [CrossRef] [PubMed]

35. Groen-blokhuis, M.M.; Franic, S.; Van Beijsterveldt, C.E.M.; De Geus, E.; Bartels, M.; Davies, G.E.; Ehli, E.A.; Xiao, X.; Scheet, P.A.; Althoff, R.; et al. Neuropsychiatric Genetics A Prospective Study of the Effects of Breastfeeding and FADS2 Polymorphisms on Cognition and Hyperactivity/Attention Problems. *Am. J. Med. Genet.* **2013**, *162*, 457–465. [CrossRef] [PubMed]

36. Al-hilal, M.; Alsaleh, A.; Maniou, Z.; Lewis, F.J.; Hall, W.L.; Sanders, T.A.B.; Dell, S.D.O. Genetic variation at the FADS1-FADS2 gene locus infl uences delta-5 desaturase activity and LC-PUFA proportions after fi sh oil supplement. *J. Lipid Res.* **2013**, *54*, 542–551. [CrossRef] [PubMed]

37. Gillingham, L.G.; Harding, S.V.; Rideout, T.C.; Yurkova, N.; Cunnane, S.C.; Eck, P.K.; Jones, P.J.H. Dietary oils and FADS1–FADS2 genetic variants modulate [^{13}C]α-linolenic acid metabolism and plasma fatty acid composition. *Am. J. Clin. Nutr.* **2013**, *97*, 195–207. [CrossRef] [PubMed]

38. One, Y.S.; Ido, T.K.; Inuki, T.A.; Onoda, M.S.; Chi, I.I. Genetic variants of the fatty acid desaturase gene cluster are associated with plasma LDL cholesterol levels in japanese males. *J. Nutr. Sci. Vitaminol.* **2013**, *59*, 325–335.

39. Colombo, J.; Carlson, S.E.; Cheatham, C.L.; Fitzgerald-gustafson, K.M.; Kepler, A.; Doty, T. Long Chain Polyunsaturated Fatty Acid Supplementation in Infancy Reduces Heart Rate and Positively Affects Distribution of Attention. *Pediatr. Res.* **2012**, *70*, 406–410. [CrossRef]

40. Carlson, S.E.; Colombo, J. Docosahexaenoic Acid and Arachidonic Acid Nutrition in Early Development. *Adv. Pediatr.* **2016**, *63*, 453–471. [CrossRef]

41. Uhl, O.; Fleddermann, M.; Hellmuth, C.; Demmelmair, H.; Koletzko, B. Phospholipid species in newborn and 4 month old infants after consumption of different formulas or breast milk. *PLoS ONE* **2016**, *11*, e0162040. [CrossRef]

42. Jakobik, V.; Weck, M.; Weyermann, M.; Grallert, H.; Lattka, E.; Rzehak, P. Genetic variants in the FADS gene cluster are associated with arachidonic acid concentrations of human breast milk at 1.5 and 6 mo postpartum and influence the course of milk dodecanoic, tetracosenoic, and trans-9-octadecenoic acid concentrations. *Am. J. Clin. Nutr.* **2011**, *93*, 382–391. [CrossRef]

43. Wu, Y.; Zeng, L.; Chen, X.; Xu, Y.; Ye, L.; Qin, L.; Chen, L.; Xie, L. Association of the FADS gene cluster with coronary artery disease and plasma lipid concentrations in the northern Chinese Han population. *Prostaglandins Leukot. Essent. Fat. Acids* **2017**, *117*, 11–16. [CrossRef] [PubMed]

44. Hester, A.G.; Murphy, R.C.; Uhlson, C.J.; Ivester, P.; Lee, T.C.; Sergeant, S.; Miller, L.R.; Howard, T.D.; Mathias, R.A.; Chilton, F.H. Relationship between a common variant in the fatty acid desaturase (FADS) cluster and eicosanoid generation in humans. *J. Biol. Chem.* **2014**, *289*, 22482–22489. [CrossRef] [PubMed]

45. Martinelli, N.; Girelli, D.; Malerba, G.; Guarini, P.; Illig, T.; Trabetti, E.; Sandri, M.; Friso, S.; Pizzolo, F.; Schaeffer, L.; et al. FADS genotypes and desaturase activity estimated by the ratio of arachidonic acid to linoleic acid are associated with inflammation and coronary artery disease. *Am. J. Clin. Nutr.* **2008**, *88*, 941–949. [CrossRef]

46. Zietemann, V.; Kröger, J.; Enzenbach, C.; Jansen, E.; Fritsche, A.; Weikert, C.; Boeing, H.; Schulze, M.B. Genetic variation of the FADS1 FADS2 gene cluster and n-6 PUFA composition in erythrocyte membranes in the European Prospective Investigation into Cancer and Nutrition-Potsdam study. *Br. J. Nutr.* **2010**, *104*, 1748–1759. [CrossRef]

47. Meldrum, S.J.; Li, Y.; Zhang, G.; Heaton, A.E.M.; D'Vaz, N.; Manz, J.; Reischl, E.; Koletzko, B.V.; Prescott, S.L.; Simmer, K. Can polymorphisms in the fatty acid desaturase (FADS) gene cluster alter the effects of fish oil supplementation on plasma and erythrocyte fatty acid profiles? An exploratory study. *Eur. J. Nutr.* **2017**, 1–12. [CrossRef] [PubMed]

48. Koletzko, B.; Lattka, E.; Zeilinger, S.; Illig, T.; Steer, C. Genetic variants of the fatty acid desaturase gene cluster predict amounts of red blood cell docosahexaenoic and other polyunsaturated fatty acids in pregnant women: Findings from the Avon Longitudinal Study of Parents and Children. *Am. J. Nutr.* **2011**, *93*, 211–219. [CrossRef] [PubMed]

49. Barman, M.; Nilsson, S.; Naluai, Å.T.; Sandin, A.; Wold, A.E.; Sandberg, A.S. Single nucleotide polymorphisms in the FADS gene cluster but not the ELOVL2 gene are associated with serum polyunsaturated fatty acid composition and development of allergy (in a Swedish birth cohort). *Nutrients* **2015**, *7*, 10100–10115. [CrossRef] [PubMed]

50. Ding, Z.; Liu, G.-L.; Li, X.; Chen, X.-Y.; Wu, Y.-X.; Cui, C.-C.; Zhang, X.; Yang, G.; Xie, L. Association of polyunsaturated fatty acids in breast milk with fatty acid desaturase gene polymorphisms among Chinese lactating mothers. *Prostaglandins Leukot. Essent. Fat. Acids* **2016**, *109*, 66–71. [CrossRef]

51. Hellstrand, S.; Ericson, U.; Gullberg, B.; Hedblad, B.; Orho-melander, M.; Sonestedt, E. Genetic Variation in FADS1 Has Little Effect on the Association between Dietary PUFA Intake and Cardiovascular Disease. *J. Nutr.* **2014**, *144*, 1356–1363. [CrossRef]

52. Roke, K.; Mutch, D.M. The role of FADS1/2 polymorphisms on cardiometabolic markers and fatty acid profiles in young adults consuming fish oil supplements. *Nutrients* **2014**, *6*, 2290–2304. [CrossRef]

53. Miklavcic, J.J.; Larsen, B.M.K.; Mazurak, V.C.; Scalabrin, D.M.F.; MacDonald, I.M.; Shoemaker, G.K.; Casey, L.; Van Aerde, J.E.; Clandinin, M.T. Reduction of Arachidonate Is Associated With Increase in B-Cell Activation Marker in Infants: A Randomized Trial. *J. Pediatr. Gastroenterol. Nutr.* **2017**, *64*, 446–453. [CrossRef] [PubMed]

54. Schwartz, J.; Drossard, C.; Dube, K.; Kannenberg, F.; Kunz, C.; Kalhoff, H.; Kersting, M. Dietary intake and plasma concentrations of PUFA and LC-PUFA in breastfed and formula fed infants under real-life conditions. *Eur. J. Nutr.* **2010**, *49*, 189–195. [CrossRef] [PubMed]

55. Michaelsen, K.F.; Skafte, L.; Badsberg, J.H.; Jørgensen, M. Variation in macronutrients in human bank milk: Influencing factors and implications for human milk banking. *J. Pediatr. Gastroenterol. Nutr.* **1990**, *11*, 229–239. [CrossRef]

56. Stam, J.; Sauer, P.J.; Boehm, G. Can we define an infant's need from the composition of human milk? *Am. J. Clin. Nutr.* **2013**, *98*, 521S–528S. [CrossRef]

57. Moon, Y.-A.; Hammer, R.E.; Horton, J.D. Deletion of ELOVL5 leads to fatty liver through activation of SREBP-1c in mice. *J. Lipid Res.* **2009**, *50*, 412–423. [CrossRef]

58. Hallmann, J.; Kolossa, S.; Gedrich, K.; Celis-Morales, C.; Forster, H.; O'Donovan, C.B.; Woolhead, C.; Macready, A.L.; Fallaize, R.; Marsaux, C.F.M.; et al. Predicting fatty acid profiles in blood based on food intake and the FADS1 rs174546 SNP. *Mol. Nutr. Food Res.* **2015**, *59*, 2565–2573. [CrossRef] [PubMed]

59. Field, C.J.; Van Aerde, J.E.; Robinson, L.E.; Clandinin, M.T. Effect of providing a formula supplemented with long-chain polyunsaturated fatty acids on immunity in full-term neonates. *Br. J. Nutr.* **2008**, *99*, 91–99. [CrossRef] [PubMed]

60. Ward, G.R.; Huang, Y.S.; Bobik, E.; Xing, H.C.; Mutsaers, L.; Auestad, N.; Montalto, M.; Wainwright, P. Long-chain polyunsaturated fatty acid levels in formulae influence deposition of docosahexaenoic acid and arachidonic acid in brain and red blood cells of artificially reared neonatal rats. *J. Nutr.* **1998**, *128*, 2473–2487. [CrossRef] [PubMed]

61. Janssen, C.I.F.; Zerbi, V.; Mutsaers, M.P.C.; de Jong, B.S.W.; Wiesmann, M.; Arnoldussen, I.A.C.; Geenen, B.; Heerschap, A.; Muskiet, F.A.J.; Jouni, Z.E.; et al. Impact of dietary n-3 polyunsaturated fatty acids on cognition, motor skills and hippocampal neurogenesis in developing C57BL/6J mice. *J. Nutr. Biochem.* **2015**, *26*, 24–35. [CrossRef] [PubMed]

62. Lauritzen, L.; Sørensen, L.B.; Harsløf, L.B.; Ritz, C.; Stark, K.D.; Astrup, A.; Dyssegaard, C.B.; Egelund, N.; Michaelsen, K.F.; Damsgaard, C.T. Mendelian randomization shows sex-specific associations between long-chain PUFA-related genotypes and cognitive performance in Danish schoolchildren. *Am. J. Clin. Nutr.* **2017**, *106*, 88–95. [CrossRef]

 *nutrients*

Article

The Effect of Whey and Soy Protein Isolates on Cognitive Function in Older Australians with Low Vitamin B_{12}: A Randomised Controlled Crossover Trial

Ian T. Zajac *, Danielle Herreen, Kathryn Bastiaans, Varinderpal S. Dhillon and Michael Fenech

Health & Biosecurity, Commonwealth Scientific and Industrial Research Organisation (CSIRO), 5000 Adelaide, South Australia, Australia
* Correspondence: ian.zajac@csiro.au; Tel.: +61-8-8303-8875

Received: 29 November 2018; Accepted: 17 December 2018; Published: 21 December 2018

Abstract: Whey protein isolate (WPI) is high in vitamin B_{12} and folate. These and other related markers (holotranscobalamin, methylmalonic acid and homocysteine) have been linked with cognitive health. This study explored the efficacy of WPI for improving cognitive function via delivery of vitamin B_{12}. Moderately vitamin B_{12}-deficient participants aged between 45 and 75 years ($n = 56$) were recruited into this randomised controlled crossover trial. Participants (55% female) consumed 50 g whey (WPI; active) or soy protein isolate (SPI; control) for eight weeks. Following a 16-week washout phase, they consumed the alternative supplement. Consumption of WPI significantly improved active B_{12} and folate status but did not result in direct improvements in cognitive function. However, there was evidence of improvement in reaction time ($p = 0.02$) and reasoning speed ($p = 0.04$) in the SPI condition for females. Additional analyses showed that changes in active B_{12}, HcY and folate measures during WPI treatment correlated with improvements in cognitive function (all $p < 0.05$). Results indicate that WPI itself did not result in improved cognitive function but some evidence of benefit of SPI for females was found. However, consistent with previous research, we present further evidence of a role for active B_{12}, HcY and folate in supporting cognitive improvement in adults with low B vitamin status.

Keywords: whey protein isolate; dairy; cognitive function; cobalamin; vitamin B_{12}

---

1. Introduction

The impact of cognitive decline in a dramatically ageing population is one of the biggest challenges of the future, given its impact on both the individual and society [1]. Not only is deterioration of cognitive functioning amongst the most feared aspects of growing old due to its ability to lower a person's quality of life, it is often accompanied by significant burden [2]. Accordingly, there is a growing need to identify practical methods of delaying cognitive decline to ensure independence in aged individuals.

Epidemiological and observational studies suggest a relationship exists between nutrient deficiencies and cognitive decline in ageing adults [3–5]. In the case of vitamin B_{12}, low levels (together with low folate levels) are a common cause of elevated total homocysteine (HcY), which has been shown to be an independent risk factor for cognitive decline and Alzheimer's disease [6–12]. Vitamin B_{12} deficiency is common in older adults, with prevalence increasing substantially with age [13–18]. In the UK, three surveys reported that around 1 in 20 people aged 65–74 years and at least 1 in 10 of those aged $\geq$75 years were deficient in vitamin B_{12} [18]. Moreover, an Australian survey of persons aged $\geq$50 years reported 23% of participants to have low vitamin B_{12} levels, with prevalence increasing to ~30% for men aged $\geq$70 years and women $\geq$80 years [13].

Over the past few decades, studies have produced inconsistent findings regarding vitamin B_{12} and its association with cognitive impairment in older adults [12]. A recent review of prospective cohort studies [19] found no association between serum vitamin B_{12} concentrations and cognitive decline. However, it was noted that the four studies that used alternate biomarkers of vitamin B_{12}, that is, methylmalonic acid (MMA; a marker of progressed vitamin B_{12} deficiency) and holotranscobalamin (the biologically active fraction of this vitamin and hereafter referred to as active B_{12}) did find evidence of an association with cognitive status [20–23]. Moreover, these markers of vitamin B_{12} deficiency appear to display a stronger relationship with cognition than serum B_{12} [16,24–26].

A 10-year longitudinal cohort study of 1648 individuals [20] found that a doubling of MMA was associated with a 50% faster rate of cognitive decline, with a doubling of active B_{12} associated with a 30% slower rate of decline in individuals aged over 65 years. Furthermore, a 5-year prospective study of 107 community-dwelling participants aged 61–87 years without cognitive impairment at enrolment showed that the decrease in brain volume was greatest among those with lower serum vitamin B_{12} and active B_{12} levels and higher plasma total homocysteine (HcY) and MMA levels at baseline [27]. These results are consistent with findings by Tangney et al. [28], in which high HcY and MMA levels were associated with rate of cognitive decline, along with a decrease in total brain volume.

A recent comprehensive review of the literature on the association of dairy foods and cognitive function in older adults concluded that "low-fat dairy products, when consumed regularly as part of a balanced diet, may have a number of beneficial outcomes for neuro-cognitive health during ageing" and identified the whey fraction as being particularly rich in a range of bioactives, including proline-rich polypeptides, α-lactalbumin and B_{12}, which may support brain health in diverse ways [10]. Whey is a natural by-product of the manufacture of cheese, with 100 g of dried powder providing 2.5 µg of vitamin B_{12}, equivalent to 100% of the Recommended Dietary Allowance (RDA). Acid whey powder contains high amounts of lactose (60–70 g per 100 g), which is problematic given the increased risk of lactose malabsorption in older adults [29]. Whey protein isolate (WPI), on the other hand, has less than 1 g lactose per 100 g and is also richer in vitamin B_{12} (6 µg/100 g). Additionally, WPI has an 80% lower level of sodium and 14-fold higher level of natural folate relative to acid whey powder. While WPI has already been shown to be beneficial for improving cognitive performance in the context of memory tasks in stress-vulnerable subjects [30], the bioavailability and bioefficacy of vitamin B_{12} from WPI remains relatively unexplored, particularly in relation to age-related cognitive decline.

The purpose of the present study was to consider the role that dairy-based products such as WPI may have on nutritional markers associated with age-related cognitive decline. Primary outcomes including the impact of WPI supplementation on vitamin B_{12} status and related factors in individuals with subclinical B_{12} levels at baseline have already been reported [31] but are briefly revisited herein. This study specifically explores the impact of WPI supplementation on cognitive function measured during the intervention. Furthermore, it explores the relationship between changes in various biomarkers of B_{12} status (serum B_{12}, active B_{12}, MMA and HcY) and changes in cognitive function during the course of the active intervention.

2. Materials and Methods

2.1. Screening and Recruitment of Participants

The study was advertised in local newspapers in the Adelaide metropolitan area and on the CSIRO clinic website. Interested participants contacted the clinic to arrange an appointment for screening. Fifty-six eligible participants attended an information session and read the study information sheet before providing written informed consent to participate in the study. Figure 1 provides an overview of screening and recruitment and Table 1 provides an overview of baseline characteristics of study completers. All data collection visits for this study occurred between September 2014 and June 2015.

Inclusion criteria were: healthy subjects aged between 45 and 75 years, not taking vitamin B_{12}/choline/antioxidant vitamins at doses that exceed 25% of the Recommended Dietary Allowance, subclinical vitamin B_{12} deficiency as defined previously [31] (using the following parameters: serum concentration of B_{12} in the range of 100–350 pmol/L, plasma MMA >0.20 μmol/L and serum creatinine concentration of 120 μmol/L or less) and willing to consume the quantities of WPI or SPI specified for the trial.

Exclusion criteria were: cognitive deficiencies indicated by Mini Mental State Examination Score ≤24, current smokers, people who habitually consume more than two standard alcoholic drinks per day, BMI ≥35 kg/m^2, diagnosed with diabetes and/or lactose intolerance, history of pernicious anaemia or atrophic gastritis and regular users of antacids.

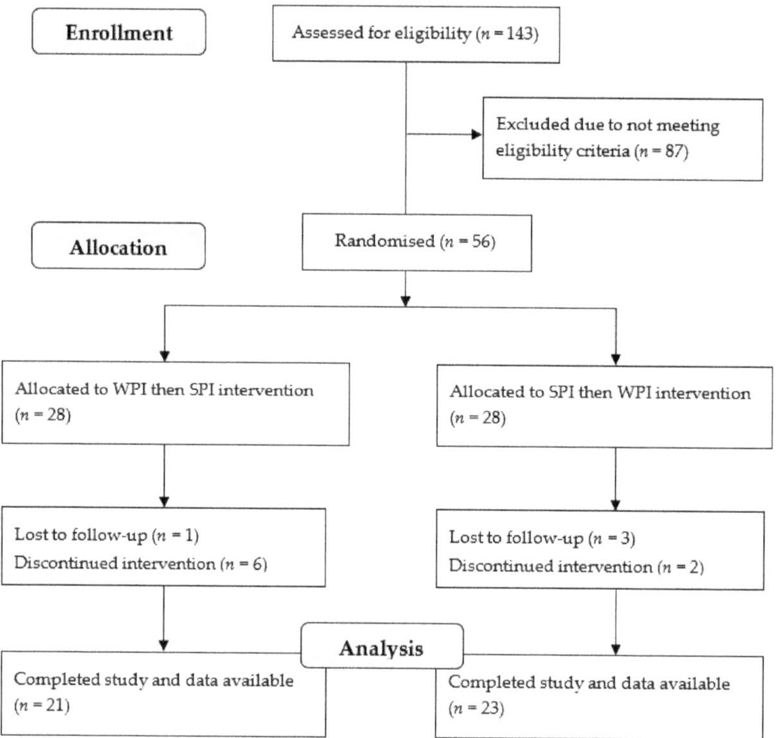

Figure 1. Consort diagram of the intervention trial.

Table 1. Baseline characteristics of participants who completed the study.

Variable	Treatment Order	
	SPI then WPI (*n* = 23)	WPI then SPI (*n* = 21)
% Female (*n*)	43% (10)	52% (11)
Mean Age in years (SD)	61.8 (1.94)	61.1 (1.78)
Mean Body Mass Index (SD)	26.8 (0.76)	26.41 (0.88)

SPI: soy protein isolate; WPI: whey protein isolate.

2.2. Intervention Design

This study was a registered (ACTRN12614000159651) randomised controlled crossover intervention, which was approved by the CSIRO Human Research Ethics Committee. Participants were randomised to daily intake of 50 g WPI or 50 g SPI (control) for eight weeks. Following an intervening

16-week washout phase, participants crossed over to the alternative supplement for the final eight weeks. The randomisation scheme was generated by the clinical trials manager using the online resource website: http://www.randomization.com. Both products were packaged in matching opaque containers with a numeric allocation number to ensure blinding of participants, researchers and clinical trial staff.

Participants were asked to maintain their habitual diet during the intervention trial but refrain from eating foods high in vitamin B_{12} (i.e., foods with vitamin B_{12} levels > 4.9 µg/100 g) and also refrain from taking supplements containing vitamin B_{12}/choline/antioxidants). Dietary restrictions included limiting consumption of vegetables or pulses (2 servings/day), fruits or juices (3 servings/day), black tea or coffee (2 cups/day), chocolate (50 g/day), wine (200 mL/day) and/or beer (375 mL/day) in order to avoid an excessive intake of antioxidants to minimise variation between subjects and groups. Participants completed 3-day food and beverage intake records at baseline and endpoint to monitor their adherence to these restrictions and were supported throughout the study by a dietitian.

The WPI or SPI was consumed daily as 25 g blended in 200 mL fruit juice or water twice a day so that the total daily intake was 50 g WPI or 50 g SPI each day. The WPI and SPI were provided in 25 g sachets (two per participant for each day of the intervention phases). A compliance checklist for use of the WPI and SPI sachets was completed by the participants.

During the washout period, participants were allowed to return to their habitual dietary habits and were required to avoid consumption of supplements containing vitamin B_{12} or foods fortified with vitamin B_{12}. Serum B_{12} was checked mid-way during the washout period. Those exceeding 350 pmol/L serum B_{12} concentration were required to restrict intake of foods rich in B_{12} until they achieved a < 350 pmol/L serum concentration before starting the second treatment phase.

2.3. Nutritional Profile of WPI and SPI

The nutritional composition of WPI and SPI is shown in Table 2. According to the manufacturer's product data, 100 g of WPI and SPI contributed 6 and 0 µg of vitamin B_{12}, respectively. Independent analysis by PathWest Laboratory Medicine WA (Nedlands, Western Australia, Australia) showed that WPI contained 5.64 µg of vitamin B_{12} per 100 g, whereas the amount of vitamin B_{12} in SPI was less than 0.34 µg/100 g.

Table 2. Nutritional composition of WPI and SPI as per manufacturer's product specifications.

Composition	Type	Amount in WPI	Amount in SPI
Protein (%)	Total proteins	93.1 (%) Dry basis	91.0% Dry basis
	β-Lactoglobulin	45%	-
	α-Lactabumin	15%	-
	GMP (glycomacropeptide)	16%	-
	Minor components	3–5%	-
	Immunoglobulins	4%	-
	Bovine serum albumin	1%	-
	Lactoferrin	0.1%	-
Fat (%)	Total fats	1%	2.5%
	Saturated fats	-	0.8%
	Polyunsaturated	-	1.6%
	Monounsaturated	-	0.6%
	Trans fatty acids	-	0.5%

Table 2. *Cont.*

Composition	Type	Amount in WPI	Amount in SPI
Carbohydrate (%)	Total carbohydrates	1.2%	1%
	Lactose	1.2%	-
	Sucrose	-	<1%
Vitamins (per 100 g)	Pantothenic acid	<1.0 mg	-
	Riboflavin (B_2)	0.32 mg	-
	B_6	0.22 mg	0.1 mg
	Thiamine (B_1)	0.12 mg	-
	α-Tocopherol	<0.1 mg	-
	Niacin (B_3)	<0.01 mg	-
	Folacin	458 µg	-
	B_{12}	6.0 µg	0.0 µg
Minerals (per 100 g)	Potassium	1005 mg	1300 mg
	Calcium	307 mg	-
	Phosphorus	210 mg	-
	Sodium	188 mg	800 mg
	Chloride	11 mg	-
Polyphenols	Isoflavones	-	10–30 mg

2.4. Biochemical Outcome Measures

Blood samples were collected at the beginning and end of each of the intervention phases and half-way through the washout phase, that is, at 0, 8, 16, 24 and 32 weeks, within the clinical research unit. Blood samples were used fresh or biobanked frozen at $-80\ ^{\circ}$C depending on the requirements of each assay. The biochemical outcome measures included in the present analyses were: serum B_{12}, active B_{12} (holotranscobalamin), methylmalonic acid (MMA), plasma total homocysteine levels (HcY) and serum folate. Complete descriptions of these measures and associated laboratory processing and results have been described elsewhere [31].

2.5. Tests of Cognitive Function

The CSIRO Cognitive Assessment Battery (C-CAB) comprised of 10 individual cognitive tasks that measure cognitive domains including processing speed, reaction time/attention, reasoning speed, verbal and numeric working memory, immediate and delayed word memory. The tasks utilised obtain latency/response time and accuracy data with performance outcomes for each measure reflecting both of these components. The test battery was programmed to run in the Inquisit version 4 environment [32] and the battery took approximately 45 minutes to complete. Participants completed the C-CAB at each of their clinic visits. The tests used to measure performance on each of the cognitive domains are now described in the order completed.

2.5.1. Immediate Word Memory

Participants were presented with a list of 15 words obtained from the MRC Psycholinguistic Database [33], which included one-to-two syllable words with high concreteness (Mean = 572.81, SD = 35.92) and meaningfulness (Mean = 681.87, SD = 83.31) ratings. Prior to presentation, participants were instructed to remember the words and were informed they will be tested immediately, as well as after approximately 20 minutes. Words were presented in the middle of the computer, one at a time, with a display time of 2-seconds and an inter-stimulus interval (ISI) of 500-millseconds (msec). Parallel word lists were used for alternate testing sessions to remove practice effects. Immediately after presentation, participants were tested using a 30 words list. The list included the original 15 words plus 15 new distractors. Words were shown one-at-a-time in the centre of the computer screen with a 500 msec ISI between the participant's response and subsequent word stimulus. Participants indicated whether the word was from the initial word list or not, by pressing 'Yes' and

'No' response keys. Raw outcome measures included mean response latency—time between stimulus onset and participant's response—and accuracy measures.

2.5.2. Processing Speed

The first task assessing processing speed was adapted from previous research [34]. A code table was presented at the top of the computer screen and comprised nine symbols arranged horizontally, to which nine digits, presented directly beneath them, were paired. For each item, one symbol was presented in the centre of the computer screen and participants responded by left clicking the mouse on its corresponding digit in a 3×3 numerical response grid positioned at the bottom of the screen. The raw outcome measure was the number of items correctly completed in 90 s.

The second task was a computerised string search task adapted from the ETS Kit of Factor Referenced Tests [35]. In this task, participants were required to decide whether two number strings, presented side-by-side, were identical and press the 'Yes' or 'No' key on the keyboard accordingly. The raw outcome measure was the number of items completed correctly in 90 s.

2.5.3. Reaction Time/Attention

Two-choice and four-choice reaction time tasks were used to measure this domain and were adapted from our previous studies [34]. To begin each trial, the participant pressed the 'H' key on a response keypad. This action triggered a cue symbol (+) of approximately 40 px which appeared in the centre of the computer screen together with two or four black bordered empty target squares (100 px) positioned 150 px each side of the cue (left and right for two-choice, with addition of top and bottom square for four-choice). After a variable ISI ranging from 1000 msec to 2500 msec, a black arrow was presented in one of the squares (arrow direction was congruent with location; pointing left, right, up or down). Participants responded by pressing the corresponding response arrow in the response keypad as quickly as possible (response keys were positioned left, right, above and below the 'H' key). Practice phases required participants to complete four and eight trials correctly. Mean response latencies for each task were subsequently estimated over 30 valid trials defined as: 1) response was correct; 2) response latency was not less than 150 ms; and 3) response latency was not greater than mean latency + 3.5 times the intra-individual standard deviation of latencies obtained during the second phase of practice trials in each task.

2.5.4. Verbal Working Memory

This was measured using a letter memory task adapted from previous studies [36]. In this task, participants were initially instructed to memorise a set of five letters presented sequentially on the computer screen for 2000 msec each with an ISI of 500 msec. Following a warning probe of 3000 msec, 31 test trials commenced and participants were required to indicate whether the test letter displayed was in the original learned set or not. In the second phase of this task, participants were required to remember a new set of eight letters which did not include the original five, followed by 31 test trials as per above. Raw outcome measures included mean response latency and accuracy.

2.5.5. Numerical Working Memory

This was measured using a number memory task adapted from previous studies [36] and was operationally equivalent to the verbal working memory task described above. However, in this task, participants were instructed to memorise sets of numbers (either five or eight) and to make yes/no decisions about numbers presented subsequently in the test phase. Raw outcome measures included mean response latency and accuracy.

2.5.6. Delayed Word Memory

Approximately 20 minutes into the cognitive battery, participants were tested again on the 15 words presented at the start of the session. The test consisted of 30 trials involving the original 15 words, plus 15 new distracters not used in the immediate recognition test. All other presentation parameters and scoring parameters were as per the immediate recognition test described above.

2.5.7. Reasoning Speed

Two tasks adapted from the odd-man-out reaction time paradigm were used to assess this domain [37]. The tests used are more complex than simple reaction time because they assess the ability to make reasoned decisions concerning distal relationships between stimuli [38]. In the first version, trials consisted of eight circles arranged horizontally with the total stimulus set being 640 px wide and 84 px high. In each set, three target circles were illuminated red and two targets were closer together in relation to the third target. Participants indicated to which side-left or right-of the two closer targets the odd-man-out (i.e., third target) was located. Participants completed two practice phases consisting of six trials each. The test-phase consisted of 40 trials and raw outcome measures included mean response latency and accuracy.

The second task is procedurally equivalent to the first task. However, in this version, trials consisted of eight mixed shapes (squares, diamonds and triangles) arranged horizontally. The target stimuli were triangles as opposed to illuminated circles and participants indicated to which side—left or right—of the two closer triangles the odd-man-out (i.e., third triangle) is located. Outcomes measures were as per above.

2.6. Sample Size, Data Processing and Analysis

The required sample size for this study was estimated using GPower. Results showed a minimum of $n = 74$ participants were required for a repeated-measures model assuming a small within-between interaction effect ($\eta^2 = 0.02$) with 90% power. For crossover designs, as used herein, this figure is halved, requiring a minimum of $n = 37$ participants. The number of recruited participants who completed this study was $n = 44$, thus satisfying statistical power requirements.

Transformations were undertaken to normalise distributions of cognitive tasks for analysis purposes. In line with previous studies [39], work rates representing the number of items completed per second were calculated. Then, these work rates were adjusted for accuracy (work rate $\times$ % correct). Finally, performance was averaged across tasks measuring the same domain (e.g., reaction time/attention) to arrive at a composite measure.

Cognitive data were analysed at the Intention-to-Treat level and therefore included all participants who commenced the trial (whether or not they withdrew). Change in cognitive function was assessed using linear mixed effects models—capable of handling missing data—and the repeated nature of the measurements were accounted for by permitting random slopes and intercepts for each participant for each treatment. Fixed effects included time, treatment and time by treatment interactions. An unstructured covariance matrix was specified for random effects and all models controlled for participant's sex, age and the order of assigned treatments. The assumption of normally distributed residuals was assessed and was satisfied for all models.

The final analyses assessed the relationship between changes in vitamin B_{12} and related parameters occurring during the WPI intervention period only and corresponding changes in cognitive function using first-order correlations adjusted for sex and age covariates. The calculation of change scores was thus based only on those who attended final follow-up ($n = 44$) and who therefore had complete data.

3. Results

Participants in this sample had a mean age of 60 years (SD = 8.5 years) and were reasonably well balanced in regard to sex (55% female). Table 3 provides average levels of cognitive performance over the course of the study as well as mean change in cognitive function for each domain. Equivalence of the groups at baseline was confirmed via no statistically significant differences between treatment arms for any measure (all $p \geq 0.42$). Descriptive statistics regarding changes in cognition show that it was generally neutral or positive.

Table 3. Means ($\pm$ standard error of the mean) for cognitive variables at baseline, endpoint and for overall change.

Cognitive Domain	Treatment	Baseline		Endpoint		Change	
		Mean	$\pm$ SE	Mean	$\pm$ SE	Mean	$\pm$ SE
Processing Speed	WPI	0.36	0.01	0.37	0.01	0.01	0.00
	SPI	0.37	0.01	0.37	0.01	0.00	0.00
Reasoning Speed	WPI	0.83	0.04	0.86	0.04	0.04	0.02
	SPI	0.79	0.04	0.88	0.04	0.09	0.02
Reaction Time/Attention	WPI	1.67	0.05	1.67	0.05	-0.01	0.02
	SPI	1.66	0.05	1.70	0.05	0.04	0.02
Numeric Working Memory	WPI	1.14	0.04	1.22	0.04	0.07	0.03
	SPI	1.18	0.05	1.22	0.04	0.04	0.02
Verbal Working Memory	WPI	0.83	0.03	0.88	0.03	0.05	0.02
	SPI	0.85	0.04	0.91	0.03	0.06	0.02
Immediate Word Memory	WPI	0.91	0.04	0.91	0.04	0.00	0.03
	SPI	0.93	0.03	0.89	0.04	-0.03	0.03
Delayed Word Memory	WPI	0.82	0.04	0.81	0.04	-0.02	0.03
	SPI	0.78	0.04	0.81	0.04	0.03	0.03

Analysis of data using mixed effects models showed significant overall time effects for reasoning speed, verbal working memory and numeric working memory (all $p \leq 0.001$). In addition, there were significant time by treatment interaction effects observed for reasoning speed ($p = 0.04$) and reaction time ($p = 0.02$). In both cases, performance increased over the duration of the intervention with a significantly greater increase observed during the SPI compared to WPI treatment. Additional three way interactions (time by treatment by sex) showed that these effects were only evident in females during SPI treatment (both $p < 0.05$). As shown in Figure 2, there was generally no change in performance for males during either treatment for either cognitive domain but significant direct changes were observed for females during SPI treatment only.

The results of this intervention on various biochemical scores have been reported in full previously [31]. However, given the presence of prior studies suggesting a relationship between alternate markers of B_{12} status and cognition, additional analyses were conducted in order to examine changes in markers of B_{12} status during the active WPI treatment phase only and their associations with observed changes in cognitive function during the same treatment period. Briefly, the results of the intervention on the parameters of interest are provided in Table 4. WPI treatment lead to significant direct changes in active B_{12} and serum folate, however changes in MMA did not quite reach significance. Correlations between changes in these biochemical measures and cognitive function during both WPI and SPI intervention periods can be seen in Table 5. These correlations were fully adjusted for age and sex covariates. As can be seen, 5/6 significant correlations emerged for the WPI intervention. Increases in in active B_{12} were associated with improvements in processing speed. Similarly, increased serum folate was associated with improvements in delayed word memory. Reductions (i.e., improvements) in homocysteine levels were related to improvements in processing speed, numeric working memory and verbal working memory. In the SPI condition, homocysteine changes were related to reasoning speed with higher levels equating to reduced performance.

We conducted the same analyses using combined B_{12} (cB_{12}), a combined indicator of corrected B_{12} status [31] and the pattern of correlations was the same as for active B_{12}.

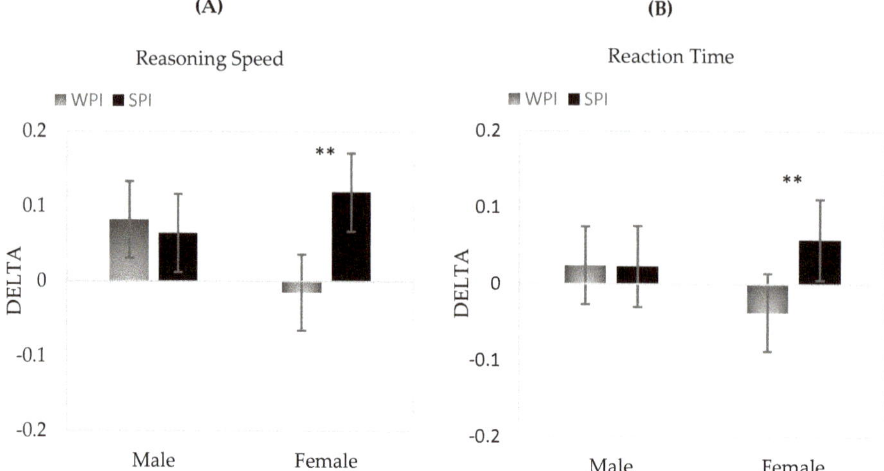

Figure 2. Mean change scores and 95% Confidence Intervals for Treatment × Gender interactions for Delta (**A**) Reasoning Speed and (**B**) Reaction Time. **$p < 0.01$.

Table 4. Baseline scores (Mean ±SE) and percentage change observed due to treatment.

	SPI		WPI		*p*-Value [a]
	Baseline	**% Change**	**Baseline**	**% Change**	
Serum B_{12} (pmol/L)	261.7 ± 8.45	↓ −9.47	260.2 ± 10.59	↓ −4.05	0.14
Active B_{12} (pmol/L)	70.16 ± 3.97	↓ −2.21	68.7 ± 4.13	↑ +19.08	<0.001
MMA (mol/L)	0.26 ± 0.021	↑ +12.24	0.26 ± 0.022	↓ −3.96	0.09
tHcy (mol/L)	11.98 ± 0.53	↑ +3.09	11.89 ± 0.55	↓ −1.78	0.14
Serum folate (nmol/L)	34.78 ± 0.26	↓ −1.05	34.45 ± 1.79	↑ +12.92	<0.01

[a] Comparison of percentage change between treatments. ↑ Increase, ↓ decrease.

Table 5. Correlations between changes in biometric measures and changes in cognitive function during each treatment period; adjusted for sex and age covariates.

		Δ Serum B_{12}	Δ Active B_{12}	Δ MMA	Δ tHcy	Δ Serum Folate
Δ Processing Speed	WPI	0.09	0.29 *	−0.03	−0.33 *	0.22
	SPI	0.24	0.06	−0.01	0.17	0.12
Δ Reasoning Speed	WPI	−0.03	0.01	−0.11	0.21	−0.10
	SPI	0.22	0.18	−0.01	−0.49 **	0.09
Δ Numeric Working Memory	WPI	0.00	0.00	0.18	−0.31 *	0.03
	SPI	0.07	0.11	−0.11	0.03	−0.04
Δ Verbal Working Memory	WPI	0.01	0.10	0.04	−0.24	−0.10
	SPI	−0.20	−0.05	0.21	0.23	0.05
Δ Reaction Time/Attention	WPI	0.03	0.13	−0.08	−0.20	0.25 *
	SPI	−0.09	−0.03	−0.03	−0.06	−0.16
Δ Immediate Word Memory	WPI	0.11	0.07	−0.11	−0.23	0.15
	SPI	0.15	0.18	−0.02	0.24	−0.14
Δ Delayed Word Memory	WPI	0.03	−0.02	0.03	−0.06	0.39 **
	SPI	0.03	0.05	−0.15	−0.03	0.12

MMA = Methylmalonic Acid; tHcy = total homocysteine. * $p < 0.05$, ** $p < 0.01$ (one-tailed).

4. Discussion

This study sought to examine the effects of WPI and SPI on cognitive function and vitamin B_{12} status in low B_{12} individuals. Cognitive function was measured for a variety of important domains using a robust cognitive assessment battery. Overall, there were no effects of WPI on cognitive function, which generally remained stable across visits during this treatment. On the contrary, SPI appeared to have benefits for the cognitive domains of reasoning speed and reaction time in females only. Given the age of participants in this sample (45–75 years) and the presence of this effect true for females only, this finding possibly reflects the impact of soy isoflavones on cognitive function in post-menopausal women [40]. A recent meta-analysis concluded that soy isoflavones exert a positive effect on a broad range of cognitive abilities in females and that this effect is robust in women as young as 60 years [41].

Despite the absence of a main treatment effect for WPI in terms of its ability to improve cognitive performance, positive findings were evident regarding changes in biometric measures of vitamin B_{12} status during WPI treatment and changes in cognitive function. Specifically, increases in active B_{12} and serum folate were associated with corresponding improvements in some cognitive domains. Reductions in homocysteine levels were also consistently related to improvements in cognitive performance, including working memory. Whilst there were no clear treatment effects linking WPI specifically to improved cognitive function, the changes observed in these key biochemical markers and the relation of these to improved cognition were consistent with the broader research literature regarding the role of B_{12}, folate and homocysteine in helping to maintain good cognitive function [12].

The finding of direct associations between changes in biochemical and cognitive measures despite the absence of any effect directly attributable to WPI likely reflects the choice of the control treatment. More specifically, although SPI is a useful comparison given its different nutritional profile to WPI in terms of levels of B-vitamins and folate, the isoflavones present can have an impact on cognitive function, most notably in females [40]. Therefore, both treatments were arguably 'active' in relation to their ability to affect cognitive performance and the absence of any interaction effect favouring WPI does not preclude potential benefits of this for improving cognition as a result of its demonstrated influence on vitamin B-status. Thus, future studies will need to consider their choice of control treatment more carefully if the primary outcome concerns cognitive function.

In addition to using a relatively active control treatment, this study employed a relatively short intervention duration (eight weeks). Prior research has proposed that interventions involving otherwise healthy participants, as used herein, require durations as long as 2-to-5 years in order to demonstrate an impact on cognitive function [42]. Thus, it has been suggested that such studies might benefit from recruiting individuals at-risk of enhanced rates of cognitive decline. Given the relationships observed herein between changes in B_{12} related markers and improved cognitive performance, future studies should consider the potential for dairy-based supplements to slow the rate of decline in individuals with quantifiable Age-Associated Memory Impairment (AAMI). These individuals are particularly interesting given the absence of underlying pathology (e.g., dementia) explaining their decline and the fact that they do not yet meet criteria for MCI [43].

The potential utility of using WPI or SPI for extended periods of time as a means of supplementing diets must be balanced against any potential negative impacts of these. For example, dairy (WPI) has been linked with self-reported gut discomfort and subsequent avoidance of such products [44]. Question marks also remain in relation to soy and its potential impact on endocrinology [45], as well as the presence of contaminants such as aluminium in soy products [46].

In conclusion, although this intervention did not demonstrate a direct effect of WPI supplementation on cognitive function, the results provide some support to the notion of vitamin B_{12} and folate status having a role in supporting cognitive health. Specifically, important vitamin B_{12} and folate related markers were improved as a result of WPI supplementation and these changes were correlated with improvements in cognitive performance. However, the findings from this study are limited to similarly aged adults with subclinical vitamin B_{12} deficiency and further research is needed to explore how these findings may relate to other groups and populations.

Author Contributions: Conceptualisation, I.T.Z., V.S.D. and M.F.; Formal analysis, I.T.Z., D.H. and K.B.; Methodology, I.T.Z.; Project administration, I.T.Z., K.B. and V.S.D.; Supervision, I.T.Z.; Writing—original draft, I.T.Z., D.H. and K.B.; Writing—review & editing, D.H., V.S.D. and M.D. All authors read and approved the final manuscript.

Funding: This research was funded by the Dairy Research Institute, USA (National Dairy Council).

Acknowledgments: We thank Anne McGuffin (Clinical Trials Manager) and Lindy Lawson (Clinical Nurse) for their role in the planning and collection of samples and other data during the intervention trial.

Conflicts of Interest: The authors declare no conflict of interest.

References

1. Williams, K.N.; Kemper, S. Interventions to Reduce Cognitive Decline in Aging. *J. Psychosoc. Nurs. Ment. Health Serv.* **2010**, *48*, 42–51. [CrossRef] [PubMed]
2. Salthouse, T.A. What and when of cognitive aging. *Curr. Dir. Psychol. Sci.* **2004**, *13*, 140–144. [CrossRef]
3. Ogawa, S. Nutritional management of older adults with cognitive decline and dementia. *Geriatr. Gerontol. Int.* **2014**, *14* (Suppl. 2), 17–22. [CrossRef] [PubMed]
4. Morris, M.C. Nutritional determinants of cognitive aging and dementia. *Proc. Nutr. Soc.* **2012**, *71*, 1–13. [CrossRef] [PubMed]
5. Solfrizzi, V.; Panza, F.; Frisardi, V.; Seripa, D.; Logroscino, G.; Imbimbo, B.P.; Pilotto, A. Diet and Alzheimer's disease risk factors or prevention: The current evidence. *Expert Rev. Neurother.* **2011**, *11*, 677–708. [CrossRef] [PubMed]
6. Duthie, S.J.; Whalley, L.J.; Collins, A.R.; Leaper, S.; Berger, K.; Deary, I.J. Homocysteine, B vitamin status, and cognitive function in the elderly. *Am. J. Clin. Nutr.* **2002**, *75*, 908–913. [CrossRef]
7. Riggs, K.M.; Spiro, A., 3rd; Tucker, K.; Rush, D. Relations of vitamin B-12, vitamin B-6, folate, and homocysteine to cognitive performance in the Normative Aging Study. *Am. J. Clin. Nutr.* **1996**, *63*, 306–314. [CrossRef] [PubMed]
8. Jensen, E.; Dehlin, O.; Erfurth, E.M.; Hagberg, B.; Samuelsson, G.; Svensson, T.; Hultberg, B. Plasma homocysteine in 80-year-olds: Relationships to medical, psychological and social variables. *Arch. Gerontol. Geriatr.* **1998**, *26*, 215–226. [CrossRef]
9. Refsum, H.; Nurk, E.; Smith, A.D.; Ueland, P.M.; Gjesdal, C.G.; Bjelland, I.; Tverdal, A.; Tell, G.S.; Nygard, O.; Vollset, S.E. The Hordaland Homocysteine Study: A community-based study of homocysteine, its determinants, and associations with disease. *J. Nutr.* **2006**, *136*, 1731S–1740S. [CrossRef]
10. Camfield, D.A.; Owen, L.; Scholey, A.B.; Pipingas, A.; Stough, C. Dairy constituents and neurocognitive health in ageing. *Br. J. Nutr.* **2011**, *106*, 159–174. [CrossRef]
11. McCracken, C.; Hudson, P.; Ellis, R.; McCaddon, A.; Medical Research Council Cognitive Function and Ageing Study. Methylmalonic acid and cognitive function in the Medical Research Council Cognitive Function and Ageing Study. *Am. J. Clin. Nutr.* **2006**, *84*, 1406–1411. [PubMed]
12. Vogel, T.; Dali-Youcef, N.; Kaltenbach, G.; Andrès, E. Homocysteine, vitamin B12, folate and cognitive functions: A systematic and critical review of the literature. *Int. J. Clin. Pract.* **2009**, *63*, 1061–1067. [CrossRef] [PubMed]
13. Flood, V.M.; Smith, W.T.; Webb, K.L.; Rochtchina, E.; Anderson, V.E.; Mitchell, P. Prevalence of low serum folate and vitamin B12 in an older Australian population. *Aust. N. Z. J. Public Health* **2006**, *30*, 38–41. [CrossRef] [PubMed]
14. Mirkazemi, C.; Peterson, G.M.; Tenni, P.C.; Jackson, S.L. Vitamin B12 deficiency in Australian residential aged care facilities. *J. Nutr. Health Aging* **2012**, *16*, 277–280. [CrossRef] [PubMed]
15. Van Asselt, D.Z.; de Groot, L.C.; van Staveren, W.A.; Blom, H.J.; Wevers, R.A.; Biemond, I.; Hoefnagels, W.H. Role of cobalamin intake and atrophic gastritis in mild cobalamin deficiency in older Dutch subjects. *Am. J. Clin. Nutr.* **1998**, *68*, 328–334. [CrossRef]
16. Lindenbaum, J.; Rosenberg, I.H.; Wilson, P.W.; Stabler, S.P.; Allen, R.H. Prevalence of cobalamin deficiency in the Framingham elderly population. *Am. J. Clin. Nutr.* **1994**, *60*, 2–11. [CrossRef] [PubMed]
17. Green, T.J.; Venn, B.J.; Skeaff, C.M.; Williams, S.M. Serum vitamin B12 concentrations and atrophic gastritis in older New Zealanders. *Eur. J. Clin. Nutr.* **2005**, *59*, 205–210. [CrossRef]

18. Clarke, R.; Grimley Evans, J.; Schneede, J.; Nexo, E.; Bates, C.; Fletcher, A.; Prentice, A.; Johnston, C.; Ueland, P.M.; Refsum, H.; et al. Vitamin B12 and folate deficiency in later life. *Age Ageing* **2004**, *33*, 34–41. [CrossRef]

19. O'Leary, F.; Allman-Farinelli, M.; Samman, S. Vitamin B(1)(2) status, cognitive decline and dementia: A systematic review of prospective cohort studies. *Br. J. Nutr.* **2012**, *108*, 1948–1961. [CrossRef]

20. Clarke, R.; Birks, J.; Nexo, E.; Ueland, P.M.; Schneede, J.; Scott, J.; Molloy, A.; Evans, J.G. Low vitamin B-12 status and risk of cognitive decline in older adults. *Am. J. Clin. Nutr.* **2007**, *86*, 1384–1391. [CrossRef]

21. Tangney, C.C.; Tang, Y.; Evans, D.A.; Morris, M.C. Biochemical indicators of vitamin B12 and folate insufficiency and cognitive decline. *Neurology* **2009**, *72*, 361–367. [CrossRef]

22. Hooshmand, B.; Solomon, A.; Kareholt, I.; Leiviska, J.; Rusanen, M.; Ahtiluoto, S.; Winblad, B.; Laatikainen, T.; Soininen, H.; Kivipelto, M. Homocysteine and holotranscobalamin and the risk of Alzheimer disease: A longitudinal study. *Neurology* **2010**, *75*, 1408–1414. [CrossRef]

23. Kivipelto, M.; Annerbo, S.; Hultdin, J.; Bäckman, L.; Viitanen, M.; Fratiglioni, L.; Lökk, J. Homocysteine and holo-transcobalamin and the risk of dementia and Alzheimers disease: A prospective study. *Eur. J. Neurol.* **2009**, *16*, 808–813. [CrossRef]

24. Clarke, R.; Sherliker, P.; Hin, H.; Nexo, E.; Hvas, A.M.; Schneede, J.; Birks, J.; Ueland, P.M.; Emmens, K.; Scott, J.M.; et al. Detection of vitamin B12 deficiency in older people by measuring vitamin B12 or the active fraction of vitamin B12, holotranscobalamin. *Clin. Chem.* **2007**, *53*, 963–970. [CrossRef] [PubMed]

25. Miller, J.W.; Garrod, M.G.; Rockwood, A.L.; Kushnir, M.M.; Allen, L.H.; Haan, M.N.; Green, R. Measurement of total vitamin B12 and holotranscobalamin, singly and in combination, in screening for metabolic vitamin B12 deficiency. *Clin. Chem.* **2006**, *52*, 278–285. [CrossRef] [PubMed]

26. Hin, H.; Clarke, R.; Sherliker, P.; Atoyebi, W.; Emmens, K.; Birks, J.; Schneede, J.; Ueland, P.M.; Nexo, E.; Scott, J.; et al. Clinical relevance of low serum vitamin B12 concentrations in older people: The Banbury B12 study. *Age Ageing* **2006**, *35*, 416–422. [CrossRef] [PubMed]

27. Vogiatzoglou, A.; Refsum, H.; Johnston, C.; Smith, S.M.; Bradley, K.M.; de Jager, C.; Budge, M.M.; Smith, A.D. Vitamin B12 status and rate of brain volume loss in community-dwelling elderly. *Neurology* **2008**, *71*, 826–832. [CrossRef]

28. Tangney, C.C.; Aggarwal, N.T.; Li, H.; Wilson, R.S.; Decarli, C.; Evans, D.A.; Morris, M.C. Vitamin B12, cognition, and brain MRI measures: A cross-sectional examination. *Neurology* **2011**, *77*, 1276–1282. [CrossRef]

29. Levitt, M.; Wilt, T.; Shaukat, A. Clinical implications of lactose malabsorption versus lactose intolerance. *J. Clin. Gastroenterol.* **2013**, *47*, 471–480. [CrossRef]

30. Markus, C.R.; Olivier, B.; de Haan, E.H. Whey protein rich in alpha-lactalbumin increases the ratio of plasma tryptophan to the sum of the other large neutral amino acids and improves cognitive performance in stress-vulnerable subjects. *Am. J. Clin. Nutr.* **2002**, *75*, 1051–1056. [CrossRef]

31. Dhillon, V.S.; Zabaras, D.; Almond, T.; Cavuoto, P.; James-Martin, G.; Fenech, M. Whey protein Isolate improves vitamin B12 and folate status in elderly Australians with subclinical deficiency of vitamin B12. *Mol. Nutr. Food Res.* **2017**, *61*, 1600915. [CrossRef] [PubMed]

32. *Inquisit 4 Computer Software*; Millisecond Software: Seattle, WA, USA, 2015.

33. Wilson, M. Mrc Psycholinguistic Database—Machine-Usable Dictionary, Version 2.00. *Behav. Res. Meththods Instrum. Comput.* **1988**, *20*, 6–10. [CrossRef]

34. Zajac, I.T.; Burns, N.R.; Nettelbeck, T. Do Purpose-Designed Auditory Tasks Measure General Speediness? *Int. J. Intell. Sci.* **2012**, *2*, 23–31. [CrossRef]

35. Ekstrom, R.B.; French, J.W.; Harman, H.H.; Derman, D. *Kit of Factor-Referenced Cognitive Tests*; Educational Testing Service: Princeton, NJ, USA, 1976.

36. Scholey, A.; Ossoukhova, A.; Owen, L.; Ibarra, A.; Pipingas, A.; He, K.; Roller, M.; Stough, C. Effects of American ginseng (*Panax quinquefolius*) on neurocognitive function: An acute, randomised, double-blind, placebo-controlled, crossover study. *Psychopharmacology* **2010**, *212*, 345–356. [CrossRef] [PubMed]

37. Jensen, A.R. Reaction time and psycyometric g. In *A Model for Intelligence*; Eysenck, H.J., Ed.; Springer: Berlin, Germany, 1982; pp. 93–132.

38. Diascro, M.N.; Brody, N. Odd-Man-out and Intelligence. *Intelligence* **1994**, *19*, 79–92. [CrossRef]

39. Danthiir, V.; Wilhelm, O.; Schulze, R.; Roberts, R.D. Factor structure and validity of paper-and-pencil measures of mental speed: Evidence for a higher-order model? *Intelligence* **2005**, *33*, 491–514. [CrossRef]

40. Kritz-Silverstein, D.; Von Muhlen, D.; Barrett-Connor, E.; Mathias, B. Isoflavones and cognitive function in older women: The SOy and Postmenopausal Health in AGing (SOPHIA) Study. *Menopause* **2003**, *10*, 196–202. [CrossRef] [PubMed]

41. Cheng, P.-F.; Chen, J.-J.; Zhou, X.-Y.; Ren, Y.-F.; Huang, W.; Zhou, J.-J.; Xie, P. Do soy isoflavones improve cognitive function in postmenopausal women? A meta-analysis. *Menopause* **2015**, *22*, 198–206. [CrossRef] [PubMed]

42. Smith, A.D.; Smith, S.M.; de Jager, C.A.; Whitbread, P.; Johnston, C.; Agacinski, G.; Oulhaj, A.; Bradley, K.M.; Jacoby, R.; Refsum, H. Homocysteine-Lowering by B Vitamins Slows the Rate of Accelerated Brain Atrophy in Mild Cognitive Impairment: A Randomized Controlled Trial. *PLoS ONE* **2010**, *5*, e12244. [CrossRef] [PubMed]

43. Lee, G.; Elashoff, D.; Di, L.; Teng, E.; Melchor, S.; Kim, J.; Lu, P.-H. Age-associated memory impairment increases risk of conversion to MCI and dementia. *Alzheimer's Dement.* **2012**, *8*, P544. [CrossRef]

44. Yantcheva, B.; Golley, S.; Topping, D.; Mohr, P. Food avoidance in an Australian adult population sample: The case of dairy products. *Public Health Nutr.* **2016**, *19*, 1616–1623. [CrossRef] [PubMed]

45. Jargin, S.V. Soy and phytoestrogens: Possible side effects. *Germ. Med. Sci.* **2014**, *12*. [CrossRef]

46. Chuchu, N.; Patel, B.; Sebastian, B.; Exley, C. The aluminium content of infant formulas remains too high. *BMC Pediatr.* **2013**, *13*, 162. [CrossRef] [PubMed]

 *nutrients*

Article

Associations of Lifestyle Behaviour and Healthy Ageing in Five Latin American and the Caribbean Countries—A 10/66 Population-Based Cohort Study

Christina Daskalopoulou [1,*], Artemis Koukounari [2], José Luis Ayuso-Mateos [3,4,5], Martin Prince [1] and A. Matthew Prina [1]

[1] Department of Health Service and Population Research, Institute of Psychiatry, Psychology and Neuroscience, King's College London, London SE5 8AF, UK; martin.prince@kcl.ac.uk (M.P.); matthew.prina@kcl.ac.uk (A.M.P.)

[2] Department of Infectious Disease Epidemiology, London School of Hygiene & Tropical Medicine, Faculty of Epidemiology and Population Health, London WC1E 7HT, UK; Artemis.Koukounari@lshtm.ac.uk

[3] Department of Psychiatry, Universidad Autónoma de Madrid, 28029 Madrid, Spain; joseluis.ayuso@uam.es

[4] Instituto de Salud Carlos III, Centro de Investigación Biomedica en Red de Salud Mental (CIBERSAM), 28029 Madrid, Spain

[5] Department of Psychiatry, Hospital Universitario de La Princesa, Instituto de Investigación Sanitaria Princesa (IIS-Princesa), 28006 Madrid, Spain

* Correspondence: christina.daskalopoulou@kcl.ac.uk; Tel.: +44-20-7848-0714

Received: 1 October 2018; Accepted: 25 October 2018; Published: 30 October 2018

Abstract: Latin American and the Caribbean countries exhibit high life expectancy and projections show that they will experience the fastest growth of older people in the following years. As people live longer, it is important to maximise the opportunity to age healthily. We aimed to examine the associations of lifestyle behaviours with healthy ageing in Cuba, Dominican Republic, Peru, Mexico and Puerto Rico, part of the 10/66 study. Residents 65 years old and over (n = 10,900) were interviewed between 2003 and 2010. In the baseline survey, we measured four healthy behaviours: Physical activity, non-smoking, moderate drinking and fruits or vegetables consumption. Healthy ageing was conceptualised within the functional ability framework over a median of 4 years follow-up. Logistic models were calculated per country and then pooled together with fixed-effects meta-analysis. People engaging in physical activity and consuming fruits or vegetables had increased odds of healthy ageing in the follow-up (OR: 2.59, 95% CI: 2.20–3.03; OR: 1.24, 95% CI: 1.06–1.44, respectively). Compared with participants engaging in none or one healthy behaviour, the ORs of participants engaging in two, three or four healthy behaviours increased in a linear way (OR: 1.60, 95% CI: 1.40–1.84; OR: 2.29, 95% CI: 1.94–2.69; OR: 2.46, 95% CI: 1.54–3.92, respectively). Our findings highlight the importance of awareness of a healthy lifestyle behaviour among older people.

Keywords: older adults; healthy ageing; nutrition; physical activity; lifestyle behaviour

---

1. Introduction

The world population is ageing and by 2050 there will be more older than younger people [1]. Even though population ageing is a human achievement, there are still many health policy challenges to be addressed [2]. Increases in life expectancy have not been followed by relevant increases in disease-free years, meaning that people are living longer but are spending more years with illness and disability [3]. Living longer but in ill health could result in severe demand of public and health care resources [4]. As people experience ageing with great heterogeneity in their health pathways [5], it is crucial to identify those behaviours that are associated with a slower and healthier ageing process.

Until recently, healthy ageing research has been complicated by a lack of common definition and measurement of it [6,7]. However, the latest report on health and ageing from the World Health Organization (WHO) provided a common framework to conceptualise healthy ageing; the functional ability framework. Healthy ageing is now defined as the process of developing and maintaining the functional ability that enables well-being in older age [8]. Healthy lifestyle behaviours (for instance, never smoking, moderate alcohol consumption, physical activity and daily consumption of fruits and vegetables) have been consistently associated with better health outcomes in older people: successful ageing [9], increases in life-years spent in good health [10], a reduced risk of mortality [11,12], and of poor cognitive function [13]. Recent systematic reviews also indicated the beneficial effects of physical activity, non-smoking and of a healthy diet to healthy ageing [2,14,15].

The latter reviews also revealed the lack of evidence in low-and-middle income countries (LMICs) even though it is estimated that by 2050, 80% of the global population that is 60 years old and over will reside there [16]. Among those countries, Latin American and the Caribbean (LAC) are the ones that have been witnessing an increased life expectancy from 1990 to 2016 [17] and projections show that in the next 15 years they are to experience the fastest increase in the number of older people [1]. Dietary risk, physical inactivity, alcohol use and smoking are among the top leading factors of disability-adjusted life years (DALYs) in the area [18].

In this study, we aimed to examine the impact of four healthy lifestyle behaviours (i.e., physical activity, non-smoking, moderate alcohol consumption and consumption of fruits and vegetables), individually and in combination, on healthy ageing and survival in a cohort of people aged 65 years old and over living in five LAC countries after a follow-up period of 4 years.

2. Materials and Methods

2.1. Study Population

The 10/66 Dementia Research Group (10/66 DRG) population-based studies of ageing and dementia have been conducted in geographically defined catchment areas in LMICs (Cuba, Dominican Republic, Peru, Venezuela, Mexico, China, India and Puerto Rico) with a sample size between 1000 and 3000 (generally 2000). In this study, we used data from the baseline survey (2003–2008), including people 65 years old and over, and the follow-up assessments 3 to 5 years later (2008–2010) from Cuba (n = 3000), Dominican Republic, Peru, Mexico and Puerto Rico (n = 2000). Detailed description of the study settings is available elsewhere [19,20]. Access to the dataset can be obtained via the official website www.alz.co.uk/1066. Local ethical committees and the ethical committee of the Institute of Psychiatry of King's College London approved the studies: Cuba-Finlay Albarran Medical Faculty of Havana Medical University Ethical Committee (January 2003, September 2006); Dominican Republic-Consejo Nacional de Bioética y Salud (CONABIOS); Mexico-Instituto Nacional de Neurología y Neurocirugía Ethics Committee (Ref: 96/07; 78/07); Peru-El Instituto de la Memoria, Depresión y Enfermedades de Riesgo (IMEDER) Ethics Committee (August 2004, May 2007); Puerto Rico-Oficina para la Protección de Participantes Humanos en Investigación (OPPHI) (Ref: FWA00005561); King's College London-Institute of Psychiatry/SLAM NHS Trust Ethical Committee Research (ECR) (2004, Ref: 076/03), King's College Research Ethics Committee (KREC) (February 2007, Ref: CREC/06/07/38).

2.2. Baseline Measures

Lifestyle behaviours were assessed based on participants' responses to specific questionnaires. Physical activity was categorised as: "very physically active", "fairly", "not very", and "not at all". Smoking status was categorised as "current", "former" and "never smoker". Alcohol consumption was assessed by asking the units of alcohol drunk per week and then categorised as "no alcohol consumption", "moderate alcohol consumption" (1–14 units per week for women; 1–21 units per week for men) and "heavy alcohol consumption" ($\geq$15 units per week for women and $\geq$21 units per week for men) [21]. We assessed dietary habits by asking the number of fruit and vegetable servings in the last 3 days. In our analyses, we compared physically active participants ("very physically active" or "fairly") with non-physically active ("not very" or "not at all"); non-smokers with ever-smokers ("current" or "former"); moderate drinkers with abstainers (no alcohol consumption) or heavy drinkers; participants with "daily consumption" of fruits and vegetables (3 or more servings the last 3 days) with "non-daily consumption" (0–2 servings the last 3 days). We also adjusted our results for socioeconomic factors: age, sex and level of education (none, some, did not complete primary, completed primary, completed secondary, tertiary).

2.3. Outcome Assessment

In accordance with the WHO definition of healthy ageing, we conceptualised healthy ageing within the functional ability framework [8]. We created a healthy ageing index by employing information from 26 health related question-indicators which were either self-reported or were provided by key informants. Items included information on daily disabilities and difficulties, on pain and sleep problems and on cognition abilities among others (Supplementary File S1).

To create a measurement model for healthy ageing, we employed the framework of Exploratory Factor Analysis (EFA) and Confirmatory Factor Analysis (CFA) [22]. We initially identified the number of factors needed to reproduce the latent (unobserved) construct of healthy ageing (EFA) and then we used a bifactor model to represent a single general construct of it (CFA) [23]. To establish model fit, we examined the comparative fit index (CFI) and root mean square error of approximation (RMSEA) with 90% confidence intervals (CI). We considered a model to have an acceptable fit when CFI $\geq$ 0.90 and RMSEA values were close or less than 0.06 [24]. We also calculated omega hierarchical (ω_H) reliability coefficient of the bifactor structure; a high ω_H > 0.80 indicates that the general factor is the dominant source of systematic variance [25]. Finally, we calculated healthy ageing (factor) scores by using the regression method [26].

We transformed scores in a scale of 0–100 with higher values indicating better health. We examined 3 categories for the outcome of interest: healthy ageing, normal ageing and death at follow-up. To create these categories, we calculated the quintiles of the baseline healthy ageing score distribution and participants belonging in the three lowest fifths (i.e., 0–67.92 scores) were characterised as "normal agers", whereas participants belonging in the two highest fifths (i.e., 67.93–100 scores) were characterised as "healthy agers".

We performed two separate logistic regression analyses: one to estimate the odds ratios (ORs) of healthy ageing (with normal ageing and death as the non-event) and another one to estimate the ORs for survival (in which death was the non-event). We did all analyses per country and we then provided a pooled result by performing fixed-effects meta-analysis. We also computed Higgin's I^2 to estimate the proportion of variability across countries that accounted mostly for heterogeneity rather than sampling error [27]. Heterogeneity up to 40% is conventionally considered negligible, whereas heterogeneity more than 50% is considered moderate to substantial [28].

We also performed sensitivity analyses to assess the robustness of our findings. Firstly, we examined whether the associations of lifestyle behaviours with healthy ageing were influenced by deaths. To do this, we excluded deaths from our sample. Secondly, we performed our analyses by including the non-dichotomised lifestyle variables. Finally, we investigated whether our findings were influenced by the way we categorised participants as healthy or normal agers and if there were differences between men and women. Our analyses were performed in STATA 14.1 and Mplus 7.4 [26].

3. Results

The EFA indicated a latent structure of four factors and the bifactor model (χ^2 = 4039.14, df = 273, *p*-value < 0.00l) exhibited good fit (CFI: 0.985; RMSEA: 0.040, 90%CI: 0.039–0.041). The psychometrically informative bifactor-derived statistics indicated that a strong percentage of the total score variance can be attributed to the general factor (ω_H = 0.85). Hence, we proceeded by considering the score of the general factor only. More information about the EFA and CFA is provided in Supplementary File S2.

In our study there were 10,900 participants in the baseline assessment. In the follow-up survey 1648 were healthy agers, 5594 normal agers, 1734 died and 1924 were either untraceable or refused to be re-interviewed. We performed analyses per country and we checked if participants included in our analyses differ from those not available in the follow-up wave (*t*-test for continuous variables, χ^2 test for categorical). We found that participants included in our analyses in general did not differ with those excluded (Supplementary File S3). There were some statistically significant (i.e., *p*-value < 0.05) differences in Cuba and Peru regarding the dietary habits between those included in the follow-up wave and those excluded. A higher percentage of participants available in the follow-up wave endorsed daily consumption of fruits and vegetables compared to those not available (i.e., 75.4% versus 66.3% in Cuba; 82.5% versus 76.2% in Peru).

Table 1 presents the characteristics of participants characterised as healthy agers and those characterised as normal agers in the follow-up. Participants in the healthy ageing group, compared with the normal ageing group, were younger at baseline, had completed at least primary level education, were mostly physically active and endorsed daily consumption of fruits and vegetables.

The associations of each lifestyle behaviour with healthy ageing and survival at the end of the follow-up period are presented in Table 2. Compared with non-physically active participants, those who were physically active at baseline had 2.59 (95% CI: 2.20–3.03) times greater odds of being healthy agers and 2.23 (95% CI: 1.98–2.52) times greater odds of survival. Compared with people consuming two or less servings of fruits and vegetables in the last three days, people consuming three or more were more likely to experience healthy ageing (OR: 1.24, 95% CI: 1.06–1.44) and to survive at the end of the follow-up (OR: 1.13, 95% CI: 0.99–1.28). Non-smoking, compared to former or current smoking, was not associated with healthy ageing (OR: 0.95, 95% CI: 0.82–1.10) but it was positively associated with survival as never smokers had 1.17 (95% CI: 1.02–1.34) times greater odds of remaining alive in the follow-up assessment. Moderate drinking, compared to high or never drinking, was not associated with healthy ageing (OR: 1.04, 95% CI: 0.82–1.30) or survival (OR: 1.11, 95% CI: 0.90–1.37). Heterogeneity was not statistically significant in most cases (*p*-value < 0.05) indicating that the variation among countries was limited. The only significant heterogeneity was observed in the associations of physical activity with healthy ageing (I^2: 61.3%; *p*-value: 0.035). Further investigation concluded that Puerto Rico was the country exhibiting the highest point estimate (i.e., OR: 6.30; 95% CI: 3.43–11.56). We also reproduced the pooled estimates by excluding this country and the new pooled estimates did not change the direction or strength of associations between physical activity and healthy ageing (i.e., OR: 2.42; 95% CI: 2.05–2.86; I^2: 0.0%, *p*-value: 0.693).

Participants who engaged in two-four healthy lifestyle behaviours at baseline, compared with those who engaged in none or one, exhibited increased odds of healthy ageing (OR: 1.72, 95% CI: 1.45–2.05) (Figure 1a). Additionally, the association between number of healthy behaviours and odds of healthy ageing was positive linear. More specifically, compared with participants engaging in none or one healthy behaviour, those engaging in two had 1.46 times (95% CI: 1.22–1.76) the odds of healthy ageing, those engaging in three had 2.00 times (95% CI: 1.65–2.42) the odds and those in four had 2.46 times (95% CI: 1.54–3.92) the odds. A similar beneficial association was also observed in surviving (Figure 1b). Participants engaging in two or more healthy behaviours had 1.83 times (95% CI: 1.61–2.08) the odds of surviving compared with participants that engaged in none or one healthy behaviour. A positive linear relationship was also observed; two healthy behaviours OR: 1.60, 95% CI: 1.40–1.84; three healthy behaviours OR: 2.29, 95% CI: 1.94–2.69; four healthy behaviours OR: 2.64, 95% CI: 1.53–4.54 (Table 3). Heterogeneity was not statistically significant (*p*-value < 0.05) indicating that the variability among countries was limited.

Table 1. Baseline characteristics of participants in the healthy and normal ageing group * in the follow-up wave.

Country	Cuba		Dominican Republic		Mexico		Peru		Puerto Rico	
Characteristic	Healthy Ageing (n = 494)	Normal Ageing (n = 1513)	Healthy Ageing (n = 234)	Normal Ageing (n = 963)	Healthy Ageing (n = 153)	Normal Ageing (n = 1309)	Healthy Ageing (n = 579)	Normal Ageing (n = 732)	Healthy Ageing (n = 188)	Normal Ageing (n = 1077)
Age (mean, SD)	70.9 ± 4.61	74.87 ± 6.65 †	70.89 ± 5.17	74.7 ± 6.97 †	69.9 ± 4.00	74.2 ± 6.35 †	71.1 ± 5.20	76.7 ± 7.21 †	72.0 ± 4.56	75.7 ± 6.65 †
Gender-Female	291 (58.9%)	1041 (68.8%) †	140 (60.1%)	689 (71.6%) †	99 (64.7%)	845 (64.6%)	343 (59.2%)	474 (64.8%) †	141 (75.0%)	728 (67.8%)
Education Level-above primary	440 (89.4%)	1099 (72.7%) †	101 (43.2%)	259 (27.1%) †	87 (57.2%)	355 (27.1%) †	493 (86.2%)	559 (77.0%) †	164 (87.2%)	846 (78.9%) †
Never smoking	244 (49.6%)	589 (56.9%) †	124 (53.0%)	523 (54.4%)	91 (59.5%)	913 (69.7%) †	471 (81.8%)	625 (85.9%) †	150 (79.8%)	789 (73.5%)
Moderate alcohol consumption	53 (10.9%)	140 (9.4%)	19 (8.3%)	52 (5.4%)	30 (19.7%)	225 (17.3%)	21 (3.7%)	30 (4.2%)	21 (11.2%)	92 (8.6%)
Physical activity (very or fairly)	425 (86.6%)	1047 (69.4%) †	188 (80.7%)	626 (65.3%) †	133 (86.9%)	841 (64.6%) †	471 (81.6%)	462 (63.6%) †	176 (93.6%)	722 (67.3%) †
Fruits/vegetable consumption (≥3 servings in the last 3 days)	409 (83.1%)	1130 (74.8%) †	138 (59.0%)	460 (48.7%) †	116 (76.3%)	931 (71.5%)	477 (83.1%)	593 (82.4%)	156 (83.0%)	807 (75.3%) †

Notes: SD: standard deviation; †: p-value of the t test or the $\chi^2 < 0.05$. * healthy ageing group includes those participants who in the follow-up assessment were in the two highest fifths of the healthy ageing score distribution (i.e. better health level) and normal ageing group those who were in the three lowest fifths; percentages in the parentheses are calculated in the non-missing cases in the covariates.

Table 2. Associations between lifestyle behaviours, healthy ageing and survival.

Country/Lifestyle Behaviour	Healthy Ageing * Adjusted OR † (95% CI)			
	Physical Activity (Very or Fairly vs. Not Very or Never)	Smoking (Never vs. Former or Current)	Drinking Behaviour (Moderate Drinking vs. Never or Heavy Drinking)	Fruits/Vegetables Consumption (≥3 Servings in the Last Three Days vs. <3)
Cuba	2.56 (1.93–3.41)	0.97 (0.77–1.24)	0.83 (0.57–1.20)	1.41 (1.06–1.87)
Dominican Republic	2.11 (1.46–3.04)	1.03 (0.75–1.40)	1.21 (0.68–2.17)	1.35 (0.99–1.83)
Mexico	2.98 (1.81–4.93)	0.58 (0.37–0.90)	1.09 (0.66–1.81)	1.11 (0.73–1.68)
Peru	2.32 (1.75–3.08)	0.90 (0.62–1.30)	0.70 (0.34–1.44)	0.92 (0.66–1.29)
Puerto Rico	6.30 (3.43–11.56)	1.30 (0.85–1.97)	1.80 (1.02–3.17)	1.44 (0.94–2.21)
Pooled Effect	2.59 (2.20–3.03)	0.95 (0.82–1.10)	1.04 (0.82–1.30)	1.24 (1.06–1.44)
I^2; p-value	61.3%; 0.035	44.7%; 0.124	38.0%; 0.168	16.6%; 0.309
	Survival * Adjusted OR † (95% CI)			
Cuba	2.29 (1.86–2.82)	1.16 (0.92–1.45)	1.04 (0.74–1.46)	1.15 (0.92–1.44)
Dominican Republic	1.82 (1.43–2.30)	1.18 (0.93–1.50)	1.36 (0.78–2.37)	0.94 (0.75–1.19)
Mexico	2.08 (1.50–2.86)	0.94 (0.61–1.44)	1.16 (0.74–1.80)	1.17 (0.84–1.64)
Peru	2.75 (1.90–3.63)	1.20 (0.70–2.04)	0.72 (0.33–1.60)	1.06 (0.66–1.73)
Puerto Rico	2.71 (2.02–3.63)	1.34 (0.95–1.89)	1.24 (0.73–2.09)	1.45 (1.08–1.96)
Pooled Effect	2.23 (1.98–2.52)	1.17 (1.02–1.34)	1.11 (0.90–1.37)	1.13 (0.99–1.28)
I^2; p-value	34.4%; 0.192	0.0%; 0.807	0.0%; 0.737	21.9%; 0.275

Notes: CI: confidence interval; OR: Odds ratio; I^2: heterogeneity. *: healthy ageing compared to normal ageing or death; survival (healthy or normal ageing) compared to death; number of participants per country: Cuba: 2557, Dominican Republic: 1608, Mexico: 1633, Peru: 1392, Puerto Rico: 1554. †: models are adjusted for age, gender, education level and all other lifestyle behaviour variables.

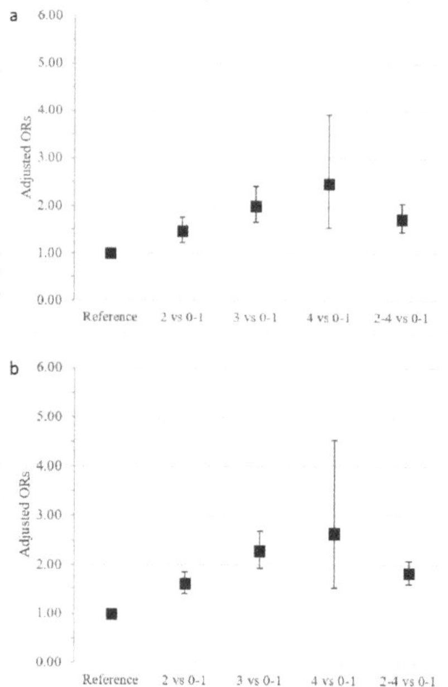

Figure 1. Associations of (**a**) healthy ageing and (**b**) survival with the number of healthy behaviours. Notes: ORs: Odds Ratios (pooled effects of the fixed-effects meta-analysis); models are adjusted for sex, age and education level and for all behaviours at baseline. Associations are given per individual number of healthy behaviours (two, three, four) and for two-four compared to no or one healthy behaviour (reference category). Error bars indicate 95% confidence intervals.

Further sensitivity analysis showed that when we excluded deaths from our sample, the associations of lifestyle behaviour with healthy ageing did not substantially change (i.e., fairly or very physically active OR: 2.28, 95% CI: 1.93–2.69; never smoking OR: 0.95, 95% CI: 0.81–1.10; moderate drinking OR: 1.05, 95% CI:0.83–1.32; daily consumption of fruits and vegetables OR: 1.22, 95% CI: 1.04–1.42). In addition, supplementary analysis showed that the higher the level of physical activity the more increased the ORs of healthy ageing compared to normal ageing or death (very physically active OR: 9.64, 95% CI: 5.62–16.55; fairly OR: 8.20, 95% CI: 4.77–14.12; not very OR: 4.20, 95% CI: 2.42–7.31; not at all: reference category). The same trend was also observed for the vegetable and fruits consumption (more than six servings OR: 1.51, 95% CI: 1.23–1.85; three to six servings OR: 1.16, 95% CI: 0.99–1.36; less than two servings: reference category). Smoking and alcohol consumption had no significant associations.

To examine whether our findings differed between men and women, we also performed our analyses separately for these two subpopulations. The observed relationships between lifestyle behaviours and healthy ageing compared to normal ageing or death did not change (Supplementary File S4). In addition, to examine whether our conclusions were influenced by the categorisation of participants as healthy agers based on the two highest quintiles of the healthy ageing score, we also performed our analyses by using a different categorisation. We performed our analyses by considering participants scoring in the three highest quintiles of the healthy ageing score distribution as healthy agers (i.e., 59.70–100 scores) and compared them against normal agers (i.e., 0–59.69 scores) or dead participants. Our conclusions regarding the protective effect of physical activity and daily consumption of fruits and vegetables did not alter (Supplementary File S5).

Table 3. Associations between number of healthy behaviours, healthy ageing and survival.

Country	Healthy Ageing *			
	Adjusted OR [†] (95% CI)			
	Number of Healthy Behaviours Compared to Zero or One			
	Two	Three	Four	Two-Four
Cuba	1.67 (1.23–2.29)	2.12 (1.54–2.92)	1.74 (0.77–3.96)	1.86 (1.39–2.49)
Dominican Republic	1.64 (1.16–2.32)	1.98 (1.32–2.96)	2.48 (0.78–7.86)	1.77 (1.28–2.43)
Mexico	1.29 (0.74–2.24)	1.31 (0.74–2.33)	2.75 (1.04–7.26)	1.35 (0.80–2.29)
Peru	0.98 (0.64–1.51)	1.64 (1.08–2.49)	1.14 (0.21–6.22)	1.31 (0.88–1.96)
Puerto Rico	1.63 (0.84–3.17)	4.21 (2.25–7.87)	4.93 (1.77–13.71)	2.92 (1.60–5.36)
Pooled Effect	1.46 (1.22–1.76)	2.00 (1.65–2.42)	2.46 (1.54–3.92)	1.72 (1.45–2.05)
I^2; *p*-value	16.0%; 0.313	53.0%; 0.074	0.0%; 0.509	31.5%; 0.211
	Survival *			
	Adjusted OR [†] (95% CI)			
Cuba	1.57 (1.24–2.00)	2.17 (1.66–2.85)	3.53 (1.21–10.28)	1.80 (1.44–2.24)
Dominican Republic	1.45 (1.13–1.88)	1.71 (1.22–2.40)	1.80 (0.48–6.72)	1.53 (1.21–1.93)
Mexico	1.79 (1.24–2.60)	2.19 (1.42–3.36)	2.13 (0.79–5.68)	1.94 (1.37–2.75)
Peru	1.52 (0.90–2.55)	2.91 (1.66–5.12)	1.80 (0.26–12.22)	2.02 (1.22–3.34)
Puerto Rico	1.84 (1.33–2.54)	3.44 (2.34–5.05)	4.11 (1.18–14.28)	2.35 (1.73–3.18)
Pooled Effect	1.60 (1.40–1.84)	2.29 (1.94–2.69)	2.64 (1.53–4.54)	1.83 (1.61–2.08)
I^2; *p*-value	0.0%; 0.790	50.3%; 0.090	0.0%; 0.840	22.0%; 0.274

Notes: CI: confidence interval; OR: Odds ratio; I^2: heterogeneity. *: healthy ageing compared to normal ageing or death; survival (healthy or normal ageing) compared to death; number of participants per country: Cuba: 2557, Dominican Republic: 1608, Mexico: 1633, Peru: 1392, Puerto Rico: 1554. [†]: models are adjusted for age, gender, education level and all other lifestyle behaviour variables.

4. Discussion

In this study, we investigated the associations of four healthy lifestyle behaviours with healthy ageing and survival from a large dataset (n = 10,900) of five LAC countries (Dominican Republic, Cuba, Peru, Puerto Rico and Mexico). Participants engaging in physical activity and in a diet with daily consumption of fruits and vegetables were individually associated with increased odds of healthy ageing and survival. In addition, we found that the more physically active the participants and the higher the number of fruits and vegetables servings, the higher the odds of ageing healthily. Never smoking and moderate alcohol consumption were not individually associated with healthy ageing but all these four behaviours in combination had a positive effect both for healthy ageing and survival.

The world's older population is increasing at a higher pace than the total population, and the region of LAC exhibits the highest growth rate of the 60-and-older population from 2015 to 2020 [29]. Although research on healthy ageing has been growing, there is still limited research available for LAC [2]. Our study is among the first reporting a beneficial association of lifestyle habits with healthy ageing or survival in a LAC dataset. These findings are in accordance with those from other studies, which examined lifestyle behaviours and adverse health outcomes in older age, in datasets from high-and-middle income countries [9,14,30]. Physical activity seems to be related with an increased healthy life expectancy by preventing many chronic diseases (i.e., arterial hypertension, diabetes mellitus type 2, store, cancer) [31], and/or age-related diseases (i.e., dementia and Alzheimer's disease) [32]. The complex biophysiological pathway between physical activity and healthy ageing and survival is yet to be solved. However, this could be partly explained by the favourable biomarker profile of physically active people; higher cell endurance, muscle tissue functionality and energy metabolism [33], reduced fat mass and adipose tissue inflammation [34]. Other studies have also reported a beneficial association of fruits and vegetables consumption with healthy ageing [35,36]. Research has shown that adherence to Mediterranean diet, which includes high consumption of fruits and vegetables, is associated with decreased risk of all-cause mortality, better health outcomes in older people and a reduced risk of frailty [37]. The high antioxidant capacity of fruits and vegetables along with their high fibre and vitamins content could possibly explain this beneficial association [38]. In addition, fruits and vegetables are sources of phytochemicals—the bioactive non-nutrient plant compound—which could contribute to the prevention or minimization of oxidative stress caused by free radicals and thus decrease the risk of chronic diseases [39]. For instance, carotenoids, polyphenols, saponins and phytoestrogens are among those phytochemicals which have shown anticarcinogenic, antioxidative and immunomodulatory effects among others [40].

In regards to moderate alcohol consumption and never smoking, our study did not indicate an individual significant association with healthy ageing. A recent meta-analysis has shown that moderate alcohol consumption seems to be positively associated with healthy ageing but in general associations are mixed [15]. The latest report on the alcohol burden of disease also did not replicate the finding that moderate consumption of alcohol is beneficial; a zero level of alcohol consumption constitutes the level that minimises health loss in the total population [41]. Nevertheless, other studies have found a protective effect of light-to-moderate alcohol consumption against functional health decline among middle-aged people [42]. However these findings should be interpreted with caution, as the beneficial effects of the low to moderate drinking groups (compared with non-drinkers) could be biased by the poor health of former drinkers that are also included in the non-drinkers group in many studies [43,44]. The non-significant result of smoking could be attributed to the limited period of follow-up time (i.e., 4 years) or the relatively older cohort. Studies, which include an older sample, are biased towards those smoking and surviving compared to those that have never smoked and died (survival bias) [45]. Previous meta-analysis also confirms that the negative effects of smoking are more pronounced in younger cohorts [15]. When considering all behaviours in combination, we found that the higher the number of healthy lifestyle behaviours that a participant engaged in during the baseline, the more increased the likelihood of healthy ageing and survival in the follow-up. This comes in agreement with previous research within a high-income setting [9].

A prolonged life longevity is accompanied by multimorbidity [46] which as a consequence poses an extra burden in health systems. As LAC area already exhibits high life expectancy and is the one with the fastest estimated ageing growth, major concerns are raised about the need for early non-pharmacological measures and interventions in older adults to promote healthy ageing. Our findings reflect the importance of healthy lifestyle behaviours and point out the public health actions that must be preserved and prioritised. Adoption of healthy behavioural habits could contribute to the improvement of well-being of the older populations in LACs. Nevertheless, other socioeconomic factors should also be considered when healthy ageing interventions are suggested. For instance, a study with data from 32 countries indicated that overall economic development of the region, public health expenditure and sanitation facilities considerably relate to healthy ageing [47].

Strengths and Limitations

Based on our knowledge, this is the first study, including a large sample, which examined healthy ageing within the functional ability framework and lifestyle behaviour in LAC countries. Our results contribute to the advance of healthy ageing knowledge in under-examined areas and especially in LMICs as current research has mostly focused on other countries [2]. The limited attrition between baseline assessment and follow-up together with our further analyses, which indicated that in most of the cases participants included in the study did not differ with those excluded, limit any potential information bias. In addition, the observed heterogeneity (I^2) was minimal in most cases, reflecting the limited variability in point estimates among countries.

In our study, we created a healthy ageing index within the functional ability framework, as it is the latest recommendation from the WHO [8]. Previous research has mostly focused on the presence or not of some disease/illness as a recent review indicated [7]. Yet, the functioning framework is considered more useful for effective public-health responses than considering specific diseases [8]. For this reason, we created this healthy ageing index by not taking into account specific conditions (i.e., depression, cardiovascular diseases, etc.). Furthermore, we did not consider other domains of healthy ageing (i.e., psychological well-being and social well-being) [48] since the healthy ageing model provided by WHO also incorporates "resilience" as an ability to maintain or improve the level of functioning during challenging periods. This suggests that older people could use psychological/physiological resources and environmental characteristics (i.e., social relationships) to maintain or improve their functional ability. However, these domains were not included in our index, as our ultimate intention was to build an index focusing on the final observed outcome -functional ability-.

Our findings should be interpreted within the context of this study limitations. Lifestyle behaviours were all assessed via self-reported questionnaires. Therefore, potential measurement error could have occurred. Furthermore, questions were too broad not allowing to holistically assess the impact of different frequencies and intensities in physical activity, or of specific fruits and vegetables that were consumed. In addition, even though interviewers underwent substantial training to ensure consistency in the way surveys were conducted in the various settings, we cannot exclude the bias of cross-cultural differences among countries in the conceptualisation of some questions. Another limitation of our study could be the quite arbitrary way of categorising participants who are in the two highest fifths of the baseline healthy ageing score distribution as healthy agers and all others as normal agers. Nevertheless, supplementary analyses (Supplementary File S5) indicated that our conclusions were not influenced by a different categorisation of participants in the health and normal ageing groups. In addition, we characterised healthy behaviour when participants were "very" or "fairly" physically active and when they consumed fruits and vegetables on a daily basis. These categorisations constitute assumptions of our study and consequently our findings should be interpreted within the limitations caused by these. The limited follow-up time inherited with these data also made it impossible to examine the impact of lifestyle behaviours in a more prolonged time. Finally, our sample included participants 65 years old and over. Hence, we were unable to assess lifestyle behaviour and examine healthy ageing within a life course perspective approach [49].

5. Conclusions

Physical activity and a diet rich in fruits and vegetables seem to be associated with healthy ageing and survival. Even though never smoking and moderate alcohol consumption were not significantly associated, once we combine all four healthy behaviours their effect is substantial. Furthermore, the more the healthy lifestyle behaviours adopted, the higher the odds of being old and maintaining well-being. Our results could help in the establishment of policies promoting a healthy lifestyle in LAC countries. Future studies evaluating physical activity, nutritional and other lifestyle interventions within the LAC area are needed to attain a better ageing process for these populations.

Supplementary Materials: The following are available online at http://www.mdpi.com/2072-6643/10/11/1593/s1. Supplementary File S1: Healthy ageing questions origin; Supplementary File S2: Development of a healthy ageing index; Supplementary File S3: Participants' (included and excluded from the analyses) baseline characteristics; Supplementary File S4: Associations between lifestyle behaviours and healthy ageing per gender; Supplementary File S5: Associations between lifestyle behaviours and healthy ageing-different categorisation of healthy ageing.

Author Contributions: C.D. made the study conception and design, analysed the data and wrote the paper; A.K. supervised the data analyses and contributed to paper writing; J.L.A.-M. contributed to paper writing; M.P. provided insight in the 10/66 cohort and contributed to paper writing; A.M.P. provided insight in the 10/66 cohort, supervised the data analyses, contributed to data interpretation and to paper writing.

Funding: The 10/66 Dementia Research Group's research has been funded by the Wellcome Trust Health Consequences of Population Change Programme (GR066133—Prevalence phase in Cuba and Brazil; GR080002—Incidence phase in Peru, Mexico, Cuba, Dominican Republic, Venezuela and China), the World Health Organization (India, Dominican Republic and China), the US Alzheimer's Association (IIRG–04–1286-Peru, Mexico and Argentina), and FONDACIT (Venezuela). On-going data collection and analysis is supported by the European Research Council (ERC-2013-ADG 340755 LIFE2YEARS1066). This work is also supported by the ATHLOS (Ageing Trajectories of Health: Longitudinal Opportunities and Synergies) project, funded by the European Union's Horizon 2020 Research and Innovation Programme under grant agreement number 635316.

Nutrients **2018**, *10*, 1593

Acknowledgments: The authors would like to thank the ATHLOS (Ageing Trajectories of Health: Longitudinal Opportunities and Synergies) project, funded by the European Union's Horizon 2020 Research and Innovation Programme under grant agreement number 635316.

Conflicts of Interest: The authors declare no conflict of interest.

References

1. United Nations. *World Population Ageing 2015*; United Nations, Department of Economic and Social Affairs, Population Division: New York, NY, USA, 2015.
2. Kralj, C.; Daskalopoulou, C.; Rodríguez-Artalejo, F.; García-Esquinas, E.; Cosco, T.D.; Prince, M.; Prina, A.M. *Healthy Ageing: A Systematic Review of Risk Factors*; ATHLOS Consortium: London, UK, 2018.
3. GBD 2013 DALYs and HALE Collaborators; Murray, C.J.; Barber, R.M.; Foreman, K.J.; Abbasoglu Ozgoren, A.; Abd-Allah, F.; Abera, S.F.; Aboyans, V.; Abraham, J.P.; Abubakar, I. Global, regional, and national disability-adjusted life years (DALYs) for 306 diseases and injuries and healthy life expectancy (HALE) for 188 countries, 1990–2013: Quantifying the epidemiological transition. *Lancet* **2015**, *386*, 2145–2191. [CrossRef]
4. Prina, A.M. Ageing, resilience and depression: Adding life to years as well as years to life. *Epidemiol. Psychiatr. Sci.* **2017**, *26*, 571–573. [CrossRef] [PubMed]
5. Lowsky, D.J.; Olshansky, S.J.; Bhattacharya, J.; Goldman, D.P. Heterogeneity in Healthy Aging. *J. Gerontol. Ser. A Boil. Sci. Med. Sci.* **2014**, *69*, 640–649. [CrossRef] [PubMed]
6. Depp, C.A.; Jeste, D.V. Definitions and predictors of successful aging: A comprehensive review of larger quantitative studies. *Am. J. Geriatr. Psychiatry Off. J. Am. Assoc. Geriatr. Psychiatry* **2006**, *14*, 6–20. [CrossRef] [PubMed]
7. Cosco, T.D.; Prina, A.M.; Perales, J.; Stephan, B.C.; Brayne, C. Operational definitions of successful aging: A systematic review. *Int. Psychogeriatr.* **2014**, *26*, 373–381. [CrossRef] [PubMed]
8. World Health Organization. *World Report on Ageing and Health*; World Health Organization: Geneva, Switzerland, 2015.
9. Sabia, S.; Singh-Manoux, A.; Hagger-Johnson, G.; Cambois, E.; Brunner, E.J.; Kivimaki, M. Influence of individual and combined healthy behaviours on successful aging. *CMAJ* **2012**, *184*, 1985–1992. [CrossRef] [PubMed]
10. Stenholm, S.; Head, J.; Kivimaki, M.; Kawachi, I.; Aalto, V.; Zins, M.; Goldberg, M.; Zaninotto, P.; Magnuson Hanson, L.; Westerlund, H.; et al. Smoking, physical inactivity and obesity as predictors of healthy and disease-free life expectancy between ages 50 and 75: A multicohort study. *Int. J. Epidemiol.* **2016**, *45*, 1260–1270. [CrossRef] [PubMed]
11. Knoops, K.T.; de Groot, L.C.; Kromhout, D.; Perrin, A.E.; Moreiras-Varela, O.; Menotti, A.; van Staveren, W.A. Mediterranean diet, lifestyle factors, and 10-year mortality in elderly European men and women: The HALE project. *JAMA* **2004**, *292*, 1433–1439. [CrossRef] [PubMed]
12. Kvaavik, E.; Batty, G.D.; Ursin, G.; Huxley, R.; Gale, C.R. Influence of individual and combined health behaviors on total and cause-specific mortality in men and women: The United Kingdom health and lifestyle survey. *Arch. Intern. Med.* **2010**, *170*, 711–718. [CrossRef] [PubMed]
13. Sabia, S.; Nabi, H.; Kivimaki, M.; Shipley, M.J.; Marmot, M.G.; Singh-Manoux, A. Health behaviors from early to late midlife as predictors of cognitive function: The Whitehall II study. *Am. J. Epidemiol.* **2009**, *170*, 428–437. [CrossRef] [PubMed]
14. Daskalopoulou, C.; Stubbs, B.; Kralj, C.; Koukounari, A.; Prince, M.; Prina, A.M. Physical activity and healthy ageing: A systematic review and meta-analysis of longitudinal cohort studies. *Ageing Res. Rev.* **2017**, *38*, 6–17. [CrossRef] [PubMed]
15. Daskalopoulou, C.; Stubbs, B.; Kralj, C.; Koukounari, A.; Prince, M.; Prina, A.M. Associations of smoking and alcohol consumption with healthy ageing: A systematic review and meta-analysis of longitudinal studies. *BMJ Open* **2018**, *8*, e019540. [CrossRef] [PubMed]
16. World Health Organization. Ageing and Health. 2015. Available online: http://www.who.int/mediacentre/factsheets/fs404/en/ (accessed on 16 March 2018).
17. Institute for Health Metrics and Evaluation. Global Burden of Disease (GBD) Data. 2016. Available online: http://ghdx.healthdata.org/gbd-results-tool (accessed on 30 August 2018).

18. Institute for Health Metrics and Evaluation; Human Development Network; The World Bank. *The Global Burden of Disease: Generating Evidence, Guiding Policy—Latin America and Caribbean Regional Edition*; IHME: Seattle, WA, USA, 2013.

19. Prince, M.; Ferri, C.P.; Acosta, D.; Albanese, E.; Arizaga, R.; Dewey, M.; Gavrilova, S.I.; Guerra, M.; Huang, Y.; Jacob, K.S.; et al. The protocols for the 10/66 dementia research group population-based research programme. *BMC Public Health* **2007**, *7*, 165. [CrossRef] [PubMed]

20. Prina, A.M.; Acosta, D.; Acosta, I.; Guerra, M.; Huang, Y.; Jotheeswaran, A.T.; Jimenez-Velazquez, I.Z.; Liu, Z.; Llibre Rodriguez, J.J.; Salas, A.; et al. Cohort Profile: The 10/66 study. *Int. J. Epidemiol.* **2017**, *46*, 406–406i. [CrossRef] [PubMed]

21. Royal College of Physicians. *Alcohol Guidelines*; Authority of the House of Commons; House of Commons Science and Technology Committee: London, UK, 2011.

22. Brown, T.A. *Confirmatory Factor Analysis for Applied Research*, 2nd ed.; The Guilford Press: New York, NY, USA, 2015.

23. Reise, S.P. Invited Paper: The Rediscovery of Bifactor Measurement Models. *Multivar. Behav. Res.* **2012**, *47*, 667–696. [CrossRef] [PubMed]

24. Hair, J.F.; Black, W.C.; Babin, B.J.; Anderson, R.E. *Multivariate Data Analysis*, 7th ed.; Pearson: New York, NY, USA, 2010.

25. Rodriguez, A.; Reise, S.P.; Haviland, M.G. Evaluating bifactor models: Calculating and interpreting statistical indices. *Psychol. Methods* **2016**, *21*, 137–150. [CrossRef] [PubMed]

26. Muthén, L.K.; Muthén, B.O. *Mplus User's Guide*, 7th ed.; Muthén & Muthén: Los Angeles, CA, USA, 1998–2015.

27. Higgins, J.P.; Thompson, S.G.; Deeks, J.J.; Altman, D.G. Measuring inconsistency in meta-analyses. *BMJ* **2003**, *327*, 557–560. [CrossRef] [PubMed]

28. Deeks, J.J.; Higgins, J.P.T.; Altman, D.G. Analysing data and undertaking meta-analyses. In *Cochrane Handbook for Systematic Reviews of Interventions*; Higgins, J.P.T., Green, S., Eds.; Version 5.1.0; The Cochrane Collaboration, 2011; Available online: www.handbook.cochrane.org (accessed on 22 October 2018).

29. National Academies of Sciences, Engineering, Medicine. Strengthening the Scientific Foundation for Policymaking to Meet the Challenges of Aging in Latin America and the Caribbean: Summary of a Workshop. Committee for the Workshop on Strengthening the Scientific Foundation, Caribbean. In *Committee on Population, Division of Behavioral and Social Sciences and Education*; Kinsella, K., Steering, R., Eds.; The National Academies Press: Washington, DC, USA, 2015.

30. Ford, E.S.; Zhao, G.; Tsai, J.; Li, C. Low-risk lifestyle behaviors and all-cause mortality: Findings from the National Health and Nutrition Examination Survey III Mortality Study. *Am. J. Public Health* **2011**, *101*, 1922–1929. [CrossRef] [PubMed]

31. Reimers, C.D.; Knapp, G.; Reimers, A.K. Does physical activity increase life expectancy? A review of the literature. *J. Aging Res.* **2012**, *2012*, 243958. [CrossRef] [PubMed]

32. Reiner, M.; Niermann, C.; Jekauc, D.; Woll, A. Long-term health benefits of physical activity–a systematic review of longitudinal studies. *BMC Public Health* **2013**, *13*, 813. [CrossRef] [PubMed]

33. Radak, Z.; Hart, N.; Sarga, L.; Koltai, E.; Atalay, M.; Ohno, H.; Boldogh, I. Exercise plays a preventive role against Alzheimer's disease. *J. Alzheimer's Dis.* **2010**, *20*, 777–783. [CrossRef] [PubMed]

34. Woods, J.A.; Wilund, K.R.; Martin, S.A.; Kistler, B.M. Exercise, Inflammation and Aging. *Aging Dis.* **2012**, *3*, 130–140. [PubMed]

35. Hodge, A.M.; O'Dea, K.; English, D.R.; Giles, G.G.; Flicker, L. Dietary patterns as predictors of successful ageing. *J. Nutr. Health Aging* **2014**, *18*, 221–227. [CrossRef] [PubMed]

36. Kiefte-de Jong, J.C.; Mathers, J.C.; Franco, O.H. Nutrition and healthy ageing: The key ingredients. *Proc. Nutr. Soc.* **2014**, *73*, 249–259. [CrossRef] [PubMed]

37. Sofi, F.; Abbate, R.; Gensini, G.F.; Casini, A. Accruing evidence on benefits of adherence to the Mediterranean diet on health: An updated systematic review and meta-analysis. *Am. J. Clin. Nutr.* **2010**, *92*, 1189–1196. [CrossRef] [PubMed]

38. Slavin, J.L.; Lloyd, B. Health Benefits of Fruits and Vegetables. *Adv. Nutr.* **2012**, *3*, 506–516. [CrossRef] [PubMed]

39. Liu, R.H. Health benefits of fruit and vegetables are from additive and synergistic combinations of phytochemicals. *Am. J. Clin. Nutr.* **2003**, *78* (Suppl. 3), 517s–520s. [CrossRef] [PubMed]

40. Leitzmann, C. Characteristics and Health Benefits of Phytochemicals. *Complement. Med. Res.* **2016**, *23*, 69–74. [CrossRef] [PubMed]

41. Griswold, M.G.; Fullman, N.; Hawley, C.; Arian, N.; Zimsen, S.R.M.; Tymeson, H.D.; Venkateswaran, V.; Tapp, A.D.; Forouzanfar, M.H.; Salama, J.S.; et al. Alcohol use and burden for 195 countries and territories, 1990–2016, a systematic analysis for the Global Burden of Disease Study 2016. *Lancet* **2018**, *392*, 1015–1035. [CrossRef]

42. Chen, L.Y.; Hardy, C.L. Alcohol Consumption and Health Status in Older Adults: A Longitudinal Analysis. *J. Aging Health* **2009**, *21*, 824–847. [CrossRef] [PubMed]

43. Hu, Y.; Pikhart, H.; Malyutina, S.; Pajak, A.; Kubinova, R.; Nikitin, Y.; Peasey, A.; Marmot, M.; Bobak, M. Alcohol consumption and physical functioning among middle-aged and older adults in Central and Eastern Europe: Results from the HAPIEE study. *Age Ageing* **2015**, *44*, 84–89. [CrossRef] [PubMed]

44. Stockwell, T.; Greer, A.; Fillmore, K.; Chikritzhs, T.; Zeisser, C. How good is the science? *BMJ* **2012**, *344*, e2294. [CrossRef] [PubMed]

45. Hernan, M.A.; Alonso, A.; Logroscino, G. Cigarette smoking and dementia: Potential selection bias in the elderly. *Epidemiology* **2008**, *19*, 448–450. [CrossRef] [PubMed]

46. Divo, M.J.; Martinez, C.H.; Mannino, D.M. Ageing and the epidemiology of multimorbidity. *Eur. Respir. J.* **2014**, *44*, 1055–1068. [CrossRef] [PubMed]

47. Kim, J.I.; Kim, G. Factors affecting the survival probability of becoming a centenarian for those aged 70, based on the human mortality database: Income, health expenditure, telephone, and sanitation. *BMC Geriatr.* **2014**, *14*, 113. [CrossRef] [PubMed]

48. Lu, W.; Pikhart, H.; Sacker, A. Domains and Measurements of Healthy Aging in Epidemiological Studies: A Review. *Gerontologist* **2018**, gny029. [CrossRef] [PubMed]

49. Kuh, D.; Richards, M.; Cooper, R.; Hardy, R.; Ben-Shlomo, Y. Life course epidemiology, ageing research, and maturing cohort studies: A dynamic combination for understanding healthy ageing. In *A Life Course Approach to Healthy Ageing*; Oxford University Press: Oxford, UK, 2014.

 *nutrients*

Review

Cytoplasmic and Mitochondrial NADPH-Coupled Redox Systems in the Regulation of Aging

Patrick C. Bradshaw

Department of Biomedical Sciences, James H. Quillen College of Medicine, East Tennessee State University, Johnson City, TN 37614, USA; bradshawp@etsu.edu; Tel.: +1-423-439-4767

Received: 30 January 2019; Accepted: 21 February 2019; Published: 27 February 2019

Abstract: The reduced form of nicotinamide adenine dinucleotide phosphate (NADPH) protects against redox stress by providing reducing equivalents to antioxidants such as glutathione and thioredoxin. NADPH levels decline with aging in several tissues, but whether this is a major driving force for the aging process has not been well established. Global or neural overexpression of several cytoplasmic enzymes that synthesize NADPH have been shown to extend lifespan in model organisms such as *Drosophila* suggesting a positive relationship between cytoplasmic NADPH levels and longevity. Mitochondrial NADPH plays an important role in the protection against redox stress and cell death and mitochondrial NADPH-utilizing thioredoxin reductase 2 levels correlate with species longevity in cells from rodents and primates. Mitochondrial NADPH shuttles allow for some NADPH flux between the cytoplasm and mitochondria. Since a decline of nicotinamide adenine dinucleotide (NAD^+) is linked with aging and because $NADP^+$ is exclusively synthesized from NAD^+ by cytoplasmic and mitochondrial NAD^+ kinases, a decline in the cytoplasmic or mitochondrial NADPH pool may also contribute to the aging process. Therefore pro-longevity therapies should aim to maintain the levels of both NAD^+ and NADPH in aging tissues.

Keywords: NADPH; aging; redox; lifespan; longevity; mitochondrial; NADP; nicotinamide adenine dinucleotide phosphate; glutathione; thioredoxin; *Drosophila*; *C. elegans*

1. NADPH and the Redox Theories of Aging

The reduced form of nicotinamide adenine dinucleotide phosphate, NADPH, protects cells from redox stress and is required for the synthesis of fatty acids, cholesterol, and deoxynucleotides. NADPH is also required for steroid biosynthesis and protects cells from toxic metabolites by providing reducing power to cytochrome P450 enzymes. NADPH levels decrease with age [1,2] due to both aging-related loss of nicotinamide adenine dinucleotide (NAD^+) [3], a precursor for its synthesis, and to the aging related increase in the $[NADP^+]/[NADPH]$ ratio that occurs as a result of oxidative stress due to increased mitochondrial electron transport chain dysfunction [4] and inflammation [5]. Therapies that increase NAD^+ levels combined with therapies that increase NADPH levels could greatly benefit subjects with aging-related disorders. This type of combination therapy has already been shown to be better than either monotherapy alone in an animal model of ischemic stroke [6]. The major objectives of this review are to discuss the mechanisms contributing to the aging-related loss of NADPH, to detail the mechanisms regulating the cytoplasmic and mitochondrial $[NADP^+]/[NADPH]$, and to give an update on enzymes of the NADPH-powered redox systems that have been identified to modulate lifespan.

The mitochondrial free radical theory of aging (MFRTA) has been a leading theory of aging for nearly 50 years [7] and states that increased levels of mitochondrial reactive oxygen species (ROS) stimulate aging. It therefore suggests that antioxidants and NADPH, which is used for the recycling of many antioxidants, will protect cells from mitochondrial generated ROS to delay tissue and organismal aging. Over the past two decades data has shown that ROS-mediated signaling events in young and healthy organisms can lead to lifespan extension, instead of lifespan decline as was originally proposed by the MFRTA [8]. There are also many experimental results that appear to refute the MFRTA [9]. For example, overexpression of antioxidant enzymes, such as superoxide dismutase (SOD) [10] or catalase [11], or administration of antioxidant supplements, such as vitamins C (ascorbate) [12] or E (tocopherols) [13], do not extend lifespan in several model organisms, although there are exceptions to this statement [14–16], the most notable being expression of mitochondrial-targeted catalase extending the lifespan of mice [17].

The fact that many supplemented antioxidants or overexpression of SODs or catalase do not generally extend lifespan could be explained if the activities of the NADPH-fueled redox systems are limiting factors for the protection against lifespan-shortening redox stress and that the activities of SODs and catalase are not limiting. However, it is acknowledged that SOD1 or SOD2 must first detoxify the superoxide produced from the mitochondrial electron transport chain (ETC) to H_2O_2 before the NADPH-linked redox systems can detoxify the H_2O_2 to water. In addition it is evident that SOD2 plays an essential role in ROS detoxification in newborns due to the death of SOD2 knockout mice shortly after birth [18]. NADPH is required for optimal catalase activity as it binds and stabilizes the catalase tetramer [19]. Antioxidant vitamins such as vitamins C or E that rely upon NADPH to be recycled to their active reduced forms would be expected to provide little longevity benefit in aged organisms where NADPH levels have declined to such an extent to prevent their efficient recycling. The Redox Stress Theory of Aging [20] and the Redox Theory of Aging [21] suggest that lifespan is regulated by redox changes, including alterations in the [NADP$^+$]/[NADPH], GSSG/GSH (oxidized glutathione disulfide/reduced glutathione), and oxidized thioredoxin/reduced thioredoxin ratios. These theories also highlight the ability of redox changes to influence aging by not only altering oxidative damage, but by altering cell signaling. Examples of this include the activation of the lifespan-extending transcriptional regulators DAF-16/FOXO and SKN-1/Nrf2 in *C. elegans* by compounds that stimulate ROS production [22,23].

Much data obtained over the past two decades greatly support the MFRTA-derived redox-based theories of aging including the strong negative correlation between the rate of mitochondrial superoxide generation and lifespan in closely related species [24], the strong positive correlation between phospholipid fatty acid saturation levels and lifespan, and the negative correlation between the frequency of cysteine residues in mitochondrial electron transport chain transmembrane spanning regions and lifespan [25]. The higher fatty acid saturation in longer lived species likely evolved to prevent the ROS-mediated oxidation of fatty acid double bonds [26], while the depletion of mitochondrial inner transmembrane cysteine residues likely evolved to prevent thiyl radical formation and potentially lifespan shortening protein crosslinking that can occur when superoxide reacts with protein sulfhydryl groups [27]. The mitochondrial inner membrane is enriched with the phospholipid cardiolipin, which is essential for ETC function and ADP/ATP transport and due to its high degree of fatty acid unsaturation is especially vulnerable to ROS-mediated damage [28].

2. Loss of NAD$^+$ as a Major Cause for Loss of NADPH With Aging

One cause for the aging-related loss of NADPH and increase in oxidative stress with aging is the decrease in the levels of cellular NAD$^+$ [3], the immediate precursor for the synthesis of NADP$^+$ by NAD$^+$ kinases. NAD$^+$ levels decline with aging in mammals for several reasons, one of which is the aging-related decrease in the salvage pathway of NAD$^+$ synthesis as a result of decreased expression of nicotinamide phosphoribosyl transferase (NAMPT) [29], a committed step in this pathway. There is also an increase in NAD$^+$ degradation with aging. The decreased NAD$^+$ levels may be a cause of sirtuin protein deacetylase-dependent [30,31] or sirtuin-independent alterations in mitochondrial ETC activity that results in increased ROS production and increased nuclear DNA damage that activates poly-ADP-ribose polymerase (PARP) in several aged tissues including liver, heart, kidney, and lung [32]. This PARP activation together with the aging-related increase in expression and activity of the NAD$^+$ and NADP$^+$ hydrolyzing enzyme CD38 [33] lead to increased hydrolysis of NAD$^+$ and NADP$^+$ in aged tissues. CD38 was shown to have greater activity (6-fold lower K_m and 2-fold higher V_{max}) using $NADP^+$ as a substrate than NAD^+ [34,35]. PARP activation also leads to decreased NADPH levels as PARP inhibits hexokinase, the first enzyme of glycolysis also required for glucose flux into the NADPH-generating pentose phosphate pathway (PPP) [36].

In brain, SARM1 is another NADase that contributes to the loss of NAD^+ under pathological conditions [37]. But whether or not SARM1 is activated in aged brain has yet to be studied in mammals. There are no homologs of CD38 present in the genomes of the aging models *Drosophila* or *C. elegans*, but both genomes encode homologs of PARP and SARM1 [38]. Expression of the *Drosophila* homolog of SARM1 increased during aging or mitochondrial ETC inhibition and was shown to play a role in inducing a pro-inflammatory state [39]. $NADP^+$ phosphatase activities, resulting in the degradation of $NADP^+$ to NAD^+, have also been observed in rat liver mitochondrial and Golgi extracts [40,41], but the proteins responsible these activities or any aging-related changes in enzyme activity levels have yet to be identified.

Nematodes and insects, as with other invertebrates, lack NAMPT homologs and the two-step NAD^+ salvage pathway present in vertebrates, but instead possess a four-step salvage pathway. In this pathway, nicotinamide is first deaminated to nicotinic acid by a nicotinamidase and then the 3-step Preiss–Handler pathway for NAD^+ salvage synthesis from nicotinic acid is employed [42,43]. In addition, like mammals, *C. elegans* can synthesize NAD^+ through a de novo pathway from tryptophan [44]. The *Drosophila* genome lacks a 3-hydroxyanthanilic acid dioxygenase gene encoding an enzyme that synthesizes quinolinic acid in the de novo NAD^+ synthesis pathway [45]. Therefore *Drosophila* appears to strictly rely upon the salvage pathway for NAD^+ synthesis.

Several lines of evidence suggest that loss of NAD^+ and NADPH may play a role in the aging of *Drosophila* and *C. elegans*. The NAD^+/NADH ratio declines with aging in both species [46,47], while in *C. elegans* this loss was driven almost entirely by a decrease in NAD^+ levels. Since greater than 80% of pyridine nucleotides are bound to proteins and not free in solution [48], the (free) [NAD^+]/[NADH] as measured by the pyruvate/lactate ratio is a much better functional indicator than measuring total levels. In aging *C. elegans* a loss of (free) [NAD^+]/[NADH] also paralleled the loss in total NAD^+/NADH [46]. Addition of NAD^+ (10 µM–1 mM) [49], nicotinamide (0.2 mM), or nicotinamide riboside (0.5 mM) [50] to the culture medium extended lifespan in *C. elegans*, even though the *C. elegans* genome apparently lacks a nicotinamide riboside kinase gene to convert nicotinamide riboside to nicotinamide mononucleotide. But *C. elegans* possesses homologs to purine nucleoside phosphorylase (PNP) and methylthioadenosine phosphorylase (MTAP) that can likely catabolize nicotinamide riboside to nicotinamide for synthesis of NAD^+ synthesis through the salvage pathway [51]

The decreased nuclear NAD^+ levels with aging in mouse muscle lead to stabilization of hypoxia inducible factor-1α (HIF-1α) and decreased c-Myc-induced expression of mitochondrial targeted genes, including mitochondrial transcription factor A (TFAM), required for mitochondrial ETC function [52]. Decreased ETC complex I function with aging increases mitochondrial NADH and decreases mitochondrial NAD^+ levels, which likely results in decreased phosphorylation of NAD^+ to $NADP^+$ by mitochondrial NAD^+ kinase 2 (NADK2). Decreased ETC function with aging also reduces the proton-motive force across the mitochondrial inner membrane, which decreases mitochondrial nicotinamide nucleotide transhydrogenase (NNT)-mediated synthesis of NADPH and NAD^+ from $NADP^+$ and NADH in the matrix space.

The increase in cytoplasmic and mitochondrial [$NADP^+$]/[NADPH] with aging decreases the activities of cytoplasmic and mitochondrial glutathione reductase (GSR) and thioredoxin reductases (cytoplasmic TXNRD1 and mitochondrial TXNRD2), which lead to more oxidized states of the glutathione redox couple (GSSG/GSH) and thioredoxins (cytoplasmic TXN and mitochondrial TXN2), resulting in decreased activities of glutathione peroxidases (GPX) and peroxiredoxins (also called thioredoxin peroxidases). See Figure 1 for an overview of the cytoplasmic and mitochondrial NADPH-linked redox systems. This oxidation of the NADPH-linked redox systems with aging also causes the oxidation of glutaredoxins, ascorbate (vitamin C), and tocopherols (vitamin E) [53] and eventually leads to oxidation of lipids, nucleic acids, and amino acids and the cell and tissue dysfunction that accompanies the aging process.

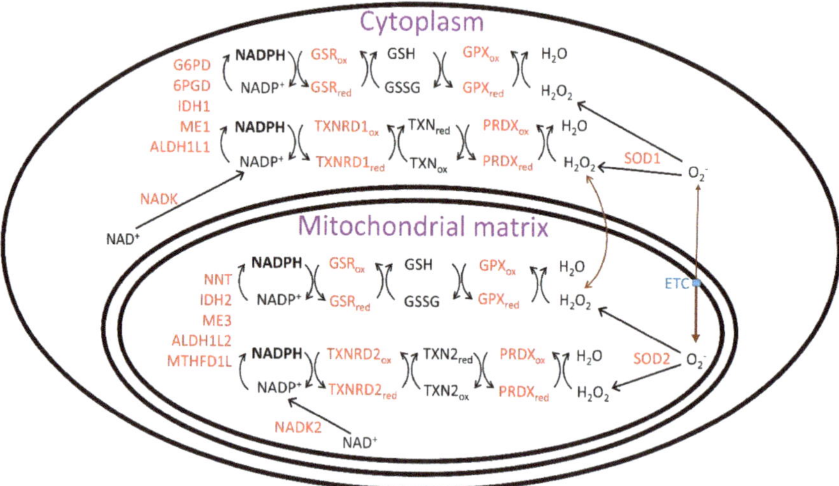

Figure 1. Cytoplasmic and mitochondrial reduced nicotinamide adenine dinucleotide phosphate (NADPH)-linked redox systems. NADK: nicotinamide adenine dinucleotide (NAD$^+$) kinase; ETC: electron transport chain; GSR: glutathione reductase; GSH: reduced glutathione; GSSG: oxidized glutathione disulfide; GPX: glutathione peroxidase; TXN: thioredoxin; TXNRD: thioredoxin reductase; PRDX: peroxiredoxin; H$_2$O$_2$: hydrogen peroxide; O$_2^-$: superoxide; G6PD: glucose-6-phosphate dehydrogenase; 6PGD: 6-phosphogluconate dehydrogenase; IDH: isocitrate dehydrogenase; ME: malic enzyme ALDH: aldehyde dehydrogenase; MTHFD: methylenetetrahydrofolate dehydrogenase; NNT: nicotinamide nucleotide transhydrogenase.

3. Aging is Associated with Redox Imbalance

All multicellular organisms that have been studied thus far, with the possible exception of extremely long-lived naked mole rats [54], show oxidizing redox changes with aging accompanied by oxidative damage to macromolecules. Redox stress appears to contribute to almost all aging-related disorders including diabetes, heart disease, and neurodegenerative disease. The [NADP$^+$]/[NADPH] redox couple is the most important redox couple in the fight against aging-induced cellular oxidation. The redox potentials of the free cytoplasmic and mitochondrial [NADP$^+$]/[NADPH] are roughly −400 to −20 mV and are the most negative redox potentials in the cell [55,56]. The cytoplasmic and the mitochondrial [NAD$^+$]/[NADH] couples [57], which are roughly 99.9% and 87% oxidized, respectively, in healthy cells [56,58] become more reduced with aging, while the cytoplasmic and most likely also the mitochondrial [NADP$^+$]/[NADPH] couple, which are both roughly 99% reduced in healthy cells [56,58] become more oxidized with aging. For example, the energy of the cytoplasmic NADP$^+$/NADPH couple became 35 mV less negative with aging corresponding to a 23% loss of brain NADPH levels from 10 to 21 months of age in mice [2]. The initial mechanisms involved in this redox imbalance remain unclear, but at least in the brain the aging-induced redox imbalances may result from the decreased expression of NAMPT and NNT [2].

The GSH/GSSG ratio in cells depends heavily on the [NADP$^+$]/[NADPH], but appears not to be in equilibrium due to the moderate expression level of glutathione reductase, the enzyme that links these couples. The GSH/GSSG is approximately 20:1 to 100:1 (95–99% reduced) [59], while during times of oxidative stress this ratio may drop to 5:1 (80% reduced) or even 1:1 (50% reduced) [60]. Cytoplasmic and mitochondrial GSH levels are 1 to 10 mM, two to three orders of magnitude higher than cytoplasmic and mitochondrial [NADPH] [61,62], due in part to the fact that the vast majority of cellular NADP(H) is bound and not free [63]. In addition in rat liver 59% of the total cellular NADP(H) was reported to be present in mitochondria [58], although mitochondria only make up 10–15% of the cell volume. Therefore it may take many hours to days for cells to completely recover from oxidation of GSH if the cells are able to recover at all [64].

Opposite to the oxidizing effects of aging on the plasma GSSG/GSH and cystine/cysteine ratios [65], the $NADP^+$/NADPH ratio in human plasma has been reported to become more reduced with aging due to increased levels of NADPH, even though $NADP^+$ levels declined [66]. Inhibiting cellular export or stimulating cellular uptake of different forms of vitamin B3 (niacin) may be a strategy to maintain cell and tissue NADPH levels with aging. In this regard supplementation with nicotinamide riboside has been shown to be beneficial to increase plasma NAD^+ levels [67]. No reports of plasma membrane transport of NADP(H) could be found, but intraperitoneal or tail vain injection of NADPH in mice was shown to increase the levels of NADPH in plasma and brain, respectively [68].

In *Drosophila* the $NADP^+$/NADPH increased in late adulthood in parallel with the decline in the NAD^+/NADH [47]. To the best of our knowledge total $NADP^+$ and NADPH levels have not been measured in young and aged *C. elegans*. The isocitrate/alpha-ketoglutarate ratio as a measure of the cytoplasmic $[NADP^+]$/[NADPH] in young and old worms has also yet to be determined. As only one malic enzyme gene is present in *C. elegans* and it contains a mitochondrial targeting signal [69], measurement of the nematode malate/pyruvate ratio will likely not be indicative of the cytoplasmic $[NADP^+]$/[NADPH] as it is in mammals, which possess a cytoplasmic $NADP^+$-dependent malic enzyme (ME1) [56].

4. Reductive Stress Can Influence the Rate of Aging

Although the study of oxidative stress is common, the study of reductive stress is somewhat rare, especially as it pertains to aging. The terms oxidative and reductive stress are somewhat unclear [70] as they do not specifically refer to the $[NADP^+]$/[NADPH] or $[NAD^+]$/[NADH] couple. Most investigators utilize the term oxidative stress to refer to a state where the $[NADP^+]$/[NADPH], GSSG/GSH, and protein cysteine thiols are more oxidized than the normal healthy condition and reductive stress to mean that these redox couples are more reduced than normal.

Quantification of oxidative stress in lysed cell, tissue, or isolated mitochondrial extracts is frequently performed by measurements of GSSG/GSH. The cytoplasmic or mitochondrial $[NADP^+]$/[NADPH] is slightly more difficult to measure as one must either use a fluorescent probe in intact cells [71] or measure the ratios of specific metabolites that are in equilibrium with the $[NADP^+]$/[NADPH] in lysed cells or tissues [56,58]. Lysed cell measurements of total $NADP^+$/NADPH (or NAD^+/NADH) are not indicative of the free levels as bound nucleotides are released into solution upon lysis [56].

The oxidation state of protein thiols in live cells can be measured using the genetically encoded fluorescent probes roGFP or roGFP2 [72]. High levels of glutaredoxins can equilibrate roGFP or roGFP2 with the redox state of the glutathione pool. Therefore, a roGFP2-glutaredoxin fusion protein was developed and expressed in *Drosophila* to measure the GSSG/GSH with aging [73]. This probe showed a stable, highly reduced cytoplasmic GSSG/GSH with aging, while there was heterogeneous oxidation of the mitochondrial GSSG/GSH in both young and aged fat body, hemocytes, and Malpighian tubules, but a stable reduced GSSG/GSH in young and aged mitochondria from muscles and gut enterocytes [74]. Feeding N-acetylcysteine (NAC) to flies led to reductive stress characterized by slightly decreased cytoplasmic H_2O_2 in gut and increased mitochondrial H_2O_2 in fat, gut, and Malpighian tubules [74].

Further elegant studies using cytoplasmic and mitochondrial targeted roGFP have shown that increasing the reduction of cytoplasmic GSSG/GSH and protein thiols to induce reductive stress as measured by cytoplasmic roGFP leads to production of ROS in mitochondria and oxidative stress as evidenced by oxidation of mitochondrial roGFP [75]. Reducing power in the cytoplasm can be transmitted to the mitochondrial matrix by both the direct transport of GSH into the mitochondrial matrix [76,77] as well as by mitochondrial NADPH shuttles that can transfer NADPH reducing equivalents [78]. A major NADPH shuttle system uses cytoplasmic isocitrate dehydrogenase 1 (IDH1), the mitochondrial tricarboxylate (citrate) carrier, and mitochondrial isocitrate dehydrogenase 2 (IDH2). Once the reducing equivalents are transported into the matrix space, the reduced [NADP$^+$]/NADPH] in the matrix space activates glutathione reductase to reduce the GSSG/GSH and activates thioredoxin reductase to reduce thioredoxin. However, once the mitochondrial GSSG/GSH and thioredoxin pools become highly reduced, superoxide and hydrogen peroxide are generated by the mitochondrial glutathione reductase and thioredoxin reductase enzymes as flavins in these two enzymes produce ROS in the absence of their GSSG and oxidized thioredoxin substrates [79].

Adding 2-deoxyglucose to *C. elegans* culture medium or knocking down expression of the glycolytic enzyme glucose-6-phosphate isomerase, *gpi-1*, to inhibit glycolysis led to lifespan extension [23]. The mechanism of lifespan extension has not been fully established, but the authors showed a transient increase in respiration and ROS production following inhibition of glycolysis and showed that lifespan extension could be blocked by supplemented antioxidants [23]. Therefore, they concluded that a ROS signal mediates lifespan extension. However, others have suggested that the addition of antioxidants may have blocked lifespan extension by inducing reductive stress instead of by quenching ROS [80]. It has also been shown that inhibiting glycolysis stimulates metabolic flux through the PPP to increase NADPH production [81], so the lifespan extension by glycolytic inhibition may also have been due in part to a reduced cytoplasmic [NADP$^+$]/[NADPH] and increased redox defenses that followed the transient increase in ROS production. Studies using yeast where the glycolytic enzyme triose phosphate isomerase was deleted showed a reduced NADP$^+$/NADPH ratio likely due to increased PPP flux, but decreased replicative lifespan [82], perhaps in part due to reductive stress.

In mice reductive stress due to increased NADPH production by the PPP enzyme glucose-6-phosphate dehydrogenase (G6PD) has been shown to be associated with protein aggregation and cardiomyopathy [83]. Reductive stress could also potentially lower the already low levels of cytoplasmic [NADP$^+$] (1–100 nM) to decrease the activity of enzymes that rely upon it to drive metabolism. As cytoplasmic oxidative stress appears to be much more prevalent than reductive stress, especially in aged tissues, it does not appear likely that cytoplasmic reductive stress is a cause of mitochondrial oxidative stress in aging tissues or major aging-related diseases.

5. Could the Low Levels of Cytoplasmic Glutathione Peroxidase Activity in Long-Lived Species Function in the Preservation of [NADP$^+$]/[NADPH] and GSSG/GSH with Aging to Promote Longevity?

Since the redox theories of aging are gaining traction, more detailed studies of the roles of the cytoplasmic and mitochondrial NADPH-glutathione reductase and NADPH-thioredoxin reductase systems in aging are needed. There is also a dearth of knowledge on the mechanisms that cause the mitochondrial ETC of short-lived species to generate more superoxide than that of long-lived species and the mechanisms through which the ETC generates increased levels of superoxide with age [8]. Although compounds that block mitochondrial superoxide production from complex I [84] and complex III [85] without affecting oxidative phosphorylation have recently been identified as potential therapeutics, compounds that stabilize the cytoplasmic and mitochondrial [NADP$^+$]/[NADPH] redox couples to maintain redox defense systems with aging are also desperately needed for clinical trials for aging-related diseases.

Naked mole rats have a mean lifespan over 30 years, at least ten times longer than similarly sized mice, and surprisingly have 70-fold less cellular glutathione peroxidase activity than mice [86]. Glutathione peroxidase detoxifies hydrogen peroxide to water through the oxidation of GSH. Therefore, it is possible that increases in specific types of ROS, such as specific types of peroxides in the cytoplasm, are compatible with increased longevity [86]. An alternative to this hypothesis is that the NADPH-thioredoxin reductase-thioredoxin-peroxiredoxin detoxification system for peroxides is upregulated in naked mole rats to compensate for the low activity of the NADPH-glutathione reductase-GSH-glutathione peroxidase system.

The extremely low glutathione peroxidase activity of naked mole rats may even contribute to their extreme longevity through the preservation of reduced cytoplasmic GSSG/GSH and [NADP$^+$]/[NADPH]. If this is true, the maintained high energy of the [NADP$^+$]/[NADPH] couple can then be used to stimulate longevity through increasing the activities of NADPH-utilizing enzymes such as glutathione reductase and thioredoxin reductase to combat other types of redox stress, such as that induced by peroxynitrite [87], which if left unchecked could decrease longevity. The increased lifespan of mice heterozygous for glutathione peroxidase-4 [88] and the strong inverse relationship between cellular glutathione peroxidase activity and lifespan in seven different rodent species [86] support a role for maintaining reduced cytoplasmic GSSG/GSH and [NADP$^+$]/[NADPH] for increased longevity. If only cytoplasmic, but not mitochondrial glutathione peroxidase activity is low in long-lived species, mitochondrial shuttles and glutathione transporters could be used for the increased detoxification of mitochondrial peroxides as a mechanism responsible for the longevity of these species.

Unexpectedly, long lived *Drosophila* strains did not have more GSH than short-lived strains [89]. Likewise, in rodents and in birds there is no correlation between GSSG/GSH in liver and longevity and even a negative relationship was found between liver GSH and longevity [90]. This data suggests that the cytoplasmic NADPH-glutathione reductase-GSH redox system does not limit longevity. As possible compensation for their low GSH levels, liver from naked mole rats contain much higher levels of the cytoplasmically localized thioredoxin reductase (TXNRD1) and major peroxiredoxin (PRDX1) than short-lived guinea pigs [91]. This may be an indication of higher Nrf2 transcriptional activity in naked mole rats [92].

As mentioned above there is also a negative correlation between mitochondrial ETC-produced superoxide production and longevity [8]. Therefore, long-lived species appear to have evolved more efficient ETC machinery that plays an important role in maintaining a reduced intracellular environment. Consistent with this, it was also reported that naked mole rats have greatly decreased levels of proteins for ETC complex I, an important source of ROS, compared to short-lived guinea pigs [91]. However, another report indicated no difference in ETC complex I activity between naked mole rats and mice, but decreased ETC complex II activity in naked mole rats [93]. It was also shown that isolated mitochondria from naked mole rats have a greatly enhanced ability to detoxify hydrogen peroxide compared to those from mice, largely due to higher activity of the mitochondrial NADPH-glutathione reductase-GSH-glutathione peroxidase system [93]

To help resolve some of the current controversies in the field, future studies could focus on measurements of [NADP$^+$]/[NADPH] and GSSG/GSH in hypothalamus and white adipose tissue in closely related species with different maximal lifespans. These tissues appear have a greater influence on organismal aging than tissues or cells where past measurements have been made [94,95]. Measurements using mitochondria isolated from these tissues would also be important, as only roughly 10–15% of cellular GSH is present in mitochondria [96]. Importantly NADPH has been shown to bind to the 39 kD subunit of ETC complex I to stabilize the complex [97,98] and ETC complex I activity declines with aging [99,100].

6. The NADPH-Thioredoxin Reductase Redox System and Aging

Thioredoxin reductases use NADPH to reduce oxidized thioredoxin. In the cytoplasm TXNRD1 reduces TXN and in mitochondria TXNRD2 reduces TXN2 [101]. Double heterozygous knock out mice (TXN$^{+/-}$/TXN2$^{+/-}$) were surprisingly long-lived [102] suggesting that TXN or TXN2 levels do not limit lifespan, although follow up studies should be performed to determine if compensatory reductions of other redox couples such as the NADP$^+$/NADPH or GSSG/GSH are responsible for the increased longevity. Transgenic TXN or TXN2 mice show increased mean, but not maximal lifespan [102,103]. Overexpression of TXN and TXN2 together surprisingly led to decreased lifespan [101] suggesting that redox signaling occurs between the cytoplasmic and mitochondrial compartments.

Therapies that maintain a reduced [NADP$^+$]/[NADPH] in the cytoplasm or mitochondria of aged organisms, but without inducing excessive reductive stress are promising longevity therapies. Longevity studies using mice overexpressing TXNRD1 or TXNRD2 would aid our understanding of the role that these proteins play in mammalian longevity. Mitochondrial TXNRD2 is an especially promising candidate to overexpress or deplete in mice for aging studies because protein levels and activities in fibroblasts positively correlate with longevity across multiple rodent and primate species [104]. Overexpression of the *Drosophila* homolog of TXNRD2, Trxr-2, extended median lifespan [104], while overexpression of the mostly cytoplasmic, but partially mitochondrial *Drosophila* thioredoxin reductase Trxr-1 did not extend lifespan during normoxia [105], but it did during hyperoxia [106]. Further evidence for a role of the mitochondrial TXNRD2-TXN2 system in enhanced longevity is that growth hormone downregulates the expression of these two genes in liver and growth-hormone deficient Ames dwarf mice are long-lived [107].

In rat brain mitochondria that naturally lack catalase, roughly 75% of the hydrogen peroxide generated was shown to be detoxified by the NADPH-TXNRD2-TXN2-peroxiredoxin system, while only 25% was detoxified by the NADPH-glutathione reductase-GSH-glutathione peroxidase system [108]. This result was quite surprising given that GSH is present at roughly 100-1,000 fold higher levels than Txn2 in mitochondria [109]. In addition to reducing oxidized thioredoxin, thioredoxin reductases can reduce dehydroascorbate [110], ubiquinone [111], cytochrome c [112], lipoamide, lipoic acid, [113], and lipid hydroperoxides [114], which also may play a role in the regulation of lifespan.

Knockdown of *C. elegans* cytoplasmic thioredoxin reductase *trxr-1* [115,116] or mitochondrial thioredoxin reductase *trxr-2* [117] by RNA interference decreased lifespan. However, deletion of *trxr-1* did not affect lifespan [117], while deletion of *trx-2* decreased lifespan at 25 °C, a temperature at the high end of the normal culture range, but not at 20 °C, the most typical culture temperature [117]. TRXR-2 was found to be protective against beta-amyloid toxicity in a *C. elegans* Alzheimer's disease model [118], while TRXR-1 was found to be protective against 6-hydroxydopamine toxicity in a nematode Parkinson's disease model [119].

C. elegans peroxiredoxin *prdx-2*, homologous to human PRDX1, was required for the lifespan extension that occurs in *daf-2* insulin receptor-deficient worms [120] or when metformin was added to the culture medium [121]. The *prdx-2* gene was also required for the longevity that was induced by exposing *C. elegans* larvae to a 25 °C temperature when the culture temperature was otherwise kept at 15 °C [122]. However PRDX-2 did not become oxidized during the normal aging process in *C. elegans* [123], so its role in lifespan extension could be independent of redox changes.

Perhaps the redox enzyme with the strongest connection with longevity in invertebrates is methionine sulfoxide reductase A (MSRA). During oxidative stress, methionine is oxidized to methionine sulfoxide. MSRA catalyzes the reduction of methionine sulfoxide back to methionine. But MSRA must be reduced by the NADPH-thioredoxin reductase-thioredoxin system to be recycled for use [124]. There are two types of genes that encode isoforms of methione sulfoxide reductase, A and B, that each reduces one of the two isomers of methione sulfoxide. MSRA has been more closely linked with lifespan extension and uses both free methionine sulfoxide and the methionine sulfoxide present in proteins as substrates, while MSRB can only the use methionine-sulfoxide present in proteins [125]. MSRA is present in the cytoplasm, mitochondria and nucleus [126], while there are 3 genes in mammals encoding MSRBs, where the gene products localize to the cytoplasm, mitochondria, nucleus, and ER [126].

Knockdown of the *C. elegans* homolog of MSRA called *msra-1* decreased lifespan [127], likely due to an increase in protein unfolding and proteotoxicity when methionine sulfoxide cannot be repaired. *Msra-1* is induced by the pro-longevity transcription factor DAF-16/FOXO that becomes activated when insulin signaling is blocked. Knockdown of *msra-1* in long-lived *daf-2* insulin receptor mutant worms decreased lifespan. *Drosophila* that simultaneously lack both MSRA and MSRB are short-lived [128], while overexpression of the *Drosophila* homolog of MSRA, mostly in neurons, extended lifespan [129] and stimulated nuclear translocation of dFOXO [130]. However, in mice overexpression of MSRA in either the cytoplasm or mitochondria did not extend lifespan [131]. This may be due to the fact that *Drosophila* MSRA lacks methionine oxidase activity while mammalian MSRA contains methionine oxidase activity [132].

7. The NADPH-Glutathione Reductase Redox System and Aging

By analyzing data from multiple rodent and primate species, a positive association between glutathione reductase (GSR) protein levels (and activity) and lifespan in fibroblasts was found [104]. But a negative correlation between brain glutathione reductase activity and lifespan was found in another study [133]. *Drosophila* lacks a glutathione reductase gene and its function is replaced by the two thioredoxin reductase genes [134]. However, *C. elegans* studies of the NADPH-glutathione reductase redox system and aging have been performed. Like mammals *C. elegans* contain only one glutathione reductase gene *gsr-1* that generates both cytoplasmic and mitochondrial protein variants [135]. Lifespan extension of 18% occurred following overexpression [136]. Knockout of *gsr-1* was lethal as it is required for molting [137]. Transgenic overexpression of the cytoplasmic isoform but not the mitochondrial isoform rescued lethality [135]. Knockdown of *gsr-1* by RNAi resulted in decreased lifespan and increased sensitivity to the superoxide generator juglone. Decreased GSR-1 levels may decrease lifespan in part through decreasing autophagy and increasing proteotoxicity [138].

Global low level overexpression of the catalytic subunit of glutamate-cysteine ligase (GCL$_c$), the rate-limiting step in glutathione synthesis, or high level overexpression in neurons has also been shown to increase lifespan in *Drosophila* by up to 50%, while overexpression of the regulatory subunit of GCL increased lifespan by up to 24% [139,140]. Administering a recombinant glutaredoxin (small protein reduced by GSH) to *C. elegans* culture medium was shown to extend lifespan dependent upon upregulation of the pro-longevity transcriptional regulator heat shock factor-1 (HSF-1) [141]. Deletion of either of the yeast glutaredoxin genes Grx1 or Grx2 decreased chronological longevity by increasing ROS production that activated Ras-protein kinase A signaling. Therefore, altered redox status and glutaredoxin function modulates lifespan through the activation of longevity regulating signaling pathways.

Surprisingly, young naked mole rats have been shown to have a more oxidized liver GSSG/GSH than mice [142], even though the naked mole rats live much longer. But the naked mole rats did not show increased levels of oxidative damage with aging as was shown in mice [54]. In mice and rats, important redox couples have been shown to be maintained at distinct, non-equilibrium potentials in different subcellular compartments [143]. The mitochondrial [NADP$^+$]/[[NADPH] couple has been estimated to be around −419 mV compared to −400 mV for the cytoplasmic couple [58]. The mitochondrial GSSG/GSH couple was estimated to be between −280 and -330 mV [144–146], while the cytoplasmic couple was estimated to be between −200 and −260 mV [145]. So the existing data on GSSG/GSH ratios cannot be used to estimate cytoplasmic or mitochondrial [NADP$^+$]/[NADPH]. Therefore, the cytoplasmic and/or mitochondrial [NADP$^+$]/[NADPH] may not decline with aging in naked mole rats as they do in mice to drive longevity through maintaining thioredoxin and glutathione reductase activities in one or both of these compartments. Since several markers of oxidative damage, such as urinary isoprostanes (marker of lipid peroxidation) and liver malondialdehyde (marker of lipid damage) are higher in aged naked mole rats than in aged mice [54], naked mole rat redox potential is likely only maintained with aging in one or more specific subcellular compartments and not the entire cell as a whole.

Identifying the redox events that drive the aging process is a high priority. Measurements using isolated mitochondria may be required if the redox events that regulate aging are mostly confined to this organelle [107]. Some data suggest that the mitochondrial [NADP$^+$]/[NADPH] redox systems may play a role in longevity regulation, but further studies manipulating the mitochondrial [NADP$^+$]/[NADPH], GSSG/GSH, and oxidized thioredoxin/reduced thioredoxin ratios and monitoring lifespan are required to obtain a better understanding of how the activity of the NADPH-powered mitochondrial redox systems influence lifespan.

8. Evidence for the Regulation of Longevity by the NADPH-Powered Redox Systems in *Drosophila*

Data supporting the importance of the NADPH-powered redox networks in the regulation of *Drosophila* longevity include overexpression of glucose-6-phosphate dehydrogenase (G6PD), an NADPH-generating enzyme of the oxidative PPP, increasing lifespan by up to 50% [147]. Expression from a neuronal-specific promoter was able to extend lifespan by up to 40%. Longer-lived strains of *Drosophila* were shown to have higher G6PD activity than shorter-lived strains [148]. Knockdown of ribose-5-phosphate isomerase, an enzyme of the non-oxidative portion of the PPP, increased G6PD and NADPH levels and lifespan by over 30% [149]. Overexpression of the NADPH-generating cytoplasmic malic enzyme (Men) [150] extended lifespan in *Drosophila* as well [151,152]. Overexpression of Men, either throughout lifespan or only during larval development, was able to stimulate longevity, while adulthood only expression in the fat body did not extend lifespan. Surprisingly ROS production was increased when overexpression occurred during larval development that increased stress resistance during adulthood. The data suggest that cytoplasmic reductive stress may have been responsible for the increased mitochondrial ROS production during larval development resulting in the lifespan extension. Men, cytoplasmic isocitrate dehydrogenase (Idh), and G6PD have been shown to form a network maintaining cytoplasmic NADPH levels stable with Men having a slightly larger role than either of the other two enzymes [153].

It will be important to determine if overexpression of G6PD only during larval development increases cytoplasmic reductive stress and mitochondrial ROS production during this time period to increase lifespan. Also, it will be important to determine if overexpression of Pug/Pugilist, a cytoplasmic NADPH-generating methylenetetrahydrofolate dehydrogenase enzyme [154,155] extends lifespan. However, these results may be difficult to interpret as targeted overexpression of Nmdmc, a mitochondrial NADH-generating methylenetetrahydrofolate dehydrogenase gene, in the fat body also extended lifespan [156]. Lastly, much information could be gained by determining the effects of overexpression of *Drosophila* mitochondrial NADPH-generating malic enzyme (Men-b) on lifespan. The *Drosophila* genome also contains two other uncharacterized malic enzyme-like genes Menl-1 and Menl-2 whose gene products also likely have a mitochondrial localization [157].

Further data linking NADPH levels with increased lifespan in *Drosophila* include the association between longevity and a specific allele of the cytoplasmic NADPH-generating isocitrate dehydrogenase gene Idh [158] and flies that were selected for a long lifespan showing increased activities of several NADPH-generating enzymes [159]. Interestingly, *Drosophila* lacks homologs to mammalian mitochondrial NADPH-generating isocitrate dehydrogenase (IDH2) and NADPH-generating mitochondrial nicotinamide nucleotide transhydrogenase (NNT), while homologs of both are present in *C. elegans*. However, a proteomics experiment identified the mostly cytoplasmic Idh protein as being present in *Drosophila* mitochondria [160]. Therefore, *Drosophila* likely depends heavily on malic enzyme (Men-b) and the mitochondrial localized fraction of Idh for mitochondrial NADPH production.

There are seven peroxiredoxin genes in *Drosophila*. Overexpression of three different *Drosophila* peroxiredoxin (thioredoxin peroxidase) genes has been shown to extend lifespan. When expressed in neurons dPrx1/Jafrac1, homologous to mammalian PRDX2, increased lifespan up to 29% [161]. *Drosophila* dPrx5 is endogenously localized to multiple subcellular compartments including the cytoplasm, nucleus, and mitochondria. Global overexpression increased lifespan by more than 30% [162]. Overexpression of dPrx5 specifically in mitochondria also extended lifespan [163]. dPrx3 is exclusively expressed in mitochondria. Although dPrx3 overexpression did not lead to lifespan extension, a dPrx3/dPrx5 double knockout showed premature aging phenotypes [164,165]. dPrx4 is endogenously localized to the ER, and moderate global overexpression or high level overexpression in neurons increased lifespan by greater than 30% [166]. Therefore, the NADPH-thioredoxin reductase-thioredoxin-peroxiredoxin network plays an important role in regulating aging in *Drosophila*.

9. Evidence for the Regulation of Longevity by the NADPH-Powered Redox Systems in Mammals

The study of NADPH synthesis in mammalian cells and tissues is complex as it synthesis is shared among at least 5 cytoplasmic enzymes including G6PD, 6PGD (6-phosphogluconate dehydrogenase), IDH1 (isocitrate dehydrogenase 1), ME1 (malic enzyme 1), and ALDH1L1 (10-formyltetrahydrofolate dehydrogenase) and 5 mitochondrial enzymes NNT, IDH2, ME3 (malic enzyme 3), ALDH1L2 (10-formyltetrahydrofolate dehydrogenase), and MTHFD1L (methylenetetrahydrofolate dehydrogenase 1 like) [167] (Figure 1). ALDH1L1, ALDH1L2, and MTHFD1L play a role in one-carbon and folate metabolism [167], and are an important source of NADPH in proliferating cells [168], where ribonucleotide reductase-mediated deoxynucleotide synthesis utilizes reducing equivalents from the NADPH-thioredoxin reductase-thioredoxin redox system. Overexpression of G6PD increased the mean lifespan of female mice [169].

Levels and activities of NADPH generating enzymes are altered by aging. For example, G6PD levels were shown to decline with aging in mouse brain [170], rat liver [171,172] and human erythrocytes [173], but not lymphocytes [174]. Activity of the PPP enzyme 6PGD was shown to decline by 26% with aging due to lysine oxidation, while malic enzyme activity declined 36% with aging due to histidine oxidation [175]. Activities of IDH1 and IDH2 decreased with aging in rat kidney, but increased with aging in rat testes [176]. Both of these changes were prevented by lifespan-extending calorie restriction (CR). The C57B/6J strain of mice was identified to contain a mutation in the NNT gene, but this strain has a longer lifespan than many other inbred strains. So it appears that the mice are able to compensate for the loss of NNT function and synthesize sufficient NADPH for proper mitochondrial function, at least under normal, non-stressed conditions [177].

Reducing the cytoplasmic $[NADP^+]/[NADPH]$ may not be beneficial for maintaining a reduced cellular redox environment in all cell types as NADPH oxidase enzymes use NADPH to produce superoxide, and sometimes also hydrogen peroxide [178]. For example, NADPH oxidase Nox2 complexes are present at high levels in macrophages and microglia and function to increase superoxide production. Most NADPH oxidase isoforms are not constitutively active, but require a signal such as increased Ca^{2+} levels for activation. However, the Nox4 isoform is widely expressed, constitutively active, and present in intracellular membranes such as those of mitochondria, ER, and the nucleus. Nox4 produces mostly hydrogen peroxide that is membrane permeable instead of superoxide that is not [179]. Nox4 has a K_m for NADPH of 55 μM [180], while human glutathione reductase has a K_m for NADPH of ~9 μM [181], and human cytoplasmic thioredoxin reductase (TXNRD1) has a K_m for NADPH of 18 μM. Cytoplasmic NADPH has been measured to be at a concentration of 3 μM in transformed cells, while mitochondrial NADPH was measured to be 37 μM in the same cells [61]. Therefore Nox4 activity is likely to be low under normal conditions, but would increase if Nox4 localized to the inner mitochondrial membrane with the active site facing the matrix space where increased levels of NADPH are present.

Stress increases Nox4 levels and increased Nox4 levels can lead to ETC complex I instability [182]. In addition dominant-negative Nox4 expression caused cytoplasmic reductive stress and increased mitochondrial ROS production, while Nox4 overexpression led to cytoplasmic oxidative stress. Although Nox4 appears to regulate mitochondrial ETC function and ROS production in heart and endothelial cells [183,184], Nox4 knockout mice have a normal lifespan [185]. This is likely due to the low levels of mitochondrial Nox4 in many tissues under normal, healthy conditions [186]. But increased Nox4 expression has been shown to contribute to hydrogen peroxide-induced senescence of endothelial cells [183].

10. Evidence for the Regulation of Longevity by the NADPH-Powered Redox Systems in Budding Yeast

When *S. cerevisiae* cells are cultured with glucose as the carbon source the majority of cytoplasmic NADPH was produced by the PPP enzyme Zwf1 (G6PD) and acetaldehyde dehydrogenase Ald6 [187], although the PPP 6PGD enzymes Gnd1 and Gnd2 also contributed to a certain extent [188]. When the yeast cells were cultured using a nonfermentable carbon source instead of glucose, levels of cytoplasmic isocitrate dehydrogenase Idp2 increased and it became an important source of cytoplasmic NADPH [189]. Deletion of Ald6 was shown to increase the chronological longevity of yeast [190].

Most mitochondrial NADPH in yeast is generated by the mitochondrial NADH kinase Pos5. Pos5 also contains an NAD$^+$ kinase activity and the Pos5 NAD$^+$ kinase activity followed by the activity of the NADPH-generating Ald4 acetaldehyde dehydrogenase, homologous to human ALDH2, was also shown to play a role in mitochondrial NADPH production. Deletion of Ald4 in yeast increased chronological lifespan [191]. Unlike multicellular eukaryotes, yeast mitochondrial malic enzyme Mae1 and mitochondrial NADP$^+$-dependent isocitrate dehydrogenase Idp1 only appeared to play minor roles in mitochondrial NADPH generation [187]. Like *Drosophila*, budding yeast lack a homolog to mammalian mitochondrial NNT. In yeast, mitochondrial transporters of NAD$^+$ (Ndt1 and Ndt2) provide the vast majority of NAD$^+$ for mitochondrial NADP(H) synthesis. Ndt1/Ndt2 double deletion mutants showed decreased mitochondrial and cellular NAD(H) levels and increased chronological lifespan, while Ndt1 overexpression resulted in increased mitochondrial and cellular NAD(H) levels and decreased chronological lifespan [192].

In chronologically aged *S. cerevisiae* cells, NADPH levels and levels of reduced cytoplasmic thioredoxin reductase drop before the majority of cysteine oxidation in proteins and before increased ROS production, while anti-aging dietary restriction (DR) delayed the oxidation of NADPH and thioredoxin reductase and extended lifespan [193]. Therefore, the drop in NADPH levels could be a key driving force in postmitotic cell aging that leads to redox stress. Consistent with this data, deletion of Zwf1 [194] or the cytoplasmic thioredoxin reductase Trr1 decreased chronological longevity when using a strain of the α haploid mating type (MATα), which may be a result of the inverse relationship between thioredoxin reductase activity and the activity of longevity controlling TORC1 kinase [195]. The chronologically short-lived MATα Zwf1 deletion strain had both increased NADP$^+$/NADPH and increased NAD$^+$/NADH demonstrating that the pro-aging effects of NADPH depletion were dominant over the pro-longevity effects of increased NAD$^+$/NADH in this context [194]. However, deletion of the Zwf1 gene increased chronological lifespan when a strain of the a mating type (MATa) was used [196]. Since deletion of NADPH-generating enzymes in yeast generally led to increased longevity, a lifespan-extending signaling pathway appears to be activated by increased NADP$^+$/NADPH. Other studies in yeast have shown that the heat shock-induced increase in replicative longevity was due to a transient increase in superoxide levels that signaled for the redirection of metabolic flux from glycolysis to the PPP to increase NADPH synthesis and GSH levels [197].

There are three different NADPH-generating isocitrate dehydrogenase genes in *S. cerevisiae*. Idp1 is mitochondrial, Idp2 is cytoplasmic as mentioned above, and Idp3 is peroxisomal. Deletion of each of these 3 genes increased replicative lifespan in a MATα strain [198,199], while deletion of Idp2 or Idp3 decreased replicative lifespan in a MATa strain. Deletion of Idp1 increased chronological longevity in a MATa strain [196], while deletion of Idp2 decreased chronological longevity in a MATa strain [191]. As a comparison, deletions of Idh1 or Idh2 subunits of the mitochondrial NADH-generating isocitrate dehydrogenase complex led to increased replicative lifespans when using either MATa or MATα strains [198,199]. Since these replicative lifespan analyses were performed on glucose-containing media, where the NADPH-generating isocitrate dehydrogenases do not greatly control NADPH levels, the effects on lifespan are likely due to changes in citric acid cycle flux that may alter the NAD$^+$/NADH. Studies like these using a nonfermentable carbon source instead of glucose would be more informative regarding the influence of [NADP$^+$]/[NADPH] on yeast replicative lifespan.

11. Evidence for the Regulation of Longevity by NADPH-Powered Redox Systems in *C. Elegans* Nematodes

The most commonly studied model of lifespan extension in *C. elegans* is the *daf-2* insulin receptor-deficient strain [200], which requires the DAF-16/FOXO transcriptional regulator and to a lesser extent the SKN-1/Nrf2 transcriptional regulator for the enhanced longevity [201]. The PPP enzyme GSPD-1/G6PD and the cytoplasmic NADP$^+$-dependent isocitrate dehydrogenase IDH-1 likely provide the vast majority of cytoplasmic NADPH in *C. elegans*, while the PPP enzyme T25B9.9/6PGD also likely contributes. Knockdown of *gspd-1* in an *idh-1* mutant strain led to severe growth, locomotion, and molting defects, although individual RNAi-mediated knockdown of *gspd-1* or individual knockout of *idh-1* had no effect on these phenotypes. The growth defect in *gspd-1* and *idh-1* double deficient animals was possibly mediated by alterations in amino acid metabolism as a result of NADPH depletion [202]. Decreased glutamate levels were found in these NADPH-deficient worms that may have been due to decreased glutamate dehydrogenase (GDH-1) activity, which relies upon NAD(P)H as a cofactor. However, GDH-1 is predicted to be localized to mitochondria [69], where GDH-1 would be less affected by GSPD-1 and IDH-1 deficiency than if it were localized to the cytoplasm. Folate metabolism could also contribute to a minor extent to cytoplasmic NADPH levels as *C. elegans* possesses two homologs to human MTHFD1 (NADP$^+$-dependent methylenetetrahydrofolate dehydrogenase 1), K07E3.4 and *dao-3*, the latter being transcriptionally regulated by DAF-16/FOXO [203].

Long-lived *daf-2* mutants showed increased IDH-1 activity [204], GSPD-1 levels [204,205], and levels of the mitochondrial citrate carrier (K11H3.3), a component of the mitochondrial NADPH shuttle [204]. The increase in IDH-1 activity may be especially important for longevity as we and others have shown that IDH-1 protein levels (as well as cytoplasmic aconitase (ACO-1) levels) decline greatly with aging [46,206]. Therefore, decreased cytoplasmic NADPH levels or decreased mitochondrial NADPH shuttle activity could contribute to aging phenotypes.

There are likely three major enzymes that contribute to mitochondrial NADPH generation in *C. elegans*, nicotinamide nucleotide transhydrogenase-1 (NNT-1), isocitrate dehydrogenase-2 (IDH-2), and malic enzyme (MEN-1) [69,157]. However, the localization of MEN-1 has yet to be confirmed experimentally. *Nnt-1* gene expression is induced mainly in the intestine and neurons by DAF-16 in the long-lived *daf-2* mutant worms [207,208]. Since knockdown of *nnt-1* by RNAi decreased lifespan by 18–20% in both N2 control and *daf-2* mutant worms [207], NNT-1 is likely a major regulator of mitochondrial NADPH levels. NADPH-generating malic enzyme (MEN-1) levels were not altered in long-lived *daf-2* mutants [206].

The increased longevity of *C. elegans daf-2* mutants required increased expression of *icl-1* (*gei-7*) [209], the bifunctional mitochondrial glyoxylate shunt enzyme possessing isocitrate lyase and malate synthase activity [210]. Increased ICL-1 activity decreases flux through mitochondrial IDH-2 and the heterotrimeric (IDHA-1, IDHB-1, and IDHG-1/IDHG-2) mitochondrial NADH-generating isocitrate dehydrogenase. ICL-1 expression levels also correlate with longevity in *eat-2* mutants that undergo dietary restriction and long-lived mitochondrial ETC mutants [211]. However, this correlation could be uncoupled for ETC deficiency, where knockout of the transcriptional regulator *nhr-49* prevented increased *icl-1* expression, but did not prevent increased lifespan [212]. With aging there is a strong decrease in the abundance of the IDHG-2 subunit of the NADH-generating isocitrate dehydrogenase complex [206] and a decrease in the NAD$^+$/NADH [46] that would likely increase flux through IDH-2 and counter aging-induced increases in the mitochondrial NADP$^+$/NADPH.

Knockdown of the *C. elegans* mitochondrial citric acid cycle genes aconitase (*aco-2*) [213], isocitrate dehydrogenase subunit *idha-1* [213], or alpha-ketoglutarate dehydrogenase subunits *ogdh-1* [214] or *dld-1* (dihydrolipoamide dehydrogenase) [215] extended lifespan. Knockdown of these enzymes could lead to accumulation of mitochondrial citrate, transport of citrate into the cytoplasm, its conversion into isocitrate (by ACO-1) and then into α-ketoglutarate by IDH-1 to reduce cytoplasmic [NADP$^+$]/[NADPH] and increase lifespan.

C. elegans GSPD-1/G6PD may be a very important source of NADPH for removing old cuticle for molting during larval development [137]. Consistent with its important role in NADPH synthesis a *gspd-1* knockout is inviable [216]. Unexpectedly, knocking down each of four cytoplasmic PPP enzymes 6PGD (T25B9.9), transaldolase (*tald-1*), transketolase (*tkt-1*) [217], or *gspd-1* (G6PD) [205] or knocking down *idh-1* [213] extended lifespan. The lifespan extension may have been a result of upregulated expression of compensatory NADPH generating enzymes, through ROS-mediated activation of SKN-1/Nrf2 and/or through activation of the mitochondrial unfolded protein response (UPR[mt]) [217,218]. UPR[mt] is a stress response pathway that frequently, but not always, associates with lifespan extension [219] and is associated with activation of the transcriptional regulators ATFS-1 (ATF4, ATF5, and CHOP in mammals), DVE-1, and SKN-1/Nrf2 [220]. During larval development ATFS-1 was activated in ETC complex I *nuo-6* mutants that resulted in DAF-16/FOXO and HIF-1 activation and increased longevity [221].

In *Drosophila*, mutations that decreased Idh and malic enzyme (Men) activities were accompanied by a compensatory upregulation of G6PD activity [153,222], while in mice decreased G6PD activity leads to a compensatory increase in IDH1 levels [223]. G6PD is also allosterically activated by NADP$^+$ [224,225]. Perhaps due to these responses, knockdown of *C. elegans idh-1* [213], *gspd-1* (G6PD) [205] or T25B9.9 (6PGD) [217] led to lifespan extension. Knockdown of the PPP gene *tald-1*, beginning from the L1 larval stage extended lifespan, but not when knockdown was initiated during adulthood [217]. In contrast, knockdown of *gcs-1*, the rate limiting step of glutathione synthesis, had no effect on lifespan when initiated at the L1 larval stage, but extended lifespan when initiated at the L4 larval stage [226]. Low levels of the glutathione depleting compound diethyl maleate increased lifespan through a DAF-16 and SKN-1 dependent mechanism, but high diethyl maleate levels decreased lifespan [226]. These compensatory effects that modulate longevity partly driven by DAF-16 and SKN-1 make aging studies where the cytoplasmic redox state is oxidized in young individuals difficult to interpret unless detailed redox measurements are made in parallel.

A *C. elegans* RNAi screen identified 41 genes including *gspd-1*, other PPP genes, and glutathione reductase (*gsr-1*) as the strongest inducers of the SKN-1/Nrf2 target gene *gcs-1* [218]. SKN-1 is activated by oxidative stress and regulates expression of hundreds of genes. It decreases expression of insulin-like peptides to activate DAF-16/FOXO [227]. Together DAF-16 and SKN-1 induce several NADPH-generating enzymes and antioxidant enzymes. However, DAF-16 and SKN-1 activity greatly decline with aging [228], which may contribute to the increased NADP$^+$/NADPH with aging. Juglone, an oxidant that induces robust activation of the SKN-1 reporter *Pgst-4::gfp* in young worms was completely ineffective in activating the reporter by day 10 of the lifespan [228], while *daf-2* knockdown failed to extend lifespan when initiated at day 8 of adulthood or later [229]. Consistent with their important roles in longevity, *skn-1* or *daf-16* knockdown can slightly decrease normal lifespan [230].

Compensatory responses to NADPH depletion also likely occur in humans as G6PD deficiency is associated with long life [231]. As NADPH is a known activator of the class I histone deacetylases (HDACs) HDAC1 and HDAC2 [232], part of this compensatory response may involve decreased activity of class I HDACs, which extends lifespan in model organisms [233].

Some data do not support a role for NADPH levels to be limiting for lifespan. For example, deletion of the mitochondrial *nnt-1* in *C. elegans* increased the GSSG/GSH ratio, but did not decrease lifespan [234]. However, unexpectedly as referred to above, RNAi knockdown of *nnt-1* in *C. elegans* decreased lifespan by 18% [207]. The reasons for the different effects between knockout and knockdown of *nnt-1* on lifespan are not known. Completely knocking out *nnt-1* may lead to greater ROS production and activation of SKN-1 and DAF-16-mediated protective pathways than incompletely knocking it down. Not only does ROS-mediated SKN-1/Nrf2 activation induce expression of NADPH-generating enzymes to partially compensate for decreased NADPH levels [235], it also increases expression of other antioxidants including superoxide dismutase [236], catalase [237], and enzymes of glutathione metabolism [238]. Activation of SKN-1 by knocking out its endogenous inhibitor WDR-23 extended lifespan, but unexpectedly increased the GSSG/GSH ratio [239], yet another example of how the cellular GSSG/GSH ratio is not always predictive of lifespan. Determining the effects of SKN-1 activation or DAF-16 activation on cytoplasmic and mitochondrial [NADP$^+$]/[NADPH] may help to decipher the mechanisms through which each extends lifespan.

12. Calorie Restriction May Extend Lifespan In Part through the NADPH-Driven Redox Systems

Calorie restriction (CR) extends lifespan in the vast majority of eukaryotic organisms from yeasts to humans [240] and studies using mice have shown that the methionine restriction (MR) induced by the CR diet provides much of this longevity effect through decreasing mitochondrial ROS production [241] and increasing hydrogen sulfide production through the transulfuration pathway [242]. Flux from methionine degradation through the transulfuration pathway results in cysteine production. The majority of cysteine from the diet is oxidized to cystine and requires NADPH oxidation to reduce it to cysteine once it is taken up by cells [243]. Reduction of the extracellular cystine/cysteine has been shown to increase the intracellular levels of NADPH and GSH [244]. Therefore, use of cystine for cysteine synthesis decreases cellular NADPH levels and use of the transulfuration pathway for cysteine production conserves NADPH. Somewhat paradoxically, CR or MR increases expression of transulfuration pathway enzymes [245]. Long-lived Ames dwarf mice also show increased transulfuration pathway activity [246]. The increased cysteine production through this pathway likely leads to decreased cellular cystine uptake, decreased oxidation of NADPH to reduce cystine, and stabilization of the $[NADP^+]/[NADPH]$ with aging leading to lifespan extension.

Methionine synthesis in yeast requires 3 NADPH molecules. Studies have shown that a Zwf1 (G6PD) mutant yeast strain becomes a methionine auxotroph as it is not able to produce the NADPH needed for methionine synthesis [247]. Genes of methionine metabolism appear to be co-regulated with PPP enzymes [248]. Methionine auxotrophs show extended replicative [249] and chronological longevity, while methionine supplementation decreases chronological longevity [250,251]. MR may provide an optimal level of cellular NADPH for longevity by minimizing methionine oxidation to methionine sulfoxide, which requires NADPH oxidation to repair, while maintaining high enough levels of methionine where methionine synthesis does not deplete NADPH levels. Arginine and lysine synthesis in yeast also require NADPH and synthesis becomes impaired when NADPH levels are low [252,253]. But, lysine auxotrophy does not increase chronological longevity [250].

Some of the largest consumers of NADPH include the enzymes for the biosynthesis of fatty acids, cholesterol, steroids, and deoxyribonucleotides [48]. During CR, AMP kinase (AMPK) becomes activated [254], prominently in the hypothalamus [255]. AMPK phosphorylates acetyl-CoA carboxylase to inhibit fatty acid synthesis [255]. This inhibition of fatty acid synthesis leads to the activation of fatty acid beta-oxidation (except in neurons that lack these enzymes) to provide cellular energy and prevent a futile cycle of fatty acid synthesis and breakdown. The inhibition of fatty acid synthesis preserves NADPH levels and contributes to the increased levels of redox defenses associated with the longevity benefits of CR. Stimulating AMPK using the AMP analog AICAR (5-Aminoimidazole-4-carboxamide ribonucleotide) increased the cellular $NAD^+/NADH$ to activate the NAD^+-dependent sirtuin deacetylase SIRT1 [256], which deacetylates and activates the master regulator of mitochondrial gene expression PGC-1α to stimulate mitochondrial biogenesis and increase expression of SOD2 and catalase. Blocking mitochondrial fatty acid uptake with etomoxovir blocked the change in $NAD^+/NADH$ [256]. The reduced $[NADP+]/[NADPH]$ induced by AMPK activation functions together with the increased expression of SOD2 and catalase to maintain redox status in the presence of increased mitochondrial ETC function.

The effect of the anti-aging CR diet in rodents on antioxidant systems has been reviewed [257]. Overall, the majority of studies showed no overall change in mitochondrial ROS production or antioxidant enzyme levels by CR. However, GSH was increased, and the GSSG/GSH and oxidative damage were decreased in the vast majority of studies supporting the redox stress theory of aging. The Nrf2 transcriptional regulator was not required for lifespan extension by CR [258]. Two major mechanisms likely drive the reduced GSSG/GSH and decreased oxidative damage in CR mice. These include the CR-mediated increase in mitochondrial SIRT3 levels and the increased activity of FOXO transcriptional regulators due to decreased insulin signaling during CR. SIRT3 has been shown to deacetylate and activate mitochondrial IDH2 to increase mitochondrial NADPH production [259], while the FOXO3 transcriptional regulator increases PPP flux to increase cytoplasmic NADPH [260].

Neural progenitor cells from FOXO3 knockout mice showed increased NADP$^+$/NADPH and GSSG/GSH ratios due to decreased glucose uptake and PPP pathway activity. Furthermore SIRT1 overexpression, which mimics the increased expression of SIRT1 during CR in muscle and white adipose tissue [261], stimulated the formation of a FOXO3/PGC-1α complex that was shown to induce expression of the mitochondrial redox factors TXNRD2, TXN2, peroxiredoxin 3 (PRDX3), and peroxiredoxin 5 (PRDX5) [262].

CR was also shown to blunt the aging-related loss of cytoplasmic TXNRD1 and TXN protein levels in rat kidney [263]. In contrast, in rat cardiac and skeletal muscle aging and CR had no effect on TXNRD1 levels, while CR decreased the aging-related loss of TXN [264]. In muscle, mitochondria-localized TXNRD2 decreased with aging, while CR blunted this effect. Muscle mitochondrial TXN2 levels increased with aging with no effect of CR [264]. CR was also shown to decrease the skeletal muscle levels of TXNIP, an inhibitor of TXN, to enhance TXN function [265]. Extension of lifespan by DR in yeast also required the peroxiredoxin Tsa1 and increased expression of a sulfiredoxin gene to decrease the oxidation of Tsa1 with aging [266]. In *C. elegans*, lifespan extension by DR required the thioredoxin *trx-1* [267]. Global protein cysteine levels were shown not to be oxidized during aging in *Drosophila*, although 24 hours of fasting caused a major oxidation of protein cysteines [268], most likely by decreasing PPP flux and NADPH generation. However, the use of tissue-specific redox probes for H_2O_2 and GSSG/GSH were able to measure an aging-induced oxidation of the cytoplasm of midgut enterocytes [74]. Surprisingly long-lived *chico* (insulin receptor substrate) heterozygotes showed increased, not decreased midgut oxidation.

Some portion of the disease delaying effects of CR may be mediated by the increased levels of ketone bodies that occur as a result of the CR diet [55], as a ketogenic diet increased the mean lifespan of mice [269,270]. Excitingly, exogenous ketone body supplementation partially mimicked the protective redox changes induced by CR in the brain (i.e. reduced GSSG/GSH [257]) as exogenous ketone bodies were able to reduce the cytoplasmic [NADP$^+$]/[NADPH] in the cerebral cortex of 3XTgAD Alzheimer's mice, but not in their hippocampus [271]. However, oxidative damage as measured by protein carbonyl levels (2,4-dintrophenylhydrazine reactivity) and lipid peroxidation (4-hydroxynonenal levels) was decreased by ketone body treatment in the hippocampus, but not the cortex of these animals. The data suggest that hippocampal NADPH was oxidized to prevent oxidative damage in this region of the brain and this oxidation of NADPH was the likely mechanism through which no ketone body-induced change in the [NADP$^+$]/[NADPH] was found. However, the cerebral cortex of these mice may have lacked high enough NADPH-coupled redox enzyme levels to utilize the reduced [NADP$^+$]/[NADPH] for the prevention of oxidative damage. The exogenous ketone treatment also resulted in increased cytoplasmic [NAD$^+$]/[NADH] in both brain regions similar to CR and increased mitochondrial [NAD$^+$]/[NADH] only in the hippocampus [271].

13. Regulation of Aging by NADPH and the Circadian Clock

It has been clearly demonstrated in several model organisms that robust circadian rhythms are associated with longevity and that disrupting circadian oscillations in humans and other models is associated with aging and aging-related disorders [272,273]. It was found that 43% of mouse protein coding genes oscillated in expression in a circadian manner in at least one of twelve tissues studied [274], while the levels of roughly half of the mRNAs for these proteins oscillated [275] suggesting post-transcriptional events that regulate the circadian changes in the other half of these proteins. The NAD$^+$/NADH and the NADP$^+$/NADPH ratios oscillate over a 24 hour cycle [276,277] due in part to circadian regulation of NAMPT [278,279], NADK [280], and the PPP [281,282]. Therefore, loss of circadian rhythms could play an important role in the decreased NAD$^+$ [3] and NADPH [1,2] levels observed in aged tissues. Disruption of the PPP through the administration of 6-aminonicotinic acid was described to lengthen the circadian period in one study [282], while disruption of the PPP with the inhibitors diphenyleneiodonium (DPI) or dehydroepiandrosterone (DHEA) was described to alter the phase and amplitude of the circadian changes without affecting the period in another [281]. To limit the potential side effects of small molecule inhibitors, future studies could express the highly specific NADPH-oxidizing enzyme triphosphopyridine nucleotide oxidase (TPNOX) [283], an engineered mutant of *Lactobacillus brevis* NADH oxidase, in cells and target it to the mitochondria or to the cytoplasm to determine the compartment-specific role of changes in the [NADP$^+$]/[NADPH] on the circadian clock.

CR may extend longevity in part by restoring the loss of circadian rhythms. This may occur through the increased NAD$^+$/NADH and reduced NADP$^+$/NADPH conferred by CR, as dimerization of the circadian clock proteins CLOCK and BMAL1 is regulated by redox state [284]. Also, NADPH binds the circadian transcriptional regulator NPAS2 and modulates its dimerization with BMAL1 and DNA binding of the heterodimer [285,286]. In addition, circadian oscillations, especially in genes involved with metabolism, occur in *C. elegans* [287], *Drosophila* S2 cells [288], and budding yeast [289], organisms that lack essential components of the well-characterized circadian clock, suggesting an evolutionarily ancient uncharacterized circadian system regulating metabolism. It will be important to determine the influence of changes in the [NAD$^+$]/[NADH] and [NADP$^+$]/[NADPH] on this ancient circadian system.

14. NAD$^+$ Kinases as Possible Regulators of NADPH Levels and Longevity

The human cytoplasmic NAD$^+$ kinase (NADK) was identified in 2001 [290] and has been fairly well-characterized since then, including its regulation by Ca2+/calmodulin-dependent protein kinases [291] and the control of expression by the circadian clock [280]. Mice lacking NADK are not viable [292]. The *Drosophila melanogaster* genome contains two homologous genes, while the *C. elegans* genome contains no homologs of human NADK. So up until the year 2012, it was unknown how *C. elegans* obtained NADP(H). In 2012 a pivotal discovery of the human mitochondrial NAD$^+$ kinase (NADK2) was made [293,294]. The *C. elegans* genome contains two homologous uncharacterized genes, while the fruit fly genome contains one. Recently, studies using genetically encoded iNap fluorescent NADPH sensors have verified the exchange of NADPH equivalents between the cytoplasmic and mitochondrial compartments in mammalian cells [61]. Overexpression of cytoplasmic NADK resulted in a 4-fold increase in cytoplasmic [NADPH] and a 57% increase in mitochondrial [NADPH].

Like other eukaryotes, yeast rely on cytoplasmic and mitochondrial NAD(H) kinases to provide NADP(H). The two yeast cytoplasmic NAD(H) kinases, Yef1 and Utr1 appear to prefer NAD$^+$ as a substrate unlike mitochondrial Pos5, which prefers NADH [187]. In the filamentous fungus *Podospora anserine*, loss of a mitochondrial NADH kinase gene extended lifespan by apparently inducing expression of an ETC alternative oxidase that decreased ROS production [295].

15. Pharmacological Modulation of Cellular Redox Status to Prevent Cellular Senescence and Delay Aging

Methylene blue is a redox active compound that has been demonstrated to delay cellular senescence [296] and showed promise in a phase II clinical trial for Alzheimer's disease [297]. It has been shown to accept electrons from NADH or reduced flavins and pass them to ETC complex III [298] bypassing superoxide generation from ETC complexes I or II [299]. Presumably these redox properties are responsible at least in part for the many other protective cellular responses including inhibition of the NLRP3 inflammasome [300], activation of the proteasome [301], and stabilization of HSF1 [302]. Methylene blue treatment was shown to extend the lifespan of *C. elegans* [303] and female mice [304], but not *Drosophila* [305].

A screen of 2684 compounds identified the redox-active compounds violuric acid (VA) and 1-naphthoquinone-2-monoxime (N2N1) as the top hits that delayed cellular senescence [306]. These compounds were then shown to increase the lifespan of *C. elegans* and mice. VA greatly reduced the level of cellular protein disulfides by facilitating electron transfer from NADH (and NADPH to a lesser extent) to GSSG to increase GSH and the cellular NAD$^+$/NADH. VA also facilitated the transfer of electrons from NADH to H$_2$O$_2$ to form water, but did not transfer electrons from NADH to cytochrome

c. N2N1 was shown to pass electrons from NADH (and NADPH to a lesser extent) to CoQ_{10} and to cytochrome c with the NQO1 (NAD(P)H: quinone oxidoreductase 1) redox system as a major target. Consistent with these results, overexpression of NQO1 together with another NADH-oxidizing redox protein, cytochrome b_5 reductase 3 (CYB5R3), extended lifespan in mice and mimicked some phenotypes of CR [307]. However, invertebrates such as *C. elegans* lack the NQO1 system, so the longevity effects of N2N1 and methylene blue in nematodes may be due to stimulation of electron transfer from NADH to ETC complex III or to the *C. elegans* CYB5R3 homologs HPO-19 or T05H4.4. Increasing the $NAD^+/NADH$ ratio is likely key to the longevity effects of these compounds, but VA may also stabilize the $NADP^+/NADPH$ to provide longevity benefits. Compounds that facilitate the selective transfer of electrons from NADH to GSSG, H_2O_2, oxidized thioredoxin, and other electron acceptors could increase the $NAD^+/NADH$ with aging and would therefore be compounds for the potential treatment of aging-related disorders. Compounds that facilitate the selective transfer of electrons from NADPH to these electron acceptors may also be beneficial at low levels, but at high levels they also have the potential to increase the $[NADP^+]/[NADPH]$ leading to oxidative stress.

16. Summary and Future Perspectives

Evidence from model organisms strongly suggests that redox changes influence the rate of aging. However, both oxidizing and reducing changes can lead to increased longevity making redox measurements in both cytoplasmic and mitochondrial compartments especially important in when determining the mechanism for lifespan extension. In *S. cerevisiae* and *C. elegans*, deleting or knocking down NADPH-generating enzymes increased longevity more frequently than decreasing it, likely due to the active stress response pathways that have evolved in young cells to compensate for the loss of NADPH. Manipulations that lead to a reduction of the $[NADP^+]/[NADPH]$ in the cytoplasm are especially linked with increased longevity. But reducing the redox potential of the cytoplasm frequently leads to reductive stress characterized by mitochondrial oxidation and ROS production. Mitochondrial ROS production and oxidation is linked with increased longevity when it occurs in young organisms capable of mounting robust longevity enhancing stress responses. The theory that mild reductive stress in the cytoplasm during development leads to increased mitochondrial superoxide production and lifespan extension should be tested in model organisms.

Another clear trend in the literature is the protective effects of the mitochondrial NADPH-linked redox systems. It would be helpful to explore potential molecular mechanisms behind the positive association between mitochondrial thioredoxin reductase activity and increased longevity. This could be accomplished in part through overexpression studies and measurements of mitochondrial superoxide or hydrogen peroxide production, GSSG/GSH, and $[NADP^+]/[NADPH]$ measurements. However, overexpression of mitochondrial thioredoxin reductase on its own would be expected to have little effect on longevity, as was verified in studies using *Drosophila* [104], without a corresponding reduction of the mitochondrial $[NADP^+]/[NADPH]$ that drives its enzymatic activity. Since little is known regarding the effects of a more reduced mitochondrial redox environment on longevity, future important experiments will be to overexpress a mitochondrial NADPH-generating enzyme, such as Men-b or mitochondrial targeted Idh in *Drosophila*, to determine the effects of reduced mitochondrial $[NADP^+]/[NADPH]$ on lifespan.

Funding: This research was funded by The National Institutes of Health grant number AG059096.

Acknowledgments: I would like to thank the pioneers of research in the field of redox biology and aging including Dean Jones, Rajindar Sohal, and William Orr.

Conflicts of Interest: The author declares no conflict of interest. The funders had no role in the writing of the manuscript or in the decision to publish.

Abbreviations

ADP: adenosine diphosphate; AICAR, 5-aminoimidazole-4-carboxamide ribonucleotide; ALDH1L1, cytoplasmic 10-formyltetrahydrofolate dehydrogenase; ALDH1L2, mitochondrial 10-formyltetrahydrofolate dehydrogenase; AMP, adenosine monophosphate; AMPK, AMP kinase; ATP, adenosine triphosphate; CR, calorie restriction; CYB5R3, cytochrome b5 reductase 3; DAF-16, dauer formation-16; DAF-2, dauer formation-2; DHEA, dehydroepiandrosterone; DPI, diphenyleneiodonium; DR, dietary restriction; ER, endoplasmic reticulum; ETC, Electron transport chain; FOXO, forkhead box O; G6PD, glucose-6-phosphate dehydrogenase; GCL, glutamate-cysteine ligase; GDH-1, glutamate dehydrogenase-1; GPI-1, glucose-6-phosphate isomerase-1; GPX, glutathione peroxidase; GSH, glutathione; GSSG, glutathione disulfide; GSR, glutathione reductase; HDAC, histone deacetylase; HSF-1, heat shock factor-1; IDH1, isocitrate dehydrogenase 1 (cytoplasmic); IDH-1, *C. elegans* isocitrate dehydrogenase-1; IDH2, isocitrate dehydrogenase 2 (mitochondrial); IDH-2, *C. elegans* isocitrate dehydrogenase-2; Idh, *Drosophila* isocitrate dehydrogenase; ICL-1, isocitrate lyase-1; K_m, Michaelis constant; MATa, haploid mating type a; MATα, haploid mating type α; Men, *Drosophila* cytoplasmic malic enzyme; Men-b, *Drosophila* mitochondrial malic enzyme; MEN-1, *C. elegans* malic enzyme; ME1, malic enzyme 1; ME3, malic enzyme 3; MFRTA, Mitochondrial free radical theory of aging; MSRA, methionine sulfoxide reductase A; MSRB, methionine sulfoxide reductase B; MTAP, methylthioadenosine phosphorylase; MTHFD1L, methylenetetrahydrofolate dehydrogenase 1 like; mV, milliVolt; mM, milliMolar; N2N1, 1-naphthoquinone-2-monoxime; NAC, N-acetylcysteine; NAD^+, nicotinamide adenine dinucleotide (oxidized form); NADK, NAD^+ kinase; NADK2, mitochondrial NAD^+ kinase; NADH, nicotinamide adenine dinucleotide (reduced form); $NADP^+$, nicotinamide adenine dinucleotide phosphate (oxidized form); NADPH, nicotinamide adenine dinucleotide phosphate (reduced form); NAMPT, nicotinamide phosphoribosyl transferase; NQO1, NAD(P)H: quinone oxidoreductase 1; NNT, nicotinamide nucleotide transhydrogenase; Nox4, NADPH oxidase 4; Nrf2, Nuclear factor (erythroid-derived 2)-like 2; PARP, poly-ADP-ribose polymerase; 6PGD, 6-phosphogluconate dehydrogenase; PNP, purine nucleoside phosphorylase; PPP, pentose phosphate pathway; PRDX, peroxiredoxin; ROS, reactive oxygen species; SARM1, Sterile alpha and TIR motif containing 1; SKN-1, skinhead-1; SOD, superoxide dismutase; TFAM, mitochondrial transcription factor A; TPNOX, triphosphopyridine nucleotide oxidase; TRXR-1, *C. elegans* cytoplasmic thioredoxin reductase; TRXR-2, *C. elegans* mitochondrial thioredoxin reductase; TXN, cytoplasmic thioredoxin; TXN2, mitochondrial thioredoxin, TXNRD1, cytoplasmic thioredoxin reductase; TXNRD2, mitochondrial thioredoxin reductase; UPR^{mt}, mitochondrial unfolded protein response; VA, violuric acid; V_{max}, maximal velocity.

References

1. Aw, T.Y. Postnatal Changes in Pyridine Nucleotides in Rat Hepatocytes: Composition and O2 Dependence. *Pediatr. Res.* **1991**, *30*, 112–117. [CrossRef] [PubMed]

2. Ghosh, D.; Levault, K.R.; Brewer, G.J. Relative importance of redox buffers GSH and NAD(P)H in age-related neurodegeneration and Alzheimer disease-like mouse neurons. *Aging Cell* **2014**, *13*, 631–640. [CrossRef] [PubMed]

3. Zhu, X.H.; Lu, M.; Lee, B.Y.; Ugurbil, K.; Chen, W. In vivo NAD assay reveals the intracellular NAD contents and redox state in healthy human brain and their age dependences. *Proc. Natl. Acad. Sci. USA* **2015**, *112*, 2876–2881. [CrossRef] [PubMed]

4. Lenaz, G.; D'Aurelio, M.; Merlo Pich, M.; Genova, M.L.; Ventura, B.; Bovina, C.; Formiggini, G.; Parenti Castelli, G. Mitochondrial bioenergetics in aging. *Biochim. Biophys. Acta* **2000**, *1459*, 397–404. [CrossRef]

5. Baciou, L.; Masoud, R.; Souabni, H.; Serfaty, X.; Karimi, G.; Bizouarn, T.; Houee Levin, C. Phagocyte NADPH oxidase, oxidative stress and lipids: Anti- or pro ageing? *Mech. Ageing Dev.* **2018**, *172*, 30–34. [CrossRef] [PubMed]

6. Huang, Q.; Sun, M.; Li, M.; Zhang, D.; Han, F.; Wu, J.C.; Fukunaga, K.; Chen, Z.; Qin, Z.H. Combination of NAD(+) and NADPH Offers Greater Neuroprotection in Ischemic Stroke Models by Relieving Metabolic Stress. *Mol. Neurobiol.* **2018**, *55*, 6063–6075. [CrossRef] [PubMed]

7. Harman, D. The biologic clock: The mitochondria? *J. Am. Geriatr. Soc.* **1972**, *20*, 145–147. [CrossRef] [PubMed]

8. Barja, G. Updating the mitochondrial free radical theory of aging: An integrated view, key aspects, and confounding concepts. *Antioxid. Redox Signal.* **2013**, *19*, 1420–1445. [CrossRef] [PubMed]

9. Gems, D.; Doonan, R. Antioxidant defense and aging in *C. elegans*: Is the oxidative damage theory of aging wrong? *Cell Cycle* **2009**, *8*, 1681–1687. [CrossRef] [PubMed]

10. Jang, Y.C.; Perez, V.I.; Song, W.; Lustgarten, M.S.; Salmon, A.B.; Mele, J.; Qi, W.; Liu, Y.; Liang, H.; Chaudhuri, A.; et al. Overexpression of Mn superoxide dismutase does not increase life span in mice. *J. Gerontol. A Biol. Sci. Med. Sci.* **2009**, *64*, 1114–1125. [CrossRef] [PubMed]

11. Perez, V.I.; Van Remmen, H.; Bokov, A.; Epstein, C.J.; Vijg, J.; Richardson, A. The overexpression of major antioxidant enzymes does not extend the lifespan of mice. *Aging Cell* **2009**, *8*, 73–75. [CrossRef] [PubMed]

12. Selman, C.; McLaren, J.S.; Meyer, C.; Duncan, J.S.; Redman, P.; Collins, A.R.; Duthie, G.G.; Speakman, J.R. Life-long vitamin C supplementation in combination with cold exposure does not affect oxidative damage or lifespan in mice, but decreases expression of antioxidant protection genes. *Mech. Ageing Dev.* **2006**, *127*, 897–904. [CrossRef] [PubMed]

13. Ernst, I.M.; Pallauf, K.; Bendall, J.K.; Paulsen, L.; Nikolai, S.; Huebbe, P.; Roeder, T.; Rimbach, G. Vitamin E supplementation and lifespan in model organisms. *Ageing Res. Rev.* **2013**, *12*, 365–375. [CrossRef] [PubMed]

14. Desjardins, D.; Cacho-Valadez, B.; Liu, J.L.; Wang, Y.; Yee, C.; Bernard, K.; Khaki, A.; Breton, L.; Hekimi, S. Antioxidants reveal an inverted U-shaped dose-response relationship between reactive oxygen species levels and the rate of aging in Caenorhabditis elegans. *Aging Cell* **2017**, *16*, 104–112. [CrossRef] [PubMed]

15. Orr, W.C.; Sohal, R.S. Does overexpression of Cu,Zn-SOD extend life span in Drosophila melanogaster? *Exp. Gerontol.* **2003**, *38*, 227–230. [CrossRef]

16. Shibamura, A.; Ikeda, T.; Nishikawa, Y. A method for oral administration of hydrophilic substances to Caenorhabditis elegans: Effects of oral supplementation with antioxidants on the nematode lifespan. *Mech. Ageing Dev.* **2009**, *130*, 652–655. [CrossRef] [PubMed]

17. Schriner, S.E.; Linford, N.J.; Martin, G.M.; Treuting, P.; Ogburn, C.E.; Emond, M.; Coskun, P.E.; Ladiges, W.; Wolf, N.; Van Remmen, H.; et al. Extension of murine life span by overexpression of catalase targeted to mitochondria. *Science* **2005**, *308*, 1909–1911. [CrossRef] [PubMed]

18. Lebovitz, R.M.; Zhang, H.; Vogel, H.; Cartwright, J., Jr.; Dionne, L.; Lu, N.; Huang, S.; Matzuk, M.M. Neurodegeneration, myocardial injury, and perinatal death in mitochondrial superoxide dismutase-deficient mice. *Proc. Natl. Acad. Sci. USA* **1996**, *93*, 9782–9787. [CrossRef] [PubMed]

19. Kirkman, H.N.; Gaetani, G.F. Catalase: A tetrameric enzyme with four tightly bound molecules of NADPH. *Proc. Natl. Acad. Sci. USA* **1984**, *81*, 4343–4347. [CrossRef] [PubMed]

20. Sohal, R.S.; Orr, W.C. The redox stress hypothesis of aging. *Free Radic. Biol. Med.* **2012**, *52*, 539–555. [CrossRef] [PubMed]

21. Go, Y.M.; Jones, D.P. Redox theory of aging: Implications for health and disease. *Clin. Sci. (Lond. Engl. 1979)* **2017**, *131*, 1669–1688. [CrossRef]

22. Heidler, T.; Hartwig, K.; Daniel, H.; Wenzel, U. *Caenorhabditis elegans* lifespan extension caused by treatment with an orally active ROS-generator is dependent on DAF-16 and SIR-2.1. *Biogerontology* **2010**, *11*, 183–195. [CrossRef] [PubMed]

23. Schulz, T.J.; Zarse, K.; Voigt, A.; Urban, N.; Birringer, M.; Ristow, M. Glucose restriction extends *Caenorhabditis elegans* life span by inducing mitochondrial respiration and increasing oxidative stress. *Cell Metab.* **2007**, *6*, 280–293. [CrossRef] [PubMed]

24. Barja, G. Rate of generation of oxidative stress-related damage and animal longevity. *Free Radic. Biol. Med.* **2002**, *33*, 1167–1172. [CrossRef]

25. Schindeldecker, M.; Stark, M.; Behl, C.; Moosmann, B. Differential cysteine depletion in respiratory chain complexes enables the distinction of longevity from aerobicity. *Mech. Ageing Dev.* **2011**, *132*, 171–179. [CrossRef] [PubMed]

26. Pamplona, R.; Barja, G.; Portero-Otin, M. Membrane fatty acid unsaturation, protection against oxidative stress, and maximum life span: A homeoviscous-longevity adaptation? *Ann. N. Y. Acad. Sci.* **2002**, *959*, 475–490. [CrossRef] [PubMed]

27. Moosmann, B. Respiratory chain cysteine and methionine usage indicate a causal role for thiyl radicals in aging. *Exp. Gerontol.* **2011**, *46*, 164–169. [CrossRef] [PubMed]

28. Paradies, G.; Paradies, V.; Ruggiero, F.M.; Petrosillo, G. Cardiolipin and mitochondrial function in health and disease. *Antioxid. Redox Signal.* **2014**, *20*, 1925–1953. [CrossRef] [PubMed]

29. Ma, C.; Pi, C.; Yang, Y.; Lin, L.; Shi, Y.; Li, Y.; Li, Y.; He, X. Nampt Expression Decreases Age-Related Senescence in Rat Bone Marrow Mesenchymal Stem Cells by Targeting Sirt1. *PLoS ONE* **2017**, *12*, e0170930. [CrossRef] [PubMed]

30. Ahn, B.H.; Kim, H.S.; Song, S.; Lee, I.H.; Liu, J.; Vassilopoulos, A.; Deng, C.X.; Finkel, T. A role for the mitochondrial deacetylase Sirt3 in regulating energy homeostasis. *Proc. Natl. Acad. Sci. USA* **2008**, *105*, 14447–14452. [CrossRef] [PubMed]

31. Nemoto, S.; Fergusson, M.M.; Finkel, T. SIRT1 functionally interacts with the metabolic regulator and transcriptional coactivator PGC-1α. *J. Biol. Chem.* **2005**, *280*, 16456–16460. [CrossRef] [PubMed]

32. Braidy, N.; Guillemin, G.J.; Mansour, H.; Chan-Ling, T.; Poljak, A.; Grant, R. Age related changes in NAD+ metabolism oxidative stress and Sirt1 activity in wistar rats. *PLoS ONE* **2011**, *6*, e19194. [CrossRef] [PubMed]

33. Camacho-Pereira, J.; Tarrago, M.G.; Chini, C.C.S.; Nin, V.; Escande, C.; Warner, G.M.; Puranik, A.S.; Schoon, R.A.; Reid, J.M.; Galina, A.; et al. CD38 Dictates Age-Related NAD Decline and Mitochondrial Dysfunction through an SIRT3-Dependent Mechanism. *Cell Metab.* **2016**, *23*, 1127–1139. [CrossRef] [PubMed]

34. Boslett, J.; Helal, M.; Chini, E.; Zweier, J.L. Genetic deletion of CD38 confers post-ischemic myocardial protection through preserved pyridine nucleotides. *J. Mol. Cell. Cardiol.* **2018**, *118*, 81–94. [CrossRef] [PubMed]

35. Vu, C.Q.; Lu, P.J.; Chen, C.S.; Jacobson, M.K. 2′-Phospho-cyclic ADP-ribose, a calcium-mobilizing agent derived from NADP. *J. Biol. Chem.* **1996**, *271*, 4747–4754. [PubMed]

36. Andrabi, S.A.; Umanah, G.K.E.; Chang, C.; Stevens, D.A.; Karuppagounder, S.S.; Gagné, J.-P.; Poirier, G.G.; Dawson, V.L.; Dawson, T.M. Poly(ADP-ribose) polymerase-dependent energy depletion occurs through inhibition of glycolysis. *Proc. Natl. Acad. Sci. USA* **2014**, *111*, 10209–10214. [CrossRef] [PubMed]

37. Essuman, K.; Summers, D.W.; Sasaki, Y.; Mao, X.; DiAntonio, A.; Milbrandt, J. The SARM1 Toll/Interleukin-1 Receptor Domain Possesses Intrinsic NAD(+) Cleavage Activity that Promotes Pathological Axonal Degeneration. *Neuron* **2017**, *93*, 1334–1343.e5. [CrossRef]

38. Ferrero, E.; Malavasi, F. A Natural History of the Human CD38 Gene. In *Cyclic ADP-Ribose and NAADP: Structures, Metabolism and Functions*; Lee, H.C., Ed.; Springer: Boston, MA, USA, 2002; pp. 65–79.

39. Sur, M.; Dey, P.; Sarkar, A.; Bar, S.; Banerjee, D.; Bhat, S.; Mukherjee, P. Sarm1 induction and accompanying inflammatory response mediates age-dependent susceptibility to rotenone-induced neurotoxicity. *Cell Death Discov.* **2018**, *4*, 114. [CrossRef] [PubMed]

40. Navas, P.; Minnifield, N.; Sun, I.; Morre, D.J. NADP phosphatase as a marker in free-flow electrophoretic separations for cisternae of the Golgi apparatus midregion. *Biochim. Biophys. Acta* **1986**, *881*, 1–9. [CrossRef]

41. Richter, C. NADP+ phosphatase: A novel mitochondrial enzyme. *Biochem. Biophys. Res. Commun.* **1987**, *146*, 253–257. [CrossRef]

42. Balan, V.; Miller, G.S.; Kaplun, L.; Balan, K.; Chong, Z.Z.; Li, F.; Kaplun, A.; VanBerkum, M.F.; Arking, R.; Freeman, D.C.; et al. Life span extension and neuronal cell protection by Drosophila nicotinamidase. *J. Biol. Chem.* **2008**, *283*, 27810–27819. [CrossRef] [PubMed]

43. Vrablik, T.L.; Huang, L.; Lange, S.E.; Hanna-Rose, W. Nicotinamidase modulation of NAD+ biosynthesis and nicotinamide levels separately affect reproductive development and cell survival in *C. elegans*. *Development* **2009**, *136*, 3637–3646. [CrossRef] [PubMed]

44. McReynolds, M.R.; Wang, W.; Holleran, L.M.; Hanna-Rose, W. Uridine monophosphate synthetase enables eukaryotic de novo NAD+ biosynthesis from quinolinic acid. *J. Biol. Chem.* **2017**, *292*, 11147–11153. [CrossRef] [PubMed]

45. Breda, C.; Sathyasaikumar, K.V.; Sograte Idrissi, S.; Notarangelo, F.M.; Estranero, J.G.; Moore, G.G.; Green, E.W.; Kyriacou, C.P.; Schwarcz, R.; Giorgini, F. Tryptophan-2,3-dioxygenase (TDO) inhibition ameliorates neurodegeneration by modulation of kynurenine pathway metabolites. *Proc. Natl. Acad. Sci. USA* **2016**, *113*, 5435–5440. [CrossRef] [PubMed]

46. Copes, N.; Edwards, C.; Chaput, D.; Saifee, M.; Barjuca, I.; Nelson, D.; Paraggio, A.; Saad, P.; Lipps, D.; Stevens, S.M., Jr.; et al. Metabolome and proteome changes with aging in *Caenorhabditis elegans*. *Exp. Gerontol.* **2015**, *72*, 67–84. [CrossRef] [PubMed]

47. Sohal, R.S.; Arnold, L.; Orr, W.C. Effect of age on superoxide dismutase, catalase, glutathione reductase, inorganic peroxides, TBA-reactive material, GSH/GSSG, NADPH/NADP+ and NADH/NAD+ in Drosophila melanogaster. *Mech. Ageing Dev.* **1990**, *56*, 223–235. [CrossRef]

48. Pollak, N.; Dolle, C.; Ziegler, M. The power to reduce: Pyridine nucleotides—Small molecules with a multitude of functions. *Biochem. J.* **2007**, *402*, 205–218. [CrossRef] [PubMed]

49. Hashimoto, T.; Horikawa, M.; Nomura, T.; Sakamoto, K. Nicotinamide adenine dinucleotide extends the lifespan of *Caenorhabditis elegans* mediated by sir-2.1 and daf-16. *Biogerontology* **2010**, *11*, 31–43. [CrossRef]

50. Mouchiroud, L.; Houtkooper, R.H.; Moullan, N.; Katsyuba, E.; Ryu, D.; Canto, C.; Mottis, A.; Jo, Y.S.; Viswanathan, M.; Schoonjans, K.; et al. The NAD(+)/Sirtuin Pathway Modulates Longevity through Activation of Mitochondrial UPR and FOXO Signaling. *Cell* **2013**, *154*, 430–441. [CrossRef] [PubMed]

51. Belenky, P.; Christensen, K.C.; Gazzaniga, F.; Pletnev, A.A.; Brenner, C. Nicotinamide riboside and nicotinic acid riboside salvage in fungi and mammals. Quantitative basis for Urh1 and purine nucleoside phosphorylase function in NAD+ metabolism. *J. Boil. Chem.* **2009**, *284*, 158–164. [CrossRef]

52. Gomes, A.P.; Price, N.L.; Ling, A.J.; Moslehi, J.J.; Montgomery, M.K.; Rajman, L.; White, J.P.; Teodoro, J.S.; Wrann, C.D.; Hubbard, B.P.; et al. Declining NAD(+) induces a pseudohypoxic state disrupting nuclear-mitochondrial communication during aging. *Cell* **2013**, *155*, 1624–1638. [CrossRef] [PubMed]

53. Ren, X.; Zou, L.; Zhang, X.; Branco, V.; Wang, J.; Carvalho, C.; Holmgren, A.; Lu, J. Redox Signaling Mediated by Thioredoxin and Glutathione Systems in the Central Nervous System. *Antioxid. Redox Signal.* **2017**, *27*, 989–1010. [CrossRef] [PubMed]

54. Andziak, B.; Buffenstein, R. Disparate patterns of age-related changes in lipid peroxidation in long-lived naked mole-rats and shorter-lived mice. *Aging Cell* **2006**, *5*, 525–532. [CrossRef] [PubMed]

55. Veech, R.L.; Bradshaw, P.C.; Clarke, K.; Curtis, W.; Pawlosky, R.; King, M.T. Ketone bodies mimic the life span extending properties of caloric restriction. *IUBMB Life* **2017**, *69*, 305–314. [CrossRef] [PubMed]

56. Veech, R.L.; Eggleston, L.V.; Krebs, H.A. The redox state of free nicotinamide-adenine dinucleotide phosphate in the cytoplasm of rat liver. *Biochem. J.* **1969**, *115*, 609–619. [CrossRef] [PubMed]

57. Son, M.J.; Kwon, Y.; Son, T.; Cho, Y.S. Restoration of Mitochondrial NAD(+) Levels Delays Stem Cell Senescence and Facilitates Reprogramming of Aged Somatic Cells. *Stem Cells* **2016**, *34*, 2840–2851. [CrossRef] [PubMed]

58. Tischler, M.E.; Friedrichs, D.; Coll, K.; Williamson, J.R. Pyridine nucleotide distributions and enzyme mass action ratios in hepatocytes from fed and starved rats. *Arch. Biochem. Biophys.* **1977**, *184*, 222–236. [CrossRef]

59. Owen, J.B.; Butterfield, D.A. Measurement of oxidized/reduced glutathione ratio. *Methods Mol. Biol.* **2010**, *648*, 269–277. [CrossRef] [PubMed]

60. Chai, Y.C.; Ashraf, S.S.; Rokutan, K.; Johnston, R.B., Jr.; Thomas, J.A. S-thiolation of individual human neutrophil proteins including actin by stimulation of the respiratory burst: Evidence against a role for glutathione disulfide. *Arch. Biochem. Biophys.* **1994**, *310*, 273–281. [CrossRef] [PubMed]

61. Tao, R.; Zhao, Y.; Chu, H.; Wang, A.; Zhu, J.; Chen, X.; Zou, Y.; Shi, M.; Liu, R.; Su, N.; et al. Genetically encoded fluorescent sensors reveal dynamic regulation of NADPH metabolism. *Nat. Methods* **2017**, *14*, 720–728. [CrossRef] [PubMed]

62. Yap, L.P.; Garcia, J.V.; Han, D.; Cadenas, E. The energy-redox axis in aging and age-related neurodegeneration. *Adv. Drug Deliv. Rev.* **2009**, *61*, 1283–1298. [CrossRef] [PubMed]

63. Canepa, L.; Ferraris, A.M.; Miglino, M.; Gaetani, G.F. Bound and unbound pyridine dinucleotides in normal and glucose-6-phosphate dehydrogenase-deficient erythrocytes. *Biochim. Biophys. Acta* **1991**, *1074*, 101–104. [CrossRef]

64. Booty, L.M.; Gawel, J.M.; Cvetko, F.; Caldwell, S.T.; Hall, A.R.; Mulvey, J.F.; James, A.M.; Hinchy, E.C.; Prime, T.A.; Arndt, S.; et al. Selective Disruption of Mitochondrial Thiol Redox State in Cells and In Vivo. *Cell Chem. Boil.* **2019**. [CrossRef] [PubMed]

65. Jones, D.P.; Mody, V.C., Jr.; Carlson, J.L.; Lynn, M.J.; Sternberg, P., Jr. Redox analysis of human plasma allows separation of pro-oxidant events of aging from decline in antioxidant defenses. *Free Radic. Biol. Med.* **2002**, *33*, 1290–1300. [CrossRef]

66. Clement, J.; Wong, M.; Poljak, A.; Sachdev, P.; Braidy, N. The Plasma NAD(+) Metabolome Is Dysregulated in "Normal" Aging. *Rejuvenation Res.* **2018**. [CrossRef] [PubMed]

67. Dellinger, R.W.; Santos, S.R.; Morris, M.; Evans, M.; Alminana, D.; Guarente, L.; Marcotulli, E. Repeat dose NRPT (nicotinamide riboside and pterostilbene) increases NAD(+) levels in humans safely and sustainably: A randomized, double-blind, placebo-controlled study. *Npj Aging Mech. Dis.* **2017**, *3*, 17. [CrossRef] [PubMed]

68. Zhou, Y.; Wu, J.; Sheng, R.; Li, M.; Wang, Y.; Han, R.; Han, F.; Chen, Z.; Qin, Z.H. Reduced Nicotinamide Adenine Dinucleotide Phosphate Inhibits MPTP-Induced Neuroinflammation and Neurotoxicity. *Neuroscience* **2018**, *391*, 140–153. [CrossRef] [PubMed]

69. Horton, P.; Park, K.J.; Obayashi, T.; Fujita, N.; Harada, H.; Adams-Collier, C.J.; Nakai, K. WoLF PSORT: Protein localization predictor. *Nucleic Acids Res.* **2007**, *35*, W585–W587. [CrossRef] [PubMed]

70. Gutteridge, J.M.C.; Halliwell, B. Mini-Review: Oxidative stress, redox stress or redox success? *Biochem. Biophys. Res. Commun.* **2018**, *502*, 183–186. [CrossRef] [PubMed]

71. Zou, Y.; Wang, A.; Shi, M.; Chen, X.; Liu, R.; Li, T.; Zhang, C.; Zhang, Z.; Zhu, L.; Ju, Z.; et al. Analysis of redox landscapes and dynamics in living cells and in vivo using genetically encoded fluorescent sensors. *Nat. Protoc.* **2018**, *13*, 2362–2386. [CrossRef] [PubMed]

72. Hanson, G.T.; Aggeler, R.; Oglesbee, D.; Cannon, M.; Capaldi, R.A.; Tsien, R.Y.; Remington, S.J. Investigating mitochondrial redox potential with redox-sensitive green fluorescent protein indicators. *J. Biol. Chem.* **2004**, *279*, 13044–13053. [CrossRef] [PubMed]

73. Gutscher, M.; Pauleau, A.L.; Marty, L.; Brach, T.; Wabnitz, G.H.; Samstag, Y.; Meyer, A.J.; Dick, T.P. Real-time imaging of the intracellular glutathione redox potential. *Nat. Methods* **2008**, *5*, 553–559. [CrossRef] [PubMed]

74. Albrecht, S.C.; Barata, A.G.; Grosshans, J.; Teleman, A.A.; Dick, T.P. In vivo mapping of hydrogen peroxide and oxidized glutathione reveals chemical and regional specificity of redox homeostasis. *Cell Metab.* **2011**, *14*, 819–829. [CrossRef] [PubMed]

75. Zhang, H.; Limphong, P.; Pieper, J.; Liu, Q.; Rodesch, C.K.; Christians, E.; Benjamin, I.J. Glutathione-dependent reductive stress triggers mitochondrial oxidation and cytotoxicity. *FASEB J.* **2012**, *26*, 1442–1451. [CrossRef] [PubMed]

76. Booty, L.M.; King, M.S.; Thangaratnarajah, C.; Majd, H.; James, A.M.; Kunji, E.R.; Murphy, M.P. The mitochondrial dicarboxylate and 2-oxoglutarate carriers do not transport glutathione. *FEBS Lett.* **2015**, *589*, 621–628. [CrossRef] [PubMed]

77. Wadey, A.L.; Muyderman, H.; Kwek, P.T.; Sims, N.R. Mitochondrial glutathione uptake: Characterization in isolated brain mitochondria and astrocytes in culture. *J. Neurochem.* **2009**, *109* (Suppl. 1), 101–108. [CrossRef] [PubMed]

78. Blacker, T.S.; Duchen, M.R. Investigating mitochondrial redox state using NADH and NADPH autofluorescence. *Free Radic. Biol. Med.* **2016**, *100*, 53–65. [CrossRef] [PubMed]

79. Korge, P.; Calmettes, G.; Weiss, J.N. Increased reactive oxygen species production during reductive stress: The roles of mitochondrial glutathione and thioredoxin reductases. *Biochim. Biophys. Acta* **2015**, *1847*, 514–525. [CrossRef] [PubMed]

80. Ralser, M.; Benjamin, I.J. Reductive stress on life span extension in *C. elegans. BMC Res. Notes* **2008**, *1*, 19. [CrossRef] [PubMed]

81. Ralser, M.; Wamelink, M.M.; Kowald, A.; Gerisch, B.; Heeren, G.; Struys, E.A.; Klipp, E.; Jakobs, C.; Breitenbach, M.; Lehrach, H.; et al. Dynamic rerouting of the carbohydrate flux is key to counteracting oxidative stress. *J. Boil.* **2007**, *6*, 10. [CrossRef] [PubMed]

82. Breitenbach, M.; Laun, P.; Dickinson, J.R.; Klocker, A.; Rinnerthaler, M.; Dawes, I.W.; Aung-Htut, M.T.; Breitenbach-Koller, L.; Caballero, A.; Nystrom, T.; et al. The role of mitochondria in the aging processes of yeast. *Sub-Cell. Biochem.* **2012**, *57*, 55–78. [CrossRef]

83. Rajasekaran, N.S.; Connell, P.; Christians, E.S.; Yan, L.J.; Taylor, R.P.; Orosz, A.; Zhang, X.Q.; Stevenson, T.J.; Peshock, R.M.; Leopold, J.A.; et al. Human alpha B-crystallin mutation causes oxido-reductive stress and protein aggregation cardiomyopathy in mice. *Cell* **2007**, *130*, 427–439. [CrossRef] [PubMed]

84. Brand, M.D.; Goncalves, R.L.; Orr, A.L.; Vargas, L.; Gerencser, A.A.; Borch Jensen, M.; Wang, Y.T.; Melov, S.; Turk, C.N.; Matzen, J.T.; et al. Suppressors of Superoxide-H_2O_2 Production at Site IQ of Mitochondrial Complex I Protect against Stem Cell Hyperplasia and Ischemia-Reperfusion Injury. *Cell Metab.* **2016**, *24*, 582–592. [CrossRef] [PubMed]

85. Orr, A.L.; Vargas, L.; Turk, C.N.; Baaten, J.E.; Matzen, J.T.; Dardov, V.J.; Attle, S.J.; Li, J.; Quackenbush, D.C.; Goncalves, R.L.; et al. Suppressors of superoxide production from mitochondrial complex III. *Nat. Chem. Boil.* **2015**, *11*, 834–836. [CrossRef] [PubMed]

86. Lewis, K.N.; Andziak, B.; Yang, T.; Buffenstein, R. The naked mole-rat response to oxidative stress: Just deal with it. *Antioxid. Redox Signal.* **2013**, *19*, 1388–1399. [CrossRef] [PubMed]

87. Broniowska, K.A.; Diers, A.R.; Hogg, N. S-nitrosoglutathione. *Biochim. Biophys. Acta* **2013**, *1830*, 3173–3181. [CrossRef] [PubMed]

88. Ran, Q.; Liang, H.; Ikeno, Y.; Qi, W.; Prolla, T.A.; Roberts, L.J., 2nd; Wolf, N.; Van Remmen, H.; Richardson, A. Reduction in glutathione peroxidase 4 increases life span through increased sensitivity to apoptosis. *J. Gerontol. A Biol. Sci. Med. Sci.* **2007**, *62*, 932–942. [CrossRef] [PubMed]

89. Mockett, R.J.; Orr, W.C.; Rahmandar, J.J.; Sohal, B.H.; Sohal, R.S. Antioxidant status and stress resistance in long- and short-lived lines of Drosophila melanogaster. *Exp. Gerontol.* **2001**, *36*, 441–463. [CrossRef]

90. Lopez-Torres, M.; Perez-Campo, R.; Rojas, C.; Cadenas, S.; Barja, G. Maximum life span in vertebrates: Relationship with liver antioxidant enzymes, glutathione system, ascorbate, urate, sensitivity to peroxidation, true malondialdehyde, in vivo H_2O_2, and basal and maximum aerobic capacity. *Mech. Ageing Dev.* **1993**, *70*, 177–199. [CrossRef]

91. Heinze, I.; Bens, M.; Calzia, E.; Holtze, S.; Dakhovnik, O.; Sahm, A.; Kirkpatrick, J.M.; Szafranski, K.; Romanov, N.; Sama, S.N.; et al. Species comparison of liver proteomes reveals links to naked mole-rat longevity and human aging. *BMC Biol.* **2018**, *16*, 82. [CrossRef] [PubMed]

92. Hawkes, H.J.; Karlenius, T.C.; Tonissen, K.F. Regulation of the human thioredoxin gene promoter and its key substrates: A study of functional and putative regulatory elements. *Biochim. Biophys. Acta* **2014**, *1840*, 303–314. [CrossRef] [PubMed]

93. Munro, D.; Baldy, C.; Pamenter, M.E.; Treberg, J.R. The exceptional longevity of the naked mole-rat may be explained by mitochondrial antioxidant defenses. *Aging Cell* **2019**, e12916. [CrossRef] [PubMed]

94. Perez, L.M.; Pareja-Galeano, H.; Sanchis-Gomar, F.; Emanuele, E.; Lucia, A.; Galvez, B.G. 'Adipaging': Ageing and obesity share biological hallmarks related to a dysfunctional adipose tissue. *J. Physiol.* **2016**, *594*, 3187–3207. [CrossRef] [PubMed]

95. Tang, Y.; Purkayastha, S.; Cai, D. Hypothalamic microinflammation: A common basis of metabolic syndrome and aging. *Trends Neurosci.* **2015**, *38*, 36–44. [CrossRef] [PubMed]

96. Ribas, V.; García-Ruiz, C.; Fernández-Checa, J.C. Glutathione and mitochondria. *Front. Pharmacol.* **2014**, *5*, 151. [CrossRef] [PubMed]

97. Abdrakhmanova, A.; Zwicker, K.; Kerscher, S.; Zickermann, V.; Brandt, U. Tight binding of NADPH to the 39-kDa subunit of complex I is not required for catalytic activity but stabilizes the multiprotein complex. *Biochim. Biophys. Acta* **2006**, *1757*, 1676–1682. [CrossRef] [PubMed]

98. Fiedorczuk, K.; Letts, J.A.; Degliesposti, G.; Kaszuba, K.; Skehel, M.; Sazanov, L.A. Atomic structure of the entire mammalian mitochondrial complex I. *Nature* **2016**, *538*, 406–410. [CrossRef] [PubMed]

99. Lesnefsky, E.J.; Hoppel, C.L. Oxidative phosphorylation and aging. *Ageing Res. Rev.* **2006**, *5*, 402–433. [CrossRef] [PubMed]

100. Stefanatos, R.; Sanz, A. Mitochondrial complex I: A central regulator of the aging process. *Cell Cycle* **2011**, *10*, 1528–1532. [CrossRef] [PubMed]

101. Cunningham, G.M.; Flores, L.C.; Roman, M.G.; Cheng, C.; Dube, S.; Allen, C.; Valentine, J.M.; Hubbard, G.B.; Bai, Y.; Saunders, T.L.; et al. Thioredoxin overexpression in both the cytosol and mitochondria accelerates age-related disease and shortens lifespan in male C57BL/6 mice. *GeroScience* **2018**, *40*, 453–468. [CrossRef] [PubMed]

102. Cunningham, G.M.; Roman, M.G.; Flores, L.C.; Hubbard, G.B.; Salmon, A.B.; Zhang, Y.; Gelfond, J.; Ikeno, Y. The paradoxical role of thioredoxin on oxidative stress and aging. *Arch. Biochem. Biophys.* **2015**, *576*, 32–38. [CrossRef] [PubMed]

103. Flores, L.C.; Roman, M.G.; Cunningham, G.M.; Cheng, C.; Dube, S.; Allen, C.; Van Remmen, H.; Hubbard, G.B.; Saunders, T.L.; Ikeno, Y. Continuous overexpression of thioredoxin 1 enhances cancer development and does not extend maximum lifespan in male C57BL/6 mice. *Pathobiol. Aging Age Relat. Dis.* **2018**, *8*, 1533754. [CrossRef] [PubMed]

104. Pickering, A.M.; Lehr, M.; Gendron, C.M.; Pletcher, S.D.; Miller, R.A. Mitochondrial thioredoxin reductase 2 is elevated in long-lived primate as well as rodent species and extends fly mean lifespan. *Aging Cell* **2017**, *16*, 683–692. [CrossRef] [PubMed]

105. Orr, W.C.; Mockett, R.J.; Benes, J.J.; Sohal, R.S. Effects of overexpression of copper-zinc and manganese superoxide dismutases, catalase, and thioredoxin reductase genes on longevity in Drosophila melanogaster. *J. Biol. Chem.* **2003**, *278*, 26418–26422. [CrossRef] [PubMed]

106. Mockett, R.J.; Sohal, R.S.; Orr, W.C. Overexpression of glutathione reductase extends survival in transgenic Drosophila melanogaster under hyperoxia but not normoxia. *FASEB J.* **1999**, *13*, 1733–1742. [CrossRef] [PubMed]

107. Rojanathammanee, L.; Rakoczy, S.; Brown-Borg, H.M. Growth hormone alters the glutathione S-transferase and mitochondrial thioredoxin systems in long-living Ames dwarf mice. *J. Gerontol. A Biol. Sci. Med. Sci.* **2014**, *69*, 1199–1211. [CrossRef] [PubMed]

108. Drechsel, D.A.; Patel, M. Respiration-dependent H_2O_2 Removal in Brain Mitochondria via the Thioredoxin/Peroxiredoxin System. *J. Boil. Chem.* **2010**, *285*, 27850–27858. [CrossRef] [PubMed]

109. Calabrese, G.; Morgan, B.; Riemer, J. Mitochondrial Glutathione: Regulation and Functions. *Antioxid. Redox Signal.* **2017**, *27*, 1162–1177. [CrossRef] [PubMed]

110. May, J.M.; Mendiratta, S.; Hill, K.E.; Burk, R.F. Reduction of dehydroascorbate to ascorbate by the selenoenzyme thioredoxin reductase. *J. Biol. Chem.* **1997**, *272*, 22607–22610. [CrossRef] [PubMed]

111. Xia, L.; Nordman, T.; Olsson, J.M.; Damdimopoulos, A.; Bjorkhem-Bergman, L.; Nalvarte, I.; Eriksson, L.C.; Arner, E.S.; Spyrou, G.; Bjornstedt, M. The mammalian cytosolic selenoenzyme thioredoxin reductase reduces ubiquinone. A novel mechanism for defense against oxidative stress. *J. Biol. Chem.* **2003**, *278*, 2141–2146. [CrossRef] [PubMed]

112. Nalvarte, I.; Damdimopoulos, A.E.; Spyrou, G. Human mitochondrial thioredoxin reductase reduces cytochrome c and confers resistance to complex III inhibition. *Free Radic. Biol. Med.* **2004**, *36*, 1270–1278. [CrossRef] [PubMed]

113. Arner, E.S.; Nordberg, J.; Holmgren, A. Efficient reduction of lipoamide and lipoic acid by mammalian thioredoxin reductase. *Biochem. Biophys. Res. Commun.* **1996**, *225*, 268–274. [CrossRef] [PubMed]

114. Bjornstedt, M.; Hamberg, M.; Kumar, S.; Xue, J.; Holmgren, A. Human thioredoxin reductase directly reduces lipid hydroperoxides by NADPH and selenocystine strongly stimulates the reaction via catalytically generated selenols. *J. Biol. Chem.* **1995**, *270*, 11761–11764. [CrossRef] [PubMed]

115. Jee, C.; Vanoaica, L.; Lee, J.; Park, B.J.; Ahnn, J. Thioredoxin is related to life span regulation and oxidative stress response in Caenorhabditis elegans. *Genes Cells* **2005**, *10*, 1203–1210. [CrossRef] [PubMed]

116. Miranda-Vizuete, A.; Fierro Gonzalez, J.C.; Gahmon, G.; Burghoorn, J.; Navas, P.; Swoboda, P. Lifespan decrease in a *Caenorhabditis elegans* mutant lacking TRX-1, a thioredoxin expressed in ASJ sensory neurons. *FEBS Lett.* **2006**, *580*, 484–490. [CrossRef] [PubMed]

117. Li, W.; Bandyopadhyay, J.; Hwaang, H.S.; Park, B.J.; Cho, J.H.; Lee, J.I.; Ahnn, J.; Lee, S.K. Two thioredoxin reductases, trxr-1 and trxr-2, have differential physiological roles in Caenorhabditis elegans. *Mol. Cells* **2012**, *34*, 209–218. [CrossRef] [PubMed]

118. Cacho-Valadez, B.; Munoz-Lobato, F.; Pedrajas, J.R.; Cabello, J.; Fierro-Gonzalez, J.C.; Navas, P.; Swoboda, P.; Link, C.D.; Miranda-Vizuete, A. The characterization of the *Caenorhabditis elegans* mitochondrial thioredoxin system uncovers an unexpected protective role of thioredoxin reductase 2 in beta-amyloid peptide toxicity. *Antioxid. Redox Signal.* **2012**, *16*, 1384–1400. [CrossRef] [PubMed]

119. Arodin, L.; Miranda-Vizuete, A.; Swoboda, P.; Fernandes, A.P. Protective effects of the thioredoxin and glutaredoxin systems in dopamine-induced cell death. *Free Radic. Biol. Med.* **2014**, *73*, 328–336. [CrossRef] [PubMed]

120. Oláhová, M.; Veal, E.A. A peroxiredoxin, PRDX-2, is required for insulin secretion and insulin/IIS-dependent regulation of stress resistance and longevity. *Aging Cell* **2015**, *14*, 558–568. [CrossRef] [PubMed]

121. De Haes, W.; Frooninckx, L.; Van Assche, R.; Smolders, A.; Depuydt, G.; Billen, J.; Braeckman, B.P.; Schoofs, L.; Temmerman, L. Metformin promotes lifespan through mitohormesis via the peroxiredoxin PRDX-2. *Proc. Natl. Acad. Sci. USA* **2014**, *111*, E2501–E2509. [CrossRef] [PubMed]

122. Henderson, D.; Huebner, C.; Markowitz, M.; Taube, N.; Harvanek, Z.M.; Jakob, U.; Knoefler, D. Do developmental temperatures affect redox level and lifespan in *C. elegans* through upregulation of peroxiredoxin? *Redox Boil.* **2017**, *14*, 386–390. [CrossRef] [PubMed]

123. Thamsen, M.; Kumsta, C.; Li, F.; Jakob, U. Is overoxidation of peroxiredoxin physiologically significant? *Antioxid. Redox Signal.* **2011**, *14*, 725–730. [CrossRef] [PubMed]

124. Jiang, B.; Moskovitz, J. The Functions of the Mammalian Methionine Sulfoxide Reductase System and Related Diseases. *Antioxidants* **2018**, *7*, 122. [CrossRef] [PubMed]

125. Lee, B.C.; Lee, H.M.; Kim, S.; Avanesov, A.S.; Lee, A.; Chun, B.-H.; Vorbruggen, G.; Gladyshev, V.N. Expression of the methionine sulfoxide reductase lost during evolution extends Drosophila lifespan in a methionine-dependent manner. *Sci. Rep.* **2018**, *8*, 1010. [CrossRef] [PubMed]

126. Kim, H.Y.; Gladyshev, V.N. Methionine sulfoxide reductases: Selenoprotein forms and roles in antioxidant protein repair in mammals. *Biochem. J.* **2007**, *407*, 321–329. [CrossRef] [PubMed]

127. Minniti, A.N.; Cataldo, R.; Trigo, C.; Vasquez, L.; Mujica, P.; Leighton, F.; Inestrosa, N.C.; Aldunate, R. Methionine sulfoxide reductase A expression is regulated by the DAF-16/FOXO pathway in Caenorhabditis elegans. *Aging Cell* **2009**, *8*, 690–705. [CrossRef] [PubMed]

128. Bruce, L.; Singkornrat, D.; Wilson, K.; Hausman, W.; Robbins, K.; Huang, L.; Foss, K.; Binninger, D. In Vivo Effects of Methionine Sulfoxide Reductase Deficiency in Drosophila melanogaster. *Antioxidants* **2018**, *7*, 155. [CrossRef] [PubMed]

129. Ruan, H.; Tang, X.D.; Chen, M.L.; Joiner, M.L.; Sun, G.; Brot, N.; Weissbach, H.; Heinemann, S.H.; Iverson, L.; Wu, C.F.; et al. High-quality life extension by the enzyme peptide methionine sulfoxide reductase. *Proc. Natl. Acad. Sci. USA* **2002**, *99*, 2748–2753. [CrossRef] [PubMed]

130. Chung, H.; Kim, A.K.; Jung, S.A.; Kim, S.W.; Yu, K.; Lee, J.H. The Drosophila homolog of methionine sulfoxide reductase A extends lifespan and increases nuclear localization of FOXO. *FEBS Lett.* **2010**, *584*, 3609–3614. [CrossRef] [PubMed]

131. Salmon, A.B.; Kim, G.; Liu, C.; Wren, J.D.; Georgescu, C.; Richardson, A.; Levine, R.L. Effects of transgenic methionine sulfoxide reductase A (MsrA) expression on lifespan and age-dependent changes in metabolic function in mice. *Redox Boil.* **2016**, *10*, 251–256. [CrossRef] [PubMed]
132. Tarafdar, S.; Kim, G.; Levine, R.L. Drosophila methionine sulfoxide reductase A (MSRA) lacks methionine oxidase activity. *Free Radic. Biol. Med.* **2018**, *131*, 154–161. [CrossRef] [PubMed]
133. Barja, G.; Cadenas, S.; Rojas, C.; Lopez-Torres, M.; Perez-Campo, R. A decrease of free radical production near critical targets as a cause of maximum longevity in animals. *Comp. Biochem. Physiol. Part B* **1994**, *108*, 501–512. [CrossRef]
134. Kanzok, S.M.; Fechner, A.; Bauer, H.; Ulschmid, J.K.; Muller, H.M.; Botella-Munoz, J.; Schneuwly, S.; Schirmer, R.; Becker, K. Substitution of the thioredoxin system for glutathione reductase in Drosophila melanogaster. *Science* **2001**, *291*, 643–646. [CrossRef] [PubMed]
135. Mora-Lorca, J.A.; Saenz-Narciso, B.; Gaffney, C.J.; Naranjo-Galindo, F.J.; Pedrajas, J.R.; Guerrero-Gomez, D.; Dobrzynska, A.; Askjaer, P.; Szewczyk, N.J.; Cabello, J.; et al. Glutathione reductase gsr-1 is an essential gene required for *Caenorhabditis elegans* early embryonic development. *Free Radic. Biol. Med.* **2016**, *96*, 446–461. [CrossRef] [PubMed]
136. Luersen, K.; Stegehake, D.; Daniel, J.; Drescher, M.; Ajonina, I.; Ajonina, C.; Hertel, P.; Woltersdorf, C.; Liebau, E. The glutathione reductase GSR-1 determines stress tolerance and longevity in Caenorhabditis elegans. *PLoS ONE* **2013**, *8*, e60731. [CrossRef] [PubMed]
137. Stenvall, J.; Fierro-Gonzalez, J.C.; Swoboda, P.; Saamarthy, K.; Cheng, Q.; Cacho-Valadez, B.; Arner, E.S.; Persson, O.P.; Miranda-Vizuete, A.; Tuck, S. Selenoprotein TRXR-1 and GSR-1 are essential for removal of old cuticle during molting in Caenorhabditis elegans. *Proc. Natl. Acad. Sci. USA* **2011**, *108*, 1064–1069. [CrossRef] [PubMed]
138. Guerrero-Gomez, D.; Mora-Lorca, J.A.; Saenz-Narciso, B.; Naranjo-Galindo, F.J.; Munoz-Lobato, F.; Parrado-Fernandez, C.; Goikolea, J.; Cedazo-Minguez, A.; Link, C.D.; Neri, C.; et al. Loss of glutathione redox homeostasis impairs proteostasis by inhibiting autophagy-dependent protein degradation. *Cell Death Differ.* **2019**. [CrossRef] [PubMed]
139. Luchak, J.M.; Prabhudesai, L.; Sohal, R.S.; Radyuk, S.N.; Orr, W.C. Modulating longevity in Drosophila by over- and underexpression of glutamate-cysteine ligase. *Ann. N. Y. Acad. Sci.* **2007**, *1119*, 260–273. [CrossRef] [PubMed]
140. Orr, W.C.; Radyuk, S.N.; Prabhudesai, L.; Toroser, D.; Benes, J.J.; Luchak, J.M.; Mockett, R.J.; Rebrin, I.; Hubbard, J.G.; Sohal, R.S. Overexpression of glutamate-cysteine ligase extends life span in Drosophila melanogaster. *J. Biol. Chem.* **2005**, *280*, 37331–37338. [CrossRef] [PubMed]
141. Li, F.; Ma, X.; Cui, X.; Li, J.; Wang, Z. Recombinant buckwheat glutaredoxin intake increases lifespan and stress resistance via hsf-1 upregulation in Caenorhabditis elegans. *Exp. Gerontol.* **2018**, *104*, 86–97. [CrossRef] [PubMed]
142. Andziak, B.; O'Connor, T.P.; Qi, W.; DeWaal, E.M.; Pierce, A.; Chaudhuri, A.R.; Van Remmen, H.; Buffenstein, R. High oxidative damage levels in the longest-living rodent, the naked mole-rat. *Aging Cell* **2006**, *5*, 463–471. [CrossRef] [PubMed]
143. Go, Y.M.; Jones, D.P. Redox compartmentalization in eukaryotic cells. *Biochim. Biophys. Acta* **2008**, *1780*, 1273–1290. [CrossRef] [PubMed]
144. Cai, J.; Jones, D.P. Superoxide in apoptosis. Mitochondrial generation triggered by cytochrome c loss. *J. Biol. Chem.* **1998**, *273*, 11401–11404. [CrossRef] [PubMed]
145. Kemp, M.; Go, Y.M.; Jones, D.P. Nonequilibrium thermodynamics of thiol/disulfide redox systems: A perspective on redox systems biology. *Free Radic. Biol. Med.* **2008**, *44*, 921–937. [CrossRef] [PubMed]
146. Rebrin, I.; Sohal, R.S. Comparison of thiol redox state of mitochondria and homogenates of various tissues between two strains of mice with different longevities. *Exp. Gerontol.* **2004**, *39*, 1513–1519. [CrossRef] [PubMed]
147. Legan, S.K.; Rebrin, I.; Mockett, R.J.; Radyuk, S.N.; Klichko, V.I.; Sohal, R.S.; Orr, W.C. Overexpression of glucose-6-phosphate dehydrogenase extends the life span of Drosophila melanogaster. *J. Biol. Chem.* **2008**, *283*, 32492–32499. [CrossRef] [PubMed]
148. Luckinbill, L.S.; Riha, V.; Rhine, S.; Grudzien, T.A. The role of glucose-6-phosphate dehydrogenase in the evolution of longevity in Drosophila melanogaster. *Heredity* **1990**, *65 Pt 1*, 29–38. [CrossRef]

149. Wang, C.-T.; Chen, Y.-C.; Wang, Y.-Y.; Huang, M.-H.; Yen, T.-L.; Li, H.; Liang, C.-J.; Sang, T.-K.; Ciou, S.-C.; Yuh, C.-H.; et al. Reduced neuronal expression of ribose-5-phosphate isomerase enhances tolerance to oxidative stress, extends lifespan, and attenuates polyglutamine toxicity in Drosophila. *Aging Cell* **2012**, *11*, 93–103. [CrossRef] [PubMed]

150. Farkas, R.; Danis, P.; Medved'ova, L.; Mechler, B.M.; Knopp, J. Regulation of cytosolic malate dehydrogenase by juvenile hormone in Drosophila melanogaster. *Cell Biochem. Biophys.* **2002**, *37*, 37–52. [CrossRef]

151. Kim, G.H.; Lee, Y.E.; Lee, G.H.; Cho, Y.H.; Lee, Y.N.; Jang, Y.; Paik, D.; Park, J.J. Overexpression of malic enzyme in the larval stage extends Drosophila lifespan. *Biochem. Biophys. Res. Commun.* **2015**, *456*, 676–682. [CrossRef] [PubMed]

152. Paik, D.; Jang, Y.G.; Lee, Y.E.; Lee, Y.N.; Yamamoto, R.; Gee, H.Y.; Yoo, S.; Bae, E.; Min, K.J.; Tatar, M.; et al. Misexpression screen delineates novel genes controlling Drosophila lifespan. *Mech. Ageing Dev.* **2012**, *133*, 234–245. [CrossRef] [PubMed]

153. Merritt, T.J.S.; Kuczynski, C.; Sezgin, E.; Zhu, C.-T.; Kumagai, S.; Eanes, W.F. Quantifying interactions within the NADP(H) enzyme network in Drosophila melanogaster. *Genetics* **2009**, *182*, 565–574. [CrossRef] [PubMed]

154. Rong, Y.S.; Golic, K.G. A targeted gene knockout in Drosophila. *Genetics* **2001**, *157*, 1307–1312. [PubMed]

155. Tremblay, G.B.; Sohi, S.S.; Retnakaran, A.; MacKenzie, R.E. NAD-dependent methylenetetrahydrofolate dehydrogenase-methenyltetrahydrofolate cyclohydrolase is targeted to the cytoplasm in insect cell lines. *FEBS Lett.* **1995**, *368*, 177–182. [CrossRef]

156. Yu, S.; Jang, Y.; Paik, D.; Lee, E.; Park, J.J. Nmdmc overexpression extends Drosophila lifespan and reduces levels of mitochondrial reactive oxygen species. *Biochem. Biophys. Res. Commun.* **2015**, *465*, 845–850. [CrossRef] [PubMed]

157. Gaudet, P.; Livstone, M.S.; Lewis, S.E.; Thomas, P.D. Phylogenetic-based propagation of functional annotations within the Gene Ontology consortium. *Brief. Bioinform.* **2011**, *12*, 449–462. [CrossRef] [PubMed]

158. Da Cunha, G.L.; de Oliveira, A.K. Citric acid cycle: A mainstream metabolic pathway influencing life span in Drosophila melanogaster? *Exp. Gerontol.* **1996**, *31*, 705–715. [CrossRef]

159. Arking, R.; Burde, V.; Graves, K.; Hari, R.; Feldman, E.; Zeevi, A.; Soliman, S.; Saraiya, A.; Buck, S.; Vettraino, J.; et al. Forward and reverse selection for longevity in Drosophila is characterized by alteration of antioxidant gene expression and oxidative damage patterns. *Exp. Gerontol.* **2000**, *35*, 167–185. [CrossRef]

160. Alonso, J.; Rodriguez, J.M.; Baena-Lopez, L.A.; Santaren, J.F. Characterization of the Drosophila melanogaster mitochondrial proteome. *J. Proteome Res.* **2005**, *4*, 1636–1645. [CrossRef] [PubMed]

161. Lee, K.S.; Iijima-Ando, K.; Iijima, K.; Lee, W.J.; Lee, J.H.; Yu, K.; Lee, D.S. JNK/FOXO-mediated neuronal expression of fly homologue of peroxiredoxin II reduces oxidative stress and extends life span. *J. Biol. Chem.* **2009**, *284*, 29454–29461. [CrossRef] [PubMed]

162. Radyuk, S.N.; Michalak, K.; Klichko, V.I.; Benes, J.; Rebrin, I.; Sohal, R.S.; Orr, W.C. Peroxiredoxin 5 confers protection against oxidative stress and apoptosis and also promotes longevity in Drosophila. *Biochem. J.* **2009**, *419*, 437–445. [CrossRef] [PubMed]

163. Odnokoz, O.; Nakatsuka, K.; Klichko, V.I.; Nguyen, J.; Solis, L.C.; Ostling, K.; Badinloo, M.; Orr, W.C.; Radyuk, S.N. Mitochondrial peroxiredoxins are essential in regulating the relationship between Drosophila immunity and aging. *Biochim. Biophys. Acta Mol. Basis Dis.* **2017**, *1863*, 68–80. [CrossRef] [PubMed]

164. Orr, W.C.; Radyuk, S.N.; Sohal, R.S. Involvement of redox state in the aging of Drosophila melanogaster. *Antioxid. Redox Signal.* **2013**, *19*, 788–803. [CrossRef] [PubMed]

165. Radyuk, S.N.; Rebrin, I.; Klichko, V.I.; Sohal, B.H.; Michalak, K.; Benes, J.; Sohal, R.S.; Orr, W.C. Mitochondrial peroxiredoxins are critical for the maintenance of redox state and the survival of adult Drosophila. *Free Radic. Biol. Med.* **2010**, *49*, 1892–1902. [CrossRef] [PubMed]

166. Klichko, V.I.; Orr, W.C.; Radyuk, S.N. The role of peroxiredoxin 4 in inflammatory response and aging. *Biochim. Biophys. Acta* **2016**, *1862*, 265–273. [CrossRef] [PubMed]

167. Fernandez-Marcos, P.J.; Nóbrega-Pereira, S. NADPH: New oxygen for the ROS theory of aging. *Oncotarget* **2016**, *7*, 50814–50815. [CrossRef] [PubMed]

168. Fan, J.; Ye, J.; Kamphorst, J.J.; Shlomi, T.; Thompson, C.B.; Rabinowitz, J.D. Quantitative flux analysis reveals folate-dependent NADPH production. *Nature* **2014**, *510*, 298–302. [CrossRef] [PubMed]

169. Nobrega-Pereira, S.; Fernandez-Marcos, P.J.; Brioche, T.; Gomez-Cabrera, M.C.; Salvador-Pascual, A.; Flores, J.M.; Vina, J.; Serrano, M. G6PD protects from oxidative damage and improves healthspan in mice. *Nat. Commun.* **2016**, *7*, 10894. [CrossRef] [PubMed]

170. Dietzmann, K. Histochemical demonstration of enzyme activity changes in the regio postcentralis of the mouse brain under the effect of the aging process. *J. Hirnforsch.* **1981**, *22*, 405–408. [PubMed]

171. Kaliman, P.A.; Konovalova, E.O. Age-dependent characteristics of the regulation of cytoplasmic NADP+-dehydrogenases in the liver of rats on different diets. *Ukrainskii Biokhimicheskii Zhurnal (1978)* **1985**, *57*, 38–42.

172. Lemeshko, V.V.; Nikitchenko Iu, V.; Kaliman, P.A. Enzymes of the antioxidant system of rat liver during aging. *Ukrainskii Biokhimicheskii Zhurnal (1978)* **1983**, *55*, 523–528.

173. Maurya, P.K.; Kumar, P.; Chandra, P. Age-dependent detection of erythrocytes glucose-6-phosphate dehydrogenase and its correlation with oxidative stress. *Arch. Physiol. Biochem.* **2016**, *122*, 61–66. [CrossRef] [PubMed]

174. Gartler, S.M.; Hornung, S.K.; Motulsky, A.G. Effect of chronologic age on induction of cystathionine synthase, uroporphyrinogen I synthase, and glucose-6-phosphate dehydrogenase activities in lymphocytes. *Proc. Natl. Acad. Sci. USA* **1981**, *78*, 1916–1919. [CrossRef] [PubMed]

175. Machado, A.; Ayala, A.; Gordillo, E.; Revilla, E.; Santa Maria, C. Relationship between enzymatic activity loss and post-translational protein modification in aging. *Arch. Gerontol. Geriatr.* **1991**, *12*, 187–197. [CrossRef]

176. Kil, I.S.; Lee, Y.S.; Bae, Y.S.; Huh, T.L.; Park, J.W. Modulation of NADP(+)-dependent isocitrate dehydrogenase in aging. *Redox Rep.* **2004**, *9*, 271–277. [CrossRef] [PubMed]

177. Wong, N.; Blair, A.R.; Morahan, G.; Andrikopoulos, S. The deletion variant of nicotinamide nucleotide transhydrogenase (Nnt) does not affect insulin secretion or glucose tolerance. *Endocrinology* **2010**, *151*, 96–102. [CrossRef] [PubMed]

178. Serrander, L.; Cartier, L.; Bedard, K.; Banfi, B.; Lardy, B.; Plastre, O.; Sienkiewicz, A.; Fórró, L.; Schlegel, W.; Krause, K.-H. NOX4 activity is determined by mRNA levels and reveals a unique pattern of ROS generation. *Biochem. J.* **2007**, *406*, 105–114. [CrossRef] [PubMed]

179. Nisimoto, Y.; Diebold, B.A.; Cosentino-Gomes, D.; Lambeth, J.D. Nox4: A hydrogen peroxide-generating oxygen sensor. *Biochemistry* **2014**, *53*, 5111–5120. [CrossRef] [PubMed]

180. Nisimoto, Y.; Jackson, H.M.; Ogawa, H.; Kawahara, T.; Lambeth, J.D. Constitutive NADPH-dependent electron transferase activity of the Nox4 dehydrogenase domain. *Biochemistry* **2010**, *49*, 2433–2442. [CrossRef] [PubMed]

181. Worthington, D.J.; Rosemeyer, M.A. Glutathione reductase from human erythrocytes. Catalytic properties and aggregation. *Eur. J. Biochem.* **1976**, *67*, 231–238. [CrossRef] [PubMed]

182. Koziel, R.; Pircher, H.; Kratochwil, M.; Lener, B.; Hermann, M.; Dencher, N.A.; Jansen-Durr, P. Mitochondrial respiratory chain complex I is inactivated by NADPH oxidase Nox4. *Biochem. J.* **2013**, *452*, 231–239. [CrossRef] [PubMed]

183. Goy, C.; Czypiorski, P.; Altschmied, J.; Jakob, S.; Rabanter, L.L.; Brewer, A.C.; Ale-Agha, N.; Dyballa-Rukes, N.; Shah, A.M.; Haendeler, J. The imbalanced redox status in senescent endothelial cells is due to dysregulated Thioredoxin-1 and NADPH oxidase 4. *Exp. Gerontol.* **2014**, *56*, 45–52. [CrossRef] [PubMed]

184. Handy, D.E.; Loscalzo, J. Responses to reductive stress in the cardiovascular system. *Free Radic. Boil. Med.* **2017**, *109*, 114–124. [CrossRef] [PubMed]

185. Rezende, F.; Schürmann, C.; Schütz, S.; Harenkamp, S.; Herrmann, E.; Seimetz, M.; Weißmann, N.; Schröder, K. Knock out of the NADPH oxidase Nox4 has no impact on life span in mice. *Redox Boil.* **2016**, *11*, 312–314. [CrossRef] [PubMed]

186. Hirschhauser, C.; Bornbaum, J.; Reis, A.; Bohme, S.; Kaludercic, N.; Menabo, R.; Di Lisa, F.; Boengler, K.; Shah, A.M.; Schulz, R.; et al. NOX4 in Mitochondria: Yeast Two-Hybrid-Based Interaction with Complex I Without Relevance for Basal Reactive Oxygen Species? *Antioxid. Redox Signal.* **2015**, *23*, 1106–1112. [CrossRef] [PubMed]

187. Miyagi, H.; Kawai, S.; Murata, K. Two sources of mitochondrial NADPH in the yeast Saccharomyces cerevisiae. *J. Biol. Chem.* **2009**, *284*, 7553–7560. [CrossRef] [PubMed]

188. He, W.; Wang, Y.; Liu, W.; Zhou, C.Z. Crystal structure of Saccharomyces cerevisiae 6-phosphogluconate dehydrogenase Gnd1. *BMC Struct. Boil.* **2007**, *7*, 38. [CrossRef] [PubMed]

189. Minard, K.I.; McAlister-Henn, L. Sources of NADPH in yeast vary with carbon source. *J. Biol. Chem.* **2005**, *280*, 39890–39896. [CrossRef] [PubMed]

190. Garay, E.; Campos, S.E.; Gonzalez de la Cruz, J.; Gaspar, A.P.; Jinich, A.; Deluna, A. High-resolution profiling of stationary-phase survival reveals yeast longevity factors and their genetic interactions. *PLoS Genet.* **2014**, *10*, e1004168. [CrossRef] [PubMed]

191. Laschober, G.T.; Ruli, D.; Hofer, E.; Muck, C.; Carmona-Gutierrez, D.; Ring, J.; Hutter, E.; Ruckenstuhl, C.; Micutkova, L.; Brunauer, R.; et al. Identification of evolutionarily conserved genetic regulators of cellular aging. *Aging Cell* **2010**, *9*, 1084–1097. [CrossRef] [PubMed]

192. Orlandi, I.; Stamerra, G.; Vai, M. Altered Expression of Mitochondrial NAD(+) Carriers Influences Yeast Chronological Lifespan by Modulating Cytosolic and Mitochondrial Metabolism. *Front. Genet.* **2018**, *9*, 676. [CrossRef] [PubMed]

193. Brandes, N.; Tienson, H.; Lindemann, A.; Vitvitsky, V.; Reichmann, D.; Banerjee, R.; Jakob, U. Time line of redox events in aging postmitotic cells. *eLife* **2013**, *2*, e00306. [CrossRef] [PubMed]

194. Kwolek-Mirek, M.; Maslanka, R.; Molon, M. Disorders in NADPH generation via pentose phosphate pathway influence the reproductive potential of the Saccharomyces cerevisiae yeast due to changes in redox status. *J. Cell. Biochem.* **2018**. [CrossRef] [PubMed]

195. Picazo, C.; Matallana, E.; Aranda, A. Yeast thioredoxin reductase Trr1p controls TORC1-regulated processes. *Sci. Rep.* **2018**, *8*, 16500. [CrossRef] [PubMed]

196. Matecic, M.; Smith, D.L.; Pan, X.; Maqani, N.; Bekiranov, S.; Boeke, J.D.; Smith, J.S. A microarray-based genetic screen for yeast chronological aging factors. *PLoS Genet.* **2010**, *6*, e1000921. [CrossRef] [PubMed]

197. Musa, M.; Peric, M.; Bou Dib, P.; Sobocanec, S.; Saric, A.; Lovric, A.; Rudan, M.; Nikolic, A.; Milosevic, I.; Vlahovicek, K.; et al. Heat-induced longevity in budding yeast requires respiratory metabolism and glutathione recycling. *Aging* **2018**, *10*, 2407–2427. [CrossRef] [PubMed]

198. Managbanag, J.R.; Witten, T.M.; Bonchev, D.; Fox, L.A.; Tsuchiya, M.; Kennedy, B.K.; Kaeberlein, M. Shortest-path network analysis is a useful approach toward identifying genetic determinants of longevity. *PLoS ONE* **2008**, *3*, e3802. [CrossRef] [PubMed]

199. Smith, E.D.; Tsuchiya, M.; Fox, L.A.; Dang, N.; Hu, D.; Kerr, E.O.; Johnston, E.D.; Tchao, B.N.; Pak, D.N.; Welton, K.L.; et al. Quantitative evidence for conserved longevity pathways between divergent eukaryotic species. *Genome Res.* **2008**, *18*, 564–570. [CrossRef] [PubMed]

200. Kenyon, C.; Chang, J.; Gensch, E.; Rudner, A.; Tabtiang, R. A *C. elegans* mutant that lives twice as long as wild type. *Nature* **1993**, *366*, 461–464. [CrossRef] [PubMed]

201. Tullet, J.M.; Hertweck, M.; An, J.H.; Baker, J.; Hwang, J.Y.; Liu, S.; Oliveira, R.P.; Baumeister, R.; Blackwell, T.K. Direct inhibition of the longevity-promoting factor SKN-1 by insulin-like signaling in *C. elegans*. *Cell* **2008**, *132*, 1025–1038. [CrossRef] [PubMed]

202. Yang, H.C.; Yu, H.; Liu, Y.C.; Chen, T.L.; Stern, A.; Lo, S.J.; Chiu, D.T. IDH-1 deficiency induces growth defects and metabolic alterations in GSPD-1-deficient *Caenorhabditis elegans*. *J. Mol. Med.* **2019**, *366*, 461–464. [CrossRef] [PubMed]

203. Yu, H.; Larsen, P.L. DAF-16-dependent and independent expression targets of DAF-2 insulin receptor-like pathway in *Caenorhabditis elegans* include FKBPs. *J. Mol. Biol.* **2001**, *314*, 1017–1028. [CrossRef] [PubMed]

204. Depuydt, G.; Xie, F.; Petyuk, V.A.; Smolders, A.; Brewer, H.M.; Camp, D.G., 2nd; Smith, R.D.; Braeckman, B.P. LC-MS Proteomics Analysis of the Insulin/IGF-1-Deficient *Caenorhabditis elegans* daf-2(e1370) Mutant Reveals Extensive Restructuring of Intermediary Metabolism. *J. Proteome Res.* **2014**, *13*, 1938–1956. [CrossRef] [PubMed]

205. Martell, J.; Seo, Y.; Bak, D.W.; Kingsley, S.F.; Tissenbaum, H.A.; Weerapana, E. Global Cysteine-Reactivity Profiling during Impaired Insulin/IGF-1 Signaling in *C. elegans* Identifies Uncharacterized Mediators of Longevity. *Cell Chem. Boil.* **2016**, *23*, 955–966. [CrossRef] [PubMed]

206. Walther, D.M.; Kasturi, P.; Zheng, M.; Pinkert, S.; Vecchi, G.; Ciryam, P.; Morimoto, R.I.; Dobson, C.M.; Vendruscolo, M.; Mann, M.; et al. Widespread Proteome Remodeling and Aggregation in Aging *C. elegans*. *Cell* **2015**, *161*, 919–932. [CrossRef] [PubMed]

207. Pinkston-Gosse, J.; Kenyon, C. DAF-16/FOXO targets genes that regulate tumor growth in *Caenorhabditis elegans*. *Nat. Genet.* **2007**, *39*, 1403–1409. [CrossRef] [PubMed]

208. Zhang, P.; Judy, M.; Lee, S.J.; Kenyon, C. Direct and indirect gene regulation by a life-extending FOXO protein in *C. elegans*: Roles for GATA factors and lipid gene regulators. *Cell Metab.* **2013**, *17*, 85–100. [CrossRef]

209. Murphy, C.T.; McCarroll, S.A.; Bargmann, C.I.; Fraser, A.; Kamath, R.S.; Ahringer, J.; Li, H.; Kenyon, C. Genes that act downstream of DAF-16 to influence the lifespan of *Caenorhabditis elegans*. *Nature* **2003**, *424*, 277–283. [CrossRef] [PubMed]

210. Erkut, C.; Gade, V.R.; Laxman, S.; Kurzchalia, T.V. The glyoxylate shunt is essential for desiccation tolerance in *C. elegans* and budding yeast. *eLife* **2016**, *5*, e13614. [CrossRef] [PubMed]

211. Gallo, M.; Park, D.; Riddle, D.L. Increased longevity of some *C. elegans* mitochondrial mutants explained by activation of an alternative energy-producing pathway. *Mech. Ageing Dev.* **2011**, *132*, 515–518. [CrossRef]

212. Zuryn, S.; Kuang, J.; Tuck, A.; Ebert, P.R. Mitochondrial dysfunction in *Caenorhabditis elegans* causes metabolic restructuring, but this is not linked to longevity. *Mech. Ageing Dev.* **2010**, *131*, 554–561. [CrossRef] [PubMed]

213. Hamilton, B.; Dong, Y.; Shindo, M.; Liu, W.; Odell, I.; Ruvkun, G.; Lee, S.S. A systematic RNAi screen for longevity genes in *C. elegans*. *Genes Dev.* **2005**, *19*, 1544–1555. [CrossRef] [PubMed]

214. Chin, R.M.; Fu, X.; Pai, M.Y.; Vergnes, L.; Hwang, H.; Deng, G.; Diep, S.; Lomenick, B.; Meli, V.S.; Monsalve, G.C.; et al. The metabolite alpha-ketoglutarate extends lifespan by inhibiting ATP synthase and TOR. *Nature* **2014**, *510*, 397–401. [CrossRef] [PubMed]

215. Kim, Y.; Sun, H. Functional genomic approach to identify novel genes involved in the regulation of oxidative stress resistance and animal lifespan. *Aging Cell* **2007**, *6*, 489–503. [CrossRef] [PubMed]

216. Chen, T.L.; Yang, H.C.; Hung, C.Y.; Ou, M.H.; Pan, Y.Y.; Cheng, M.L.; Stern, A.; Lo, S.J.; Chiu, D.T. Impaired embryonic development in glucose-6-phosphate dehydrogenase-deficient *Caenorhabditis elegans* due to abnormal redox homeostasis induced activation of calcium-independent phospholipase and alteration of glycerophospholipid metabolism. *Cell Death Dis.* **2017**, *8*, e2545. [CrossRef] [PubMed]

217. Bennett, C.F.; Kwon, J.J.; Chen, C.; Russell, J.; Acosta, K.; Burnaevskiy, N.; Crane, M.M.; Bitto, A.; Vander Wende, H.; Simko, M.; et al. Transaldolase inhibition impairs mitochondrial respiration and induces a starvation-like longevity response in Caenorhabditis elegans. *PLoS Genet.* **2017**, *13*, e1006695. [CrossRef] [PubMed]

218. Wang, J.; Robida-Stubbs, S.; Tullet, J.M.; Rual, J.F.; Vidal, M.; Blackwell, T.K. RNAi screening implicates a SKN-1-dependent transcriptional response in stress resistance and longevity deriving from translation inhibition. *PLoS Genet.* **2010**, *6*, e1001048. [CrossRef] [PubMed]

219. Bennett, C.F.; Kaeberlein, M. The mitochondrial unfolded protein response and increased longevity: Cause, consequence, or correlation? *Exp. Gerontol.* **2014**, *56*, 142–146. [CrossRef] [PubMed]

220. Melber, A.; Haynes, C.M. UPR(mt) regulation and output: A stress response mediated by mitochondrial-nuclear communication. *Cell Res.* **2018**, *28*, 281–295. [CrossRef] [PubMed]

221. Wu, Z.; Senchuk, M.M.; Dues, D.J.; Johnson, B.K.; Cooper, J.F.; Lew, L.; Machiela, E.; Schaar, C.E.; DeJonge, H.; Blackwell, T.K.; et al. Mitochondrial unfolded protein response transcription factor ATFS-1 promotes longevity in a long-lived mitochondrial mutant through activation of stress response pathways. *BMC Biol.* **2018**, *16*, 147. [CrossRef] [PubMed]

222. Rzezniczak, T.Z.; Merritt, T.J. Interactions of NADP-reducing enzymes across varying environmental conditions: A model of biological complexity. *G3 Genes Genomes Genet.* **2012**, *2*, 1613–1623. [CrossRef] [PubMed]

223. White, K.; Kim, M.J.; Ding, D.; Han, C.; Park, H.J.; Meneses, Z.; Tanokura, M.; Linser, P.; Salvi, R.; Someya, S. G6pd Deficiency Does Not Affect the Cytosolic Glutathione or Thioredoxin Antioxidant Defense in Mouse Cochlea. *J. Neurosci.* **2017**, *37*, 5770–5781. [CrossRef] [PubMed]

224. Au, S.W.; Gover, S.; Lam, V.M.; Adams, M.J. Human glucose-6-phosphate dehydrogenase: The crystal structure reveals a structural NADP(+) molecule and provides insights into enzyme deficiency. *Structure (Lond. Engl. 1993)* **2000**, *8*, 293–303.

225. Pandolfi, P.P.; Sonati, F.; Rivi, R.; Mason, P.; Grosveld, F.; Luzzatto, L. Targeted disruption of the housekeeping gene encoding glucose 6-phosphate dehydrogenase (G6PD): G6PD is dispensable for pentose synthesis but essential for defense against oxidative stress. *EMBO J.* **1995**, *14*, 5209–5215. [CrossRef] [PubMed]

226. Urban, N.; Tsitsipatis, D.; Hausig, F.; Kreuzer, K.; Erler, K.; Stein, V.; Ristow, M.; Steinbrenner, H.; Klotz, L.O. Non-linear impact of glutathione depletion on *C. elegans* life span and stress resistance. *Redox Boil.* **2017**, *11*, 502–515. [CrossRef] [PubMed]

227. Okuyama, T.; Inoue, H.; Ookuma, S.; Satoh, T.; Kano, K.; Honjoh, S.; Hisamoto, N.; Matsumoto, K.; Nishida, E. The ERK-MAPK pathway regulates longevity through SKN-1 and insulin-like signaling in *Caenorhabditis elegans*. *J. Biol. Chem.* **2010**, *285*, 30274–30281. [CrossRef] [PubMed]

228. Przybysz, A.J.; Choe, K.P.; Roberts, L.J.; Strange, K. Increased age reduces DAF-16 and SKN-1 signaling and the hormetic response of *Caenorhabditis elegans* to the xenobiotic juglone. *Mech. Ageing Dev.* **2009**, *130*, 357–369. [CrossRef] [PubMed]

229. Dillin, A.; Crawford, D.K.; Kenyon, C. Timing requirements for insulin/IGF-1 signaling in *C. elegans*. *Science* **2002**, *298*, 830–834. [CrossRef] [PubMed]

230. Edwards, C.; Canfield, J.; Copes, N.; Rehan, M.; Lipps, D.; Bradshaw, P.C. D-β-hydroxybutyrate extends lifespan in *C. elegans*. *Aging (Albany Ny)* **2014**, *6*, 621–644.

231. Schwartz, A.G.; Pashko, L.L. Dehydroepiandrosterone, glucose-6-phosphate dehydrogenase, and longevity. *Ageing Res. Rev.* **2004**, *3*, 171–187. [CrossRef] [PubMed]

232. Vogelauer, M.; Krall, A.S.; McBrian, M.A.; Li, J.Y.; Kurdistani, S.K. Stimulation of histone deacetylase activity by metabolites of intermediary metabolism. *J. Biol. Chem.* **2012**, *287*, 32006–32016. [CrossRef] [PubMed]

233. Pasyukova, E.G.; Vaiserman, A.M. HDAC inhibitors: A new promising drug class in anti-aging research. *Mech. Ageing Dev.* **2017**, *166*, 6–15. [CrossRef] [PubMed]

234. Arkblad, E.L.; Tuck, S.; Pestov, N.B.; Dmitriev, R.I.; Kostina, M.B.; Stenvall, J.; Tranberg, M.; Rydstrom, J. A *Caenorhabditis elegans* mutant lacking functional nicotinamide nucleotide transhydrogenase displays increased sensitivity to oxidative stress. *Free Radic. Biol. Med.* **2005**, *38*, 1518–1525. [CrossRef] [PubMed]

235. Wu, K.C.; Cui, J.Y.; Klaassen, C.D. Beneficial role of Nrf2 in regulating NADPH generation and consumption. *Toxicol. Sci.* **2011**, *123*, 590–600. [CrossRef] [PubMed]

236. Yang, W.; Hekimi, S. Two modes of mitochondrial dysfunction lead independently to lifespan extension in *Caenorhabditis elegans*. *Aging Cell* **2010**, *9*, 433–447. [CrossRef] [PubMed]

237. Dues, D.J.; Schaar, C.E.; Johnson, B.K.; Bowman, M.J.; Winn, M.E.; Senchuk, M.M.; Van Raamsdonk, J.M. Uncoupling of oxidative stress resistance and lifespan in long-lived isp-1 mitochondrial mutants in *Caenorhabditis elegans*. *Free Radic. Biol. Med.* **2017**, *108*, 362–373. [CrossRef] [PubMed]

238. An, J.H.; Blackwell, T.K. SKN-1 links *C. elegans* mesendodermal specification to a conserved oxidative stress response. *Genes Dev.* **2003**, *17*, 1882–1893. [CrossRef] [PubMed]

239. Tang, L.; Choe, K.P. Characterization of skn-1/wdr-23 phenotypes in *Caenorhabditis elegans*; pleiotrophy, aging, glutathione, and interactions with other longevity pathways. *Mech. Ageing Dev.* **2015**, *149*, 88–98. [CrossRef] [PubMed]

240. Sohal, R.S.; Weindruch, R. Oxidative stress, caloric restriction, and aging. *Science* **1996**, *273*, 59–63. [CrossRef] [PubMed]

241. Sanchez-Roman, I.; Barja, G. Regulation of longevity and oxidative stress by nutritional interventions: Role of methionine restriction. *Exp. Gerontol.* **2013**, *48*, 1030–1042. [CrossRef] [PubMed]

242. Hine, C.; Mitchell, J.R. Calorie restriction and methionine restriction in control of endogenous hydrogen sulfide production by the transsulfuration pathway. *Exp. Gerontol.* **2015**, *68*, 26–32. [CrossRef] [PubMed]

243. Eriksson, S.; Prigge, J.R.; Talago, E.A.; Arner, E.S.; Schmidt, E.E. Dietary methionine can sustain cytosolic redox homeostasis in the mouse liver. *Nat. Commun.* **2015**, *6*, 6479. [CrossRef] [PubMed]

244. Ghosh, D.; Brewer, G.J. External cys/cySS redox state modification controls the intracellular redox state and neurodegeneration via Akt in aging and Alzheimer's disease mouse model neurons. *J. Alzheimers Dis.* **2014**, *42*, 313–324. [CrossRef] [PubMed]

245. McIsaac, R.S.; Lewis, K.N.; Gibney, P.A.; Buffenstein, R. From yeast to human: Exploring the comparative biology of methionine restriction in extending eukaryotic life span. *Ann. N. Y. Acad. Sci.* **2016**, *1363*, 155–170. [CrossRef] [PubMed]

246. Uthus, E.O.; Brown-Borg, H.M. Methionine flux to transsulfuration is enhanced in the long living Ames dwarf mouse. *Mech. Ageing Dev.* **2006**, *127*, 444–450. [CrossRef] [PubMed]

247. Slekar, K.H.; Kosman, D.J.; Culotta, V.C. The Yeast Copper/Zinc Superoxide Dismutase and the Pentose Phosphate Pathway Play Overlapping Roles in Oxidative Stress Protection. *J. Boil. Chem.* **1996**, *271*, 28831–28836. [CrossRef]

248. Campbell, K.; Vowinckel, J.; Keller, M.A.; Ralser, M. Methionine Metabolism Alters Oxidative Stress Resistance via the Pentose Phosphate Pathway. *Antioxid. Redox Signal.* **2016**, *24*, 543–547. [CrossRef] [PubMed]

249. Lee, B.C.; Kaya, A.; Ma, S.; Kim, G.; Gerashchenko, M.V.; Yim, S.H.; Hu, Z.; Harshman, L.G.; Gladyshev, V.N. Methionine restriction extends lifespan of Drosophila melanogaster under conditions of low amino-acid status. *Nat. Commun.* **2014**, *5*, 3592. [CrossRef] [PubMed]

250. Johnson, J.E.; Johnson, F.B. Methionine restriction activates the retrograde response and confers both stress tolerance and lifespan extension to yeast, mouse and human cells. *PLoS ONE* **2014**, *9*, e97729. [CrossRef] [PubMed]

251. Ruckenstuhl, C.; Netzberger, C.; Entfellner, I.; Carmona-Gutierrez, D.; Kickenweiz, T.; Stekovic, S.; Gleixner, C.; Schmid, C.; Klug, L.; Sorgo, A.G.; et al. Lifespan extension by methionine restriction requires autophagy-dependent vacuolar acidification. *PLoS Genet.* **2014**, *10*, e1004347. [CrossRef] [PubMed]

252. Wipf, B.; Leisinger, T. Compartmentation of arginine biosynthesis in Saccharomyces cerevisiae. *Fems Microbiol. Lett.* **1977**, *2*, 239–242. [CrossRef]

253. Zabriskie, T.M.; Jackson, M.D. Lysine biosynthesis and metabolism in fungi. *Nat. Prod. Rep.* **2000**, *17*, 85–97. [CrossRef] [PubMed]

254. Canto, C.; Auwerx, J. Calorie restriction: Is AMPK a key sensor and effector? *Physiol.* **2011**, *26*, 214–224. [CrossRef] [PubMed]

255. Iio, W.; Tokutake, Y.; Koike, H.; Matsukawa, N.; Tsukahara, T.; Chohnan, S.; Toyoda, A. Effects of chronic mild food restriction on behavior and the hypothalamic malonyl-CoA signaling pathway. *Anim. Sci. J.* **2015**, *86*, 181–188. [CrossRef] [PubMed]

256. Canto, C.; Gerhart-Hines, Z.; Feige, J.N.; Lagouge, M.; Noriega, L.; Milne, J.C.; Elliott, P.J.; Puigserver, P.; Auwerx, J. AMPK regulates energy expenditure by modulating NAD+ metabolism and SIRT1 activity. *Nature* **2009**, *458*, 1056–1060. [CrossRef] [PubMed]

257. Walsh, M.E.; Shi, Y.; Van Remmen, H. The effects of dietary restriction on oxidative stress in rodents. *Free Radic. Boil. Med.* **2014**, *66*, 88–99. [CrossRef] [PubMed]

258. Pearson, K.J.; Lewis, K.N.; Price, N.L.; Chang, J.W.; Perez, E.; Cascajo, M.V.; Tamashiro, K.L.; Poosala, S.; Csiszar, A.; Ungvari, Z.; et al. Nrf2 mediates cancer protection but not prolongevity induced by caloric restriction. *Proc. Natl. Acad. Sci. USA* **2008**, *105*, 2325–2330. [CrossRef] [PubMed]

259. Someya, S.; Yu, W.; Hallows, W.C.; Xu, J.; Vann, J.M.; Leeuwenburgh, C.; Tanokura, M.; Denu, J.M.; Prolla, T.A. Sirt3 mediates reduction of oxidative damage and prevention of age-related hearing loss under caloric restriction. *Cell* **2010**, *143*, 802–812. [CrossRef] [PubMed]

260. Yeo, H.; Lyssiotis, C.A.; Zhang, Y.; Ying, H.; Asara, J.M.; Cantley, L.C.; Paik, J.H. FoxO3 coordinates metabolic pathways to maintain redox balance in neural stem cells. *EMBO J.* **2013**, *32*, 2589–2602. [CrossRef] [PubMed]

261. Chen, D.; Bruno, J.; Easlon, E.; Lin, S.J.; Cheng, H.L.; Alt, F.W.; Guarente, L. Tissue-specific regulation of SIRT1 by calorie restriction. *Genes Dev.* **2008**, *22*, 1753–1757. [CrossRef] [PubMed]

262. Olmos, Y.; Sanchez-Gomez, F.J.; Wild, B.; Garcia-Quintans, N.; Cabezudo, S.; Lamas, S.; Monsalve, M. SirT1 regulation of antioxidant genes is dependent on the formation of a FoxO3a/PGC-1alpha complex. *Antioxid. Redox Signal.* **2013**, *19*, 1507–1521. [CrossRef] [PubMed]

263. Cho, C.G.; Kim, H.J.; Chung, S.W.; Jung, K.J.; Shim, K.H.; Yu, B.P.; Yodoi, J.; Chung, H.Y. Modulation of glutathione and thioredoxin systems by calorie restriction during the aging process. *Exp. Gerontol.* **2003**, *38*, 539–548. [CrossRef]

264. Rohrbach, S.; Gruenler, S.; Teschner, M.; Holtz, J. The thioredoxin system in aging muscle: Key role of mitochondrial thioredoxin reductase in the protective effects of caloric restriction? *Am. J. Physiol. Regul. Integr. Comp. Physiol.* **2006**, *291*, R927–R935. [CrossRef] [PubMed]

265. Johnson, M.L.; Distelmaier, K.; Lanza, I.R.; Irving, B.A.; Robinson, M.M.; Konopka, A.R.; Shulman, G.I.; Nair, K.S. Mechanism by Which Caloric Restriction Improves Insulin Sensitivity in Sedentary Obese Adults. *Diabetes* **2016**, *65*, 74–84. [CrossRef] [PubMed]

266. Molin, M.; Yang, J.; Hanzen, S.; Toledano, M.B.; Labarre, J.; Nystrom, T. Life span extension and H(2)O(2) resistance elicited by caloric restriction require the peroxiredoxin Tsa1 in Saccharomyces cerevisiae. *Mol. Cell* **2011**, *43*, 823–833. [CrossRef] [PubMed]

267. Fierro-Gonzalez, J.C.; Gonzalez-Barrios, M.; Miranda-Vizuete, A.; Swoboda, P. The thioredoxin TRX-1 regulates adult lifespan extension induced by dietary restriction in Caenorhabditis elegans. *Biochem. Biophys. Res. Commun.* **2011**, *406*, 478–482. [CrossRef] [PubMed]

268. Menger, K.E.; James, A.M.; Cocheme, H.M.; Harbour, M.E.; Chouchani, E.T.; Ding, S.; Fearnley, I.M.; Partridge, L.; Murphy, M.P. Fasting, but Not Aging, Dramatically Alters the Redox Status of Cysteine Residues on Proteins in Drosophila melanogaster. *Cell Rep.* **2015**, *11*, 1856–1865. [CrossRef] [PubMed]

269. Newman, J.C.; Covarrubias, A.J.; Zhao, M.; Yu, X.; Gut, P.; Ng, C.P.; Huang, Y.; Haldar, S.; Verdin, E. Ketogenic Diet Reduces Midlife Mortality and Improves Memory in Aging Mice. *Cell Metab.* **2017**, *26*, 547–557.e8. [CrossRef]

270. Roberts, M.N.; Wallace, M.A.; Tomilov, A.A.; Zhou, Z.; Marcotte, G.R.; Tran, D.; Perez, G.; Gutierrez-Casado, E.; Koike, S.; Knotts, T.A.; et al. A Ketogenic Diet Extends Longevity and Healthspan in Adult Mice. *Cell Metab.* **2017**, *26*, 539–546.e5. [CrossRef]

271. Pawlosky, R.J.; Kemper, M.F.; Kashiwaya, Y.; King, M.T.; Mattson, M.P.; Veech, R.L. Effects of a dietary ketone ester on hippocampal glycolytic and tricarboxylic acid cycle intermediates and amino acids in a 3xTgAD mouse model of Alzheimer's disease. *J. Neurochem.* **2017**, *141*, 195–207. [CrossRef] [PubMed]

272. De Nobrega, A.K.; Lyons, L.C. Aging and the clock: Perspective from flies to humans. *Eur. J. Neurosci.* **2018**. [CrossRef] [PubMed]

273. Froy, O. Circadian rhythms, aging, and life span in mammals. *Physiology* **2011**, *26*, 225–235. [CrossRef] [PubMed]

274. Zhang, R.; Lahens, N.F.; Ballance, H.I.; Hughes, M.E.; Hogenesch, J.B. A circadian gene expression atlas in mammals: Implications for biology and medicine. *Proc. Natl. Acad. Sci. USA* **2014**, *111*, 16219–16224. [CrossRef] [PubMed]

275. Mauvoisin, D.; Wang, J.; Jouffe, C.; Martin, E.; Atger, F.; Waridel, P.; Quadroni, M.; Gachon, F.; Naef, F. Circadian clock-dependent and -independent rhythmic proteomes implement distinct diurnal functions in mouse liver. *Proc. Natl. Acad. Sci. USA* **2014**, *111*, 167–172. [CrossRef] [PubMed]

276. Mendez, I.; Vazquez-Martinez, O.; Hernandez-Munoz, R.; Valente-Godinez, H.; Diaz-Munoz, M. Redox regulation and pro-oxidant reactions in the physiology of circadian systems. *Biochimie* **2016**, *124*, 178–186. [CrossRef] [PubMed]

277. Peek, C.B.; Affinati, A.H.; Ramsey, K.M.; Kuo, H.Y.; Yu, W.; Sena, L.A.; Ilkayeva, O.; Marcheva, B.; Kobayashi, Y.; Omura, C.; et al. Circadian clock NAD+ cycle drives mitochondrial oxidative metabolism in mice. *Science* **2013**, *342*, 1243417. [CrossRef] [PubMed]

278. Nakahata, Y.; Sahar, S.; Astarita, G.; Kaluzova, M.; Sassone-Corsi, P. Circadian control of the NAD+ salvage pathway by CLOCK-SIRT1. *Science* **2009**, *324*, 654–657. [CrossRef] [PubMed]

279. Ramsey, K.M.; Yoshino, J.; Brace, C.S.; Abrassart, D.; Kobayashi, Y.; Marcheva, B.; Hong, H.K.; Chong, J.L.; Buhr, E.D.; Lee, C.; et al. Circadian clock feedback cycle through NAMPT-mediated NAD+ biosynthesis. *Science* **2009**, *324*, 651–654. [CrossRef] [PubMed]

280. Goodman, R.P.; Calvo, S.E.; Mootha, V.K. Spatiotemporal compartmentalization of hepatic NADH and NADPH metabolism. *J. Boil. Chem.* **2018**, *293*, 7508–7516. [CrossRef] [PubMed]

281. Putker, M.; Crosby, P.; Feeney, K.A.; Hoyle, N.P.; Costa, A.S.H.; Gaude, E.; Frezza, C.; O'Neill, J.S. Mammalian Circadian Period, But Not Phase and Amplitude, Is Robust Against Redox and Metabolic Perturbations. *Antioxid. Redox Signal.* **2018**, *28*, 507–520. [CrossRef] [PubMed]

282. Rey, G.; Valekunja, U.K.; Feeney, K.A.; Wulund, L.; Milev, N.B.; Stangherlin, A.; Ansel-Bollepalli, L.; Velagapudi, V.; O'Neill, J.S.; Reddy, A.B. The Pentose Phosphate Pathway Regulates the Circadian Clock. *Cell Metab.* **2016**, *24*, 462–473. [CrossRef] [PubMed]

283. Cracan, V.; Titov, D.V.; Shen, H.; Grabarek, Z.; Mootha, V.K. A genetically encoded tool for manipulation of NADP(+)/NADPH in living cells. *Nat. Chem. Boil.* **2017**, *13*, 1088–1095. [CrossRef] [PubMed]

284. Sandbichler, A.M.; Jansen, B.; Peer, B.A.; Paulitsch, M.; Pelster, B.; Egg, M. Metabolic Plasticity Enables Circadian Adaptation to Acute Hypoxia in Zebrafish Cells. *Cell. Physiol. Biochem.* **2018**, *46*, 1159–1174. [CrossRef] [PubMed]

285. Rutter, J.; Reick, M.; Wu, L.C.; McKnight, S.L. Regulation of clock and NPAS2 DNA binding by the redox state of NAD cofactors. *Science* **2001**, *293*, 510–514. [CrossRef] [PubMed]

286. Yoshii, K.; Tajima, F.; Ishijima, S.; Sagami, I. Changes in pH and NADPH regulate the DNA binding activity of neuronal PAS domain protein 2, a mammalian circadian transcription factor. *Biochemistry* **2015**, *54*, 250–259. [CrossRef] [PubMed]

287. Goya, M.E.; Romanowski, A.; Caldart, C.S.; Benard, C.Y.; Golombek, D.A. Circadian rhythms identified in *Caenorhabditis elegans* by in vivo long-term monitoring of a bioluminescent reporter. *Proc. Natl. Acad. Sci. USA* **2016**, *113*, E7837–E7845. [CrossRef] [PubMed]

288. Rey, G.; Milev, N.B.; Valekunja, U.K.; Ch, R.; Ray, S.; Silva Dos Santos, M.; Nagy, A.D.; Antrobus, R.; MacRae, J.I.; Reddy, A.B. Metabolic oscillations on the circadian time scale in Drosophila cells lacking clock genes. *Mol. Syst. Biol.* **2018**, *14*, e8376. [CrossRef] [PubMed]

289. Causton, H.C.; Feeney, K.A.; Ziegler, C.A.; O'Neill, J.S. Metabolic Cycles in Yeast Share Features Conserved among Circadian Rhythms. *Curr. Biol.* **2015**, *25*, 1056–1062. [CrossRef] [PubMed]

290. Lerner, F.; Niere, M.; Ludwig, A.; Ziegler, M. Structural and functional characterization of human NAD kinase. *Biochem. Biophys. Res. Commun.* **2001**, *288*, 69–74. [CrossRef] [PubMed]

291. Love, N.R.; Pollak, N.; Dolle, C.; Niere, M.; Chen, Y.; Oliveri, P.; Amaya, E.; Patel, S.; Ziegler, M. NAD kinase controls animal NADP biosynthesis and is modulated via evolutionarily divergent calmodulin-dependent mechanisms. *Proc. Natl. Acad. Sci. USA* **2015**, *112*, 1386–1391. [CrossRef] [PubMed]

292. Shianna, K.V.; Marchuk, D.A.; Strand, M.K. Genomic characterization of POS5, the Saccharomyces cerevisiae mitochondrial NADH kinase. *Mitochondrion* **2006**, *6*, 94–101. [CrossRef] [PubMed]

293. Ohashi, K.; Kawai, S.; Murata, K. Identification and characterization of a human mitochondrial NAD kinase. *Nat. Commun.* **2012**, *3*, 1248. [CrossRef] [PubMed]

294. Zhang, R. MNADK, a novel liver-enriched mitochondrion-localized NAD kinase. *Boil. Open* **2013**, *2*, 432–438. [CrossRef] [PubMed]

295. El-Khoury, R.; Sainsard-Chanet, A. Deletion of the mitochondrial NADH kinase increases mitochondrial DNA stability and life span in the filamentous fungus *Podospora anserina*. *Exp. Gerontol.* **2010**, *45*, 543–549. [CrossRef] [PubMed]

296. Xiong, Z.M.; O'Donovan, M.; Sun, L.; Choi, J.Y.; Ren, M.; Cao, K. Anti-Aging Potentials of Methylene Blue for Human Skin Longevity. *Sci. Rep.* **2017**, *7*, 2475. [CrossRef] [PubMed]

297. Gura, T. Hope in Alzheimer's fight emerges from unexpected places. *Nat. Med.* **2008**, *14*, 894. [CrossRef] [PubMed]

298. Gureev, A.P.; Shaforostova, E.A.; Popov, V.N.; Starkov, A.A. Methylene blue does not bypass Complex III antimycin block in mouse brain mitochondria. *FEBS Lett.* **2019**. [CrossRef] [PubMed]

299. Wen, Y.; Li, W.; Poteet, E.C.; Xie, L.; Tan, C.; Yan, L.J.; Ju, X.; Liu, R.; Qian, H.; Marvin, M.A.; et al. Alternative mitochondrial electron transfer as a novel strategy for neuroprotection. *J. Biol. Chem.* **2011**, *286*, 16504–16515. [CrossRef] [PubMed]

300. Ahn, H.; Kang, S.G.; Yoon, S.I.; Ko, H.J.; Kim, P.H.; Hong, E.J.; An, B.S.; Lee, E.; Lee, G.S. Methylene blue inhibits NLRP3, NLRC4, AIM2, and non-canonical inflammasome activation. *Sci. Rep.* **2017**, *7*, 12409. [CrossRef] [PubMed]

301. Medina, D.X.; Caccamo, A.; Oddo, S. Methylene blue reduces abeta levels and rescues early cognitive deficit by increasing proteasome activity. *Brain Pathol.* **2011**, *21*, 140–149. [CrossRef] [PubMed]

302. Huang, C.; Wu, J.; Xu, L.; Wang, J.; Chen, Z.; Yang, R. Regulation of HSF1 protein stabilization: An updated review. *Eur. J. Pharm.* **2018**, *822*, 69–77. [CrossRef] [PubMed]

303. Copes, N. *High-Throughput Screening of Age-Related Changes in Caenorhabditis elegans*; University of South Florida: Tampa, FL, USA, 2015.

304. Harrison, D.E.; Strong, R.; Allison, D.B.; Ames, B.N.; Astle, C.M.; Atamna, H.; Fernandez, E.; Flurkey, K.; Javors, M.A.; Nadon, N.L.; et al. Acarbose, 17-α-estradiol, and nordihydroguaiaretic acid extend mouse lifespan preferentially in males. *Aging Cell* **2014**, *13*, 273–282. [CrossRef] [PubMed]

305. Massie, H.R.; Aiello, V.R.; Williams, T.R. Influence of photosensitizers and light on the life span of Drosophila. *Mech. Ageing Dev.* **1993**, *68*, 175–182. [CrossRef]

306. Vatolin, S.; Radivoyevitch, T.; Maciejewski, J.P. New drugs for pharmacological extension of replicative life span in normal and progeroid cells. *Npj Aging Mech. Dis.* **2019**, *5*, 2. [CrossRef] [PubMed]

307. Diaz-Ruiz, A.; Lanasa, M.; Garcia, J.; Mora, H.; Fan, F.; Martin-Montalvo, A.; Di Francesco, A.; Calvo-Rubio, M.; Salvador-Pascual, A.; Aon, M.A.; et al. Overexpression of CYB5R3 and NQO1, two NAD(+)-producing enzymes, mimics aspects of caloric restriction. *Aging Cell* **2018**, *17*, e12767. [CrossRef] [PubMed]

MDPI

St. Alban-Anlage 66

4052 Basel

Switzerland

Tel. +41 61 683 77 34

Fax +41 61 302 89 18

www.mdpi.com

Nutrients Editorial Office

E-mail: nutrients@mdpi.com

www.mdpi.com/journal/nutrients

www.ingramcontent.com/pod-product-compliance
Lightning Source LLC
LaVergne TN
LVHW070922170726
843515LV00005B/846